Advanced Functional Materials

Edited by Nevin Tasaltin, Paul Sunday Nnamchi and Safaa Saud

Published in London, United Kingdom

IntechOpen

Supporting open minds since 2005

Advanced Functional Materials
http://dx.doi.org/10.5772/intechopen.87879
Edited by Nevin Tasaltin, Paul Sunday Nnamchi and Safaa Saud

Contributors
Chung-Sung Yang, Wen-Chia Wu, Yan Xu, Chuan-Pei Lee, Chi-Ang Tseng, Chulsoo Woo, Bin Liang, Jonghyuk Oh, Tanvir Qureshi, Abir Al-Tabbaa, Samson Oluropo Adeosun, Margaret O. Ilomuanya, Oluwashina Philips Gbenebor, Modupeola Dada, Chiosa Cletus Odili, Pengfei Wang, Shijie Jia, Shunbin Wang, Xiaosong Lu, Jibo Yu, Xin Wang, Yuxuan Jiang, Elfed Lewis, Dilan Canan Çelikel, Paul Sunday Nnamchi, Camillus Obayi, Paul Nnamchi, Nevin Tasaltin, Cory R. Knick, Amjad Khan, Shabnam Nazir, Hamad S. Alyami, Aman Ullah, Mohamed Abdelsabour Fahmy, Mujtaba Ikram, Huma Anwar, Arslan Nadeem, Sidra Baber, Abdullah Khan Durrani, Muhammad Ikram, Asghari Maqsood, Oana Cadar, Erika Andrea Levei, Thomas Dippong, Marin Senila, Rifat Kolatoğlu, Enes Aydın, Mehtap Demir, Ahmet Yıldız, Selcan Karakuş, Elif Tüzün, Nuray Beköz Üllen, Ayben Kilislioğlu, Sana Arbab, Muhammad Aamir Iqbal, Muhammad Umer Farooq

First published in London, United Kingdom, 2020 by IntechOpen
IntechOpen is the global imprint of INTECHOPEN LIMITED, registered in England and Wales, registration number: 11086078, 5 Princes Gate Court, London, SW7 2QJ, United Kingdom
Printed in Croatia

British Library Cataloguing-in-Publication Data
A catalogue record for this book is available from the British Library

Additional hard and PDF copies can be obtained from orders@intechopen.com

Advanced Functional Materials
Edited by Nevin Tasaltin, Paul Sunday Nnamchi and Safaa Saud
p. cm.
Print ISBN 978-1-83962-479-7
Online ISBN 978-1-83962-480-3
eBook (PDF) ISBN 978-1-83962-481-0

For EU product safety concerns:
IN TECH d.o.o., Prolaz Marije Krucifikse Kozulić 3, 51000 Rijeka, Croatia, info@intechopen.com or visit our website at intechopen.com.

Meet the editors

Associate Professor Nevin Taşaltın has been working at Maltepe University, Turkey since 2018. She received her MSc. and PhD. degrees in Physics from Marmara University and Gebze Technical University between 2005 and 2010. She worked as a researcher at TUBITAK Marmara Research Center between 2003 and 2006. She worked as a postdoctoral researcher at Koç University between 2010 and 2012. She worked as a senior researcher at TUBITAK Marmara Research Center between 2012 and 2018. She has research and teaching experience in nanomaterials, sensors, photovoltaics, and supercapacitors. She has been leading and working on several research projects funded by TUBITAK and the Republic of Turkey Energy Market Regulatory Authority. She has several research articles, book chapters, and patents.

Paul Nnamchi joined the Department of Metallurgical and Materials Engineering ESUT in 2019. He obtained his PhD, Meng, and B.Eng. at Sheffield University, ESUT, Technical University of Delft in the Netherlands (TUDelft) and Enugu State University of Science and Technology, Enugu-Nigeria respectively. Paul has held scholarly appointments at Northumbria University, Sheffield University, Max Planck's Institute for Iron and Steel Research, Dusseldorf, Germany, The Netherlands Institute for Metals Research (NIMR) at TU-Delft with interdisciplinary research experience in new and emerging metallic materials: bulk metallic glasses, high entropy alloys, self-healing materials, biomaterials designs, and microstructure physics, Dr. Paul Nnamchi is responsible for the metallurgical and materials academic postgraduate programs including planning and implementation of the annual management plan. He is responsible for supervising a research group and for teaching undergraduate and postgraduate students in material science and engineering projects. His research group is interested in the development of new green materials for construction and high value manufacturing. His manufacturing research is conducted on the near-industrial scale and much of it is focused on detailed investigations of novel manufacturing routes based on the use of locally available raw materials and is conducted in close collaboration with industry. Dr. Paul Nnamchi is a Fellow of the Institute of Materials, Metals, Mining (IOM3), United Kingdom, He is a Chartered Engineer (UK), Chartered Scientist (UK), and an associate professor in the Department of Metallurgical and Materials Engineering. He has developed various research subjects such as: design and integrity characterization of biomaterial alloys; development of bio-inspired structural components; development of rapid manufacturing technologies for building structures. Dr. Nnamchi has presented his research at many international conferences and symposiums. He served as a member of Professional Member African Materials Society, MRS (2009), member of International Bio Metals Association (2015), Council for Regulation of Engineering in Nigeria COREN (2011), and member of Institute of Physics UK (2011).

Dr. Saud is a Postdoctoral and Ph.D. graduate in Mechanical/Biomaterials Engineering from the Universiti Teknologi Malaysia and his research focuses on the design and development of novel biomaterials, "Shape Memory Alloys", and their related applications. Currently, he is working as a program coordinator at the Management and Science University, Shah Alam, Malaysia. Dr. Saud is registered as a professional engineer in a number of engineering associations and societies and has more than 10 years' experience in materials engineering, manufacturing, and design. He is the author of more than 90 research articles published in well-known and highly impacted journals. He has also been awarded with a number of gold and bronze medals in many innovations and innovative national and international engineering and technology exhibitions. He is also an editor and reviewer in several well-known journals and publishers.

Contents

Preface

Advanced functional materials are used in sensors, health, concrete, textile, glasses, and pharmacy applications due to their high surface area and surfactant functional groups. Advanced functional materials with different morphologies help to adapt their application-specific properties. Researchers focus on effective platforms (reaction time, sensitivity, selectivity, stability, etc.) for different applications by varying morphological structures of advanced functional materials. Many types of advanced functional materials such as metal nanomaterials, metal/core shell nanomaterials, metal oxide semiconductors, carbon-based nanomaterials, graphene/graphene-based nanomaterials, and metal-organic frameworks have been investigated for many applications. Besides these, because biopolymers enhance homogeneous dispersion of inorganic nanoparticles into an organic matrix, they are used as stabilizers in the structure of advanced functional materials. Furthermore, in order to enhance the polymer strength, inorganic nanomaterials are used as additives. Biopolymers and additives in the same matrix play a critical role for agglomeration of the bio-nanocomposites. In recent years, due to their low cost, various morphologies, high surface area, electron transfer reactions, biodegradable bio-nanocomposites have attracted attention to the structure of advanced functional materials in much more applications.

The book *Advanced Functional Materials* has sixteen chapters that provide an overview of the detailed definition of preparation and applications of advanced functional materials. We sincerely thank our authors who have contributed with experience and knowledge to this book. Especially, our thanks go to the editorial team from IntechOpen Publishing for their assistance in preparing this book.

Nevin Taşaltın
Associate Professor,
Faculty of Engineering and Science,
Department of Electrical and Electronic Engineering,
Institute of Science,
Maltepe University,
Turkey

Department of Renewable Energy Technologies and Management,
Maltepe University,
Turkey

Dr. Safaa Saud
Faculty of Information Sciences and Engineering,
Management and Science University,
Selangor, Malaysia

Dr. Paul Sunday Nnamchi
Department of Metallurgical and Materials Engineering,
University of Nigeria,
Nsukka, Nigeria

Section 1

Advanced Functional Materials for Electronics

Chapter 1

A Novel Electrochemical Sensor for the Detection of Reactive Red Dye to Determine Water Quality

Rifat Kolatoğlu, Enes Aydın, Mehtap Demir, Ahmet Yıldız, Selcan Karakuş, Elif Tüzün, Nuray Beköz Üllen, Nevin Taşaltın and Ayben Kilislioğlu

Abstract

In this study, tragacanth gum/chitosan/ZnO nanoprism-based electrochemical sensors were prepared for sensing reactive dyes in water. To use an electrochemical sensor, a ~250 nm-sized ZnO nanoprism was synthesized via ultrasonic-assisted green synthesis method, using tragacanth gum and chitosan polymer blend as a matrix. The electrochemical properties of tragacanth gum/chitosan/ZnO nanoprisms were compared against reactive red 35, reactive yellow 15, and reactive black 194. The electrochemical measurement results indicated that prepared tragacanth gum/chitosan/ZnO nanoprism-based electrochemical sensor detected 25 ppm reactive red 35 in 1 min at room temperature. This study reveals new high-potential novel tragacanth gum/chitosan/ZnO nanoprism-based sensing material for the detection of reactive red dye-consisted wastewater with high sensitivity and short response time.

Keywords: ZnO, chitosan, nanoprism, electrochemical sensor, environmental monitoring, reactive dye, wastewater

1. Introduction

The physicochemical properties of nanomaterials make them suitable candidates for sensor applications due to their high surface area and surfactant functional groups. Nanomaterials with different morphologies help to adapt their application-specific detection properties. Therefore, researchers focus on effective detection platforms (reaction time, sensitivity, and selectivity) for the detection of aqueous reactive dyes based on different detection principles with sensors prepared with different morphological structures of nanomaterials (nanoparticles, nanowires, nanoprisms, etc.). Also, the materials in nanoscale exhibit higher dissolution and higher solubility than in microscale. Many types of nanomaterials such as metal oxide semiconductors, carbon-based nanomaterials, graphene/graphene-based nanomaterials, and metal-organic frameworks have been investigated for sensing reactive dyes in water [1–15].

Electrochemical sensing method involves the measurement of the redox transformation of reactive dye molecules upon contact with the sensing nanomaterial surface. The method mainly consists of (a) conductometric/resistive, (b) amperometric/voltammetric, and (c) impedimetric electrochemical sensing. Voltammetric sensors work based on the current difference between reference and working electrodes. This approach utilizes the measurement of current as a function of variation in the applied potential difference in terms of the oxidation or reduction of an electroresponses of an electrochemical sensor. The peak current measured during voltammetry-mediated oxidation of analyte (e.g., reactive dye) is reflected as a function of its concentration. The sensitivity of the voltammetric electrochemical sensor is defined as $(I_g - I_0)/I_0$, where I_g and I_0 are the currents while sensing film - analyte (e.g., reactive dye) is interacting and not, respectively. The voltammetric-type electrochemical sensing allows the quantification of the redox state of analyte in terms of current variations [16–18].

Transducers of the electrochemical sensors have attractive attention for preparing highly sensitive sensors. Nonuniform ohmic drop on an electrochemical transducer significantly affects the cyclic voltammogram data. The shape of the cyclic voltammograms predicts various electrochemical transducer geometries and experimental conditions [19–35].

Zinc oxide (ZnO) nanomaterials have distinct properties such as high sensitivity, high surface area, nontoxic, good compatibility, specific shape, nanosize, and correspondingly high isoelectric point. The unique and adjustable properties of ZnO nanomaterials as n-type semiconductor materials show excellent chemical and thermal stability in a wide range of applications such as solar cells, optical devices, sensors, etc. As far as the morphological perspectives of ZnO are concerned, the synthesis and production procedures also play an important role. The different parameters such as surfactant, temperature, concentration, and time are very significant for the growth of nanomaterials with different morphologies in various synthesis processes. Numerous different methods have been reported worldwide for the synthesis of ZnO nanopowder, composites, and films with good surface structure. Noteworthy techniques for the synthesis of various ZnO nanomaterials are generally deposition, wet chemical technique, sol-gel treatment, hydrothermal process, solvothermal process, and microwave techniques [36, 37].

This is the first report of the preparation and structural characterization of novel tragacanth gum/chitosan/ZnO nanoprism and investigation of the voltammetric electrochemical sensing characteristics of tragacanth gum/chitosan/ZnO nanoprisms against reactive dyes in water towards environmental monitoring. We suggest that this tragacanth gum/chitosan/ZnO nanoprism material has a great potential for future applications in high-performance, low-cost, portable, small-scale voltammetric electrochemical sensors towards forthcoming electronics.

2. Materials and methods

2.1 Materials

Tragacanth gum (T) and chitosan were purchased from Sigma-Aldrich Company. Reactive red 35 (RR35), reactive yellow 15 (RY15), and acid black 194 (AB194) were obtained from Burboya Company. Zinc nitrate hexahydrate ($Zn(NO_3)_2{\cdot}6H_2O$), glacial acetic acid (glacial 100%, pro analysis), and sodium hydroxide (NaOH) were purchased from Merck Company. MF-Millipore™ membrane filter was purchased from Merck Company. Ultrapure water was provided by a human ultrapure water system (water resistance: 18.3 Mohm) and was used for the preparation of all reactive dye solutions as analytes. All electrochemical gold transducers and voltammetric electrochemical workstation were purchased from Ebtro Electronics.

2.2 Synthesis of tragacanth gum/chitosan/ZnO nanoprism

Tragacanth gum/chitosan/ZnO nanoprisms were prepared using a green sonochemical method at 25°C (35 kHz frequency, 320 W, Sonoplus, Bandelin, Germany). In the first step, 0.1 g of tragacanth gum was dissolved in 50 ml of deionized pure water on a magnetic stirrer in 2 h at 25°C. In the second step, 0.1 g of chitosan was dissolved in 50 ml of 2% glacial acetic acid. In the third step, 25 ml of the prepared solutions was taken and mixed. In the fourth step, 0.1 M of $Zn(NO_3)_2{\cdot}6H_2O$ and 0.2 M of NaOH was prepared in deionized pure water. In the fifth stage, the solution of $Zn(NO_3)_2{\cdot}6H_2O$ was added in the solutions, and then the solution of NaOH was added dropwise under the sonicator at 25°C for 40 min (%30 amp). The solution was kept at 24 h in the dark at room temperature. Then, it was first filtered through membrane filters of 0.45 μm and 0.22 μm pore size, respectively. The final solution (viscosity: 1.28 cP and pH 4) was stored at 25°C for use in the sterile container for analysis.

2.3 Preparation and measurements of electrochemical sensors

Electrochemical gold transducers were rinsed with ultrapure water, dried with nitrogen gas, and coated with tragacanth gum/chitosan/ZnO nanoprism solution by dropping and drying the solution. The 3D model structure and the property of dyes used are shown in **Figure 1**.

The potential of the working electrode was varied linearly with time, while the reference electrode was maintained at a constant potential. The potential was

Dye	Molecular Formula	3D model structure	Molar mass (g mol^{-1})	λmax (nm)
Reactive Red 35	$C_{12}H_{18}N_3Na_3O_{14}S_4$	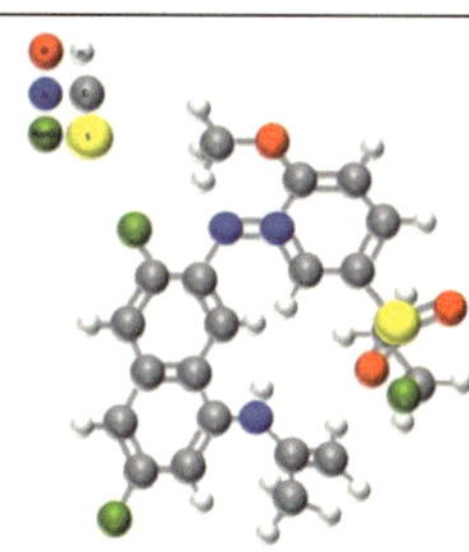	732.94	536
Reactive Yellow 15	$C_{20}H_{20}N_4Na_2O_{11}S$	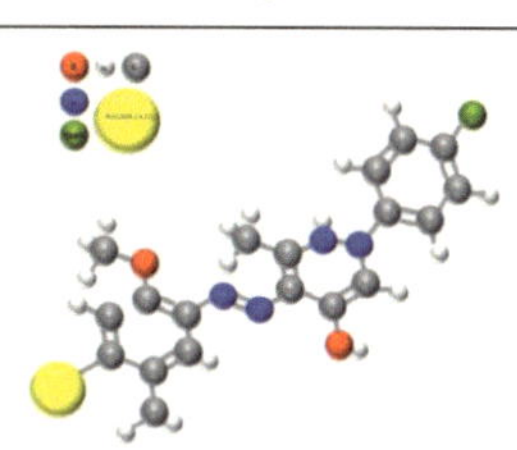	634.57	420
Reactive Black 194	$C_{20}H_{12}N_3NaO_7S$	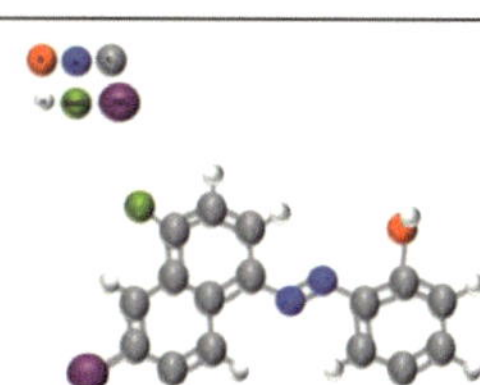	461.38	233

Figure 1.
3D model structure and property of the reactive dyes.

applied between the reference electrode and the working electrode, and the current was measured between the working electrode and the counter electrode. Ebtro voltammetric electrochemical workstation with a three-electrode configuration was used for all electrochemical tests. Cyclic voltammetry (CV) was performed in [−1, +1] V range with a scan rate of 50 mV/s.

3. Results and discussion

The structural analysis of the tragacanth gum/chitosan/ZnO nanoprism material was performed by X-ray diffraction (XRD) (**Figure 2**).

As seen in **Figure 2**, according to XRD analysis, strong peaks were observed at $2\theta = 13°, 31°$, which corresponds to the tragacanth gum/chitosan/ZnO crystalline planes. In the experiments, crystalline-structured tragacanth gum/chitosan/ZnO nanoprisms provided the advantage of obtaining a high surface area for higher interaction and reaction of sensing tragacanth gum/chitosan/ZnO nanoprism thin film on gold transducer-reactive dye with high electron mobility in terms of crystalline structure. SEM and EDX analyses of the prepared tragacanth gum/chitosan/ZnO nanoprisms were performed (**Figure 3**).

The EDX technique was employed to obtain some information on the spatial distribution of the corresponding elements. The EDX analysis of tragacanth gum/chitosan/ZnO nanoprisms provides the average percentage of zinc (Zn) and oxygen (O) at different points. All these suggest efficient preparation and presence of targeted atoms in tragacanth gum/chitosan/ZnO nanoprisms. The polymer matrix (tragacanth gum/chitosan) provides enormously large surface area for dispersion that helps ZnO to grow in the form of nanoprisms with higher reactivity for redox processing. The homogenous dispersion of ZnO in polymer matrix enhances conductivity and stability of the nanostructure. The complementary properties of tragacanth gum/chitosan/ZnO nanoprism generate a synergistic effect to enhance the electrochemical performance and provide improved charge exchange efficiency and stability during redox cycling.

Cyclic voltammetry measurements were performed to analyze the electrochemical sensor performance of tragacanth gum/chitosan/ZnO nanoprism-coated gold

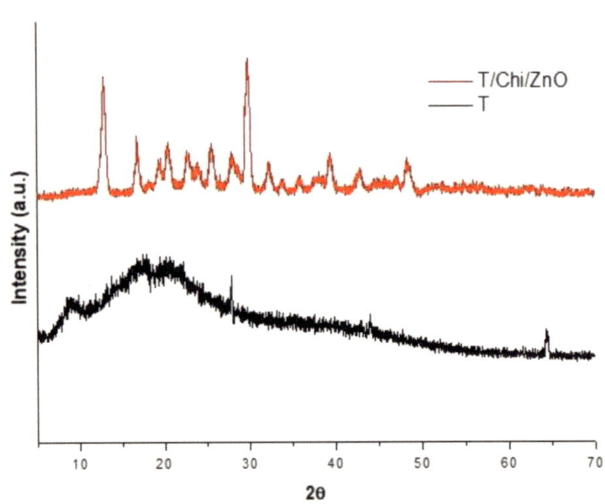

Figure 2.
XRD analysis of tragacanth gum/chitosan/ZnO nanoprisms.

DOI: http://dx.doi.org/10.5772/intechopen.92280

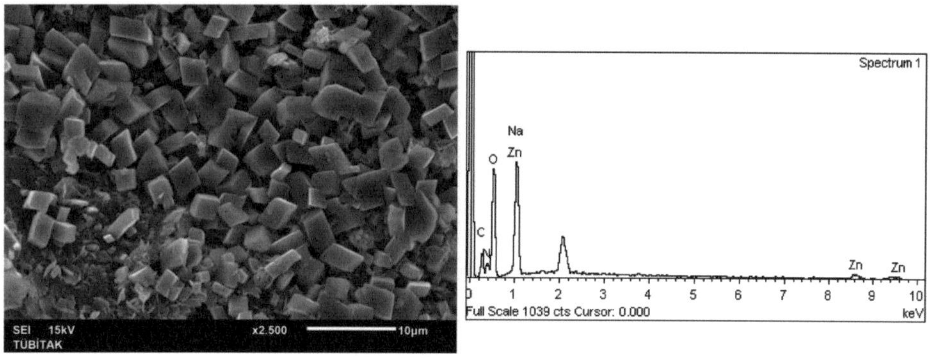

(a) (b)

Figure 3.
(a) SEM and (b) EDX analysis of tragacanth gum/chitosan/ZnO nanoprisms.

transducer. Current-voltage curves of tragacanth gum/chitosan/ZnO nanoprism-coated gold transducers against reactive red 35, reactive yellow 15, and reactive black 194 were obtained, respectively, in [−1, +10] V range with a scan rate of 50 mV/s at room temperature in real-time measurements by Ebtro voltammetric electrochemical workstation (**Figure 4**).

Figure 4 shows the comparative current-voltage curves of tragacanth gum/chitosan/ZnO nanoprism-coated gold transducers against reactive red dye, reactive yellow dye, and reactive black dye in [−1, +10] V range with a scan rate of 50 mV/s at room temperature. The measured current responses were due to either oxidation or reduction of the reactive dye analytes over the entire cycle at the surface of the bare gold transducers. The current peaks arised from redox reactions between tragacanth gum/chitosan/ZnO nanoprism and reactive red dye molecules observed. The curves showed that there are no peaks arising from reactive yellow and black dye molecules as redox reactions did not occur between tragacanth gum/chitosan/ZnO nanoprism and reactive yellow and black dye molecules. The goal of this research was to evaluate the performance of the tragacanth gum/chitosan/ZnO nanoprism-coated gold transducer of the voltammetric electrochemical sensor in discriminating different reactive dyes in water. In this context, we focused on the electrochemical sensing capability tragacanth gum/chitosan/ZnO nanoprisms against to reactive dye-consisted water. The electrochemical oxidation of reactive red dye-consisted water was observed using the scan rate of 50 mV/s at room temperature over a potential range of −0.2 to 0.8 V. In **Figure 4**, current-voltage curves of tragacanth gum/chitosan/ZnO nanoprism-coated gold transducer indicated a prominent redox peak for reactive red dye, while the other tragacanth gum/chitosan/ZnO nanoprism-coated gold transducers indicated no redox peak for reactive yellow and black dyes in water. The redox peak was attributed to a large number of SO_3^- branches of the reactive red 35 (Eqs. (1)–(3)).

$$Zn^{2+} + 2e^- \rightarrow Zn^0 \tag{1}$$

$$SO_3^{2-} + H_2O \rightarrow SO_4^{2-} + 2H^+ + 2e^- \tag{2}$$

$$Zn^{2+} + SO_3^{2-} + H_2O \rightarrow SO_4^{2-} + Zn^0 + 2H^+ \tag{3}$$

After these obtained results, sensor measurements were performed for determining reactive red dye. The different concentrations of reactive red dye were tested on tragacanth gum/chitosan/ZnO nanoprism-coated gold transducers in [−1, +10] V range with a scan rate of 50 mV/s at room temperature. Current peaks arised from redox reactions which came from between tragacanth gum/chitosan/ZnO nanoprism and reactive red dye molecules increased with increasing reactive red concentration in the range of 25–100 ppm. As the concentration of the reactive red dye molecules in the water increases, redox reactions increase the sensitivity of the sensor. Prepared tragacanth gum/chitosan/ZnO nanoprism-based electrochemical sensor detected 25 ppm reactive red dye in 1 min at room temperature.

The reproducibility of the tragacanth gum/chitosan/ZnO nanoprism-coated gold transducer was investigated by analyzing reactive red dye for four times. To ascertain the reproducibility results, the cyclic voltammetry experiments were carried out using the transducers under similar conditions. The peak currents for reactive red dye have not changed much even after a week. This showed the stability of the tragacanth gum/chitosan/ZnO nanoprism-coated gold transducer.

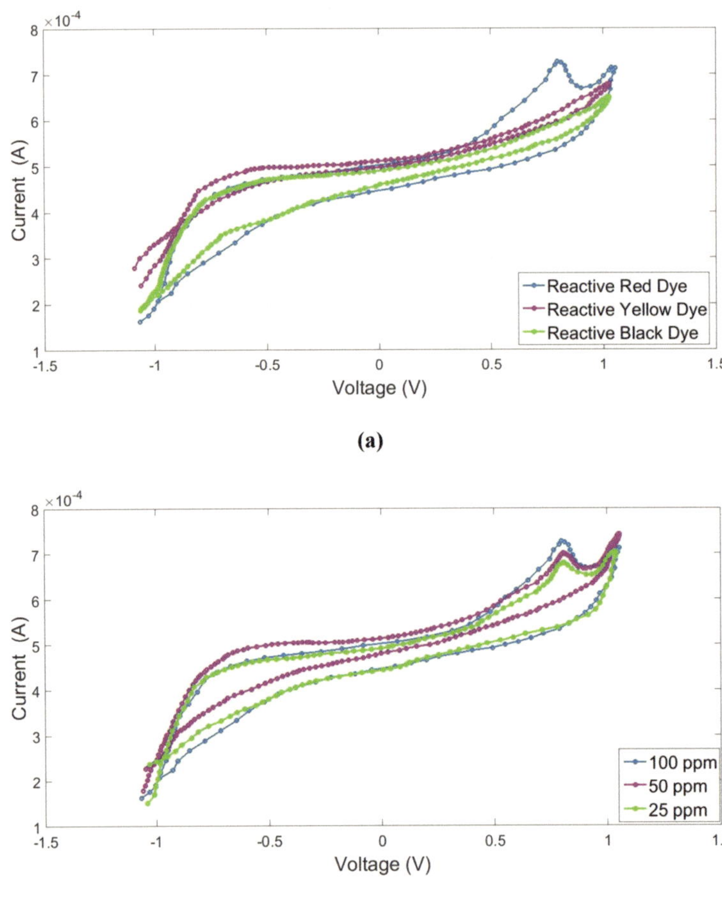

Figure 4.
Current-voltage curves of tragacanth gum/chitosan/ZnO nanoprism-coated gold transducers against (a) reactive red dye, reactive yellow dye, and reactive black dye and (b) different reactive red dye concentrations in 25–100 ppm, [−1, +10] V range with a scan rate of 50 mV/s.

4. Conclusions

In this study, for environmental monitoring of reactive dye-consisting wastewater, the novel tragacanth gum/chitosan/ZnO nanoprism-based electrochemical sensor was prepared and tested via cyclic voltammetry technique. The electrochemical measurement results indicate that prepared tragacanth gum/chitosan/ZnO nanoprism-based electrochemical sensor has a higher sensitivity against reactive red dye than reactive yellow dye and reactive black dye in water. Prepared tragacanth gum/chitosan/ZnO nanoprism-based electrochemical sensor detected 25 ppm reactive red dye in 1 min at room temperature. This study reveals new high-potential sensing material for the detection of reactive dye-consisting wastewater with high sensitivity and short response time. It is the first time that the sensing interaction of tragacanth gum/chitosan/ZnO nanoprisms and reactive red dye was explained.

Acknowledgements

This research was supported by TUBITAK Project 216M421.

Author details

Rifat Kolatoğlu[1], Enes Aydin[1,2], Mehtap Demir[1,3], Ahmet Yildiz[4], Selcan Karakuş[4], Elif Tüzün[4], Nuray Beköz Üllen[5], Nevin Taşaltın[1,6]* and Ayben Kilislioğlu[4]

1 Department of Electrical-Electronics Engineering, Sensor Technology Laboratory, Maltepe University, Istanbul, Turkey

2 ABB Electronics, Istanbul, Turkey

3 Department of Metallurgy and Material Engineering, Adiyaman University, Adıyaman, Turkey

4 Department of Chemistry, İstanbul University-Cerrahpaşa, Istanbul, Turkey

5 Department of Metallurgical and Materials Engineering, İstanbul University-Cerrahpaşa, Istanbul, Turkey

6 Department of Renewable Energy Technology, Maltepe University, Istanbul, Turkey

*Address all correspondence to: nevintasaltin@maltepe.edu.tr

References

[1] Hou SC, Zhang AY, Su M. Nanomaterials for biosensing applications. Nanomaterials. 2016;**6**:58

[2] Rowland CE, Brown CW, Delehanty JB, Medintz IL. Nanomaterial-based sensors for the detection of biological threat agents. Materials Today. 2016;**19**:464-477

[3] Saravanakumar B, Kim SJ. Growth of 2D ZnO nanowall for energy harvesting application. Journal of Physical Chemistry C. 2014;**118**:8831-8836

[4] He JH, Hsin CL, Liu J, Chen LJ, Wang ZL. Piezoelectric gated diode of a single ZnO nanowire. Advanced Materials. 2007;**19**:781-784

[5] Purusothaman Y, Alluri NR, Chandrasekhar A, Kim SJ. Elucidation of the unsymmetrical effect on the piezoelectric and semiconducting properties of Cd-doped 1D-ZnO nanorods. Journal of Materials Chemistry C. 2017;**5**:415-426

[6] Ramadoss A, Kim SJ. Facile preparation and electrochemical characterization of graphene/ZnO nanocomposite for supercapacitor applications. Materials Chemistry and Physics. 2013;**140**:405-411

[7] Brozek CK, Zhou D, Liu H, Li X, Kittilstved KR, Gamelin DR. A review on ZnO nanostructured materials: Energy, environmental and biological applications. Nano Letters. 2018;**18**:3297-3302

[8] Pullagurala VLR, Adisa IO, Rawat S, Kim B, Barrios AC, Velo IAM, et al. Finding the conditions for the beneficial use of ZnO nanoparticles towards plants—A review. Environmental Pollution. 2018;**241**:1175-1181

[9] Theerthagiri J, Chandrasekaran S, Sunitha S, Elakkiya V, Nithyadharseni P, Senthil RA, et al. Recent developments of metal oxide based heterostructures for photocatalytic applications towards environmental remediation. Journal of Solid State Chemistry. 2018;**267**:35-52

[10] Theerthagiri J, Karuppasamy K, Durai G, Rana AHS, Arunachalam P, Sangeetha K, et al. Recent advances in metal chalcogenides (MX; X = S, Se) nanostructures for electrochemical supercapacitor applications: A brief review. Nanomaterials. 2018;**8**:256

[11] Qi K, Cheng B, Yu J, Ho W. Review on the improvement of the photocatalytic and antibacterial activities of ZnO. Journal of Alloys and Compounds. 2017;**727**:792-820

[12] SoYoon S, Ramadoss A, Saravanakumar B, Kim SJ. Novel Cu/CuO/ZnO hybrid hierarchical nanostructures for non-enzymatic glucose sensor application. Journal of Electroanalytical Chemistry. 2014;**717**:90-95

[13] Alam U, Khan A, Raza W, Khan A, Bahnemann D, Muneer M. Highly efficient Y and V co-doped ZnO photocatalyst with enhanced dye sensitized visible light photocatalytic activity. Catalysis Today. 2017;**284**:169-178

[14] Mohan R, Krishnamoorthy K, Kim SJ. Enhanced photocatalytic activity of Cu-doped ZnO nanorods. Solid State Communications. 2012;**152**:375-380

[15] Theerthagiri J, Senthil RA, Senthilkumar B, Polu AR, Madhavan J, Ashokkumar M. Recent advances in MoS_2 nanostructured materials for energy and environmental applications—A review. Journal of Solid State Chemistry. 2017;**252**:43-71

[16] Wang XS, Zhang JB, He Y, Wang LY, Liu L, Wang H, et al. Porous

Nd-doped In_2O_3 nanotubes with excellent formaldehyde sensing properties. Chemical Physics Letters. 2016;**658**:319-323

[17] Stradiotto NR, Yamanaka H, Zanoni MVB. Electrochemical sensors: A powerful tool in analytical chemistry. Journal of the Brazilian Chemical Society. 2003;**14**:159-173

[18] Wang Q, Zheng JB, Zhang HF. A novel formaldehyde sensor containing AgPd alloy nanoparticles electrodeposited on an ionic liquid-chitosan composite film. Journal of Electroanalytical Chemistry. 2012;**674**:1-6

[19] Švancara I, Vytřas K, Kalcher K, Walcarius A, Wang J. Carbon paste electrodes in facts, numbers, and notes: A review on the occasion of the 50-years jubilee of carbon paste in electrochemistry and electroanalysis. Electroanalysis. 2009;**21**:7-28

[20] Zima J, Švancara I, Barek J, Vytřas K. Recent advances in electroanalysis of organic compounds at carbon paste electrodes. Critical Reviews in Analytical Chemistry. 2009;**39**:204-227

[21] Vyskocil V, Barek J. Mercury electrodes—Possibilities and limitations in environmental electroanalysis. Critical Reviews in Analytical Chemistry. 2009;**39**:189-203

[22] Stetter JR, Li J. Amperometric gas sensors—A review. Chemical Reviews. 2008;**108**:352-366

[23] Yogeswaran U, Chen SM. A review on the electrochemical sensors and biosensors composed of nanowires as sensing material. Sensors. 2008;**8**:290-313

[24] Mays DE, Hussam A. Voltammetric methods for determination and speciation of inorganic arsenic in the environment—A review. Analytica Chimica Acta. 2009;**646**:6-16

[25] Bobrowski A, Królicka A, Zarębski J. Characteristics of voltammetric determination and speciation of chromium—A review. Electroanalysis. 2009;**12**:1449-1458

[26] Amatore C, Oleinick A, Svir I. Theoretical analysis of microscopic ohmic drop effects on steady-state and transient voltammetry at the disk microelectrode: A quasi-conformal mapping modeling and simulation. Analytical Chemistry. 2008;**80**:7947-7956

[27] Amatore C, Oleinick AI, Svir I. Numerical simulation of diffusion processes at recessed disk microelectrode arrays using the quasi-conformal mapping approach. Analytical Chemistry. 2009;**81**:4397-4405

[28] Guo JD, Lindner E. Cyclic voltammograms at coplanar and shallow recessed microdisk electrode arrays: Guidelines for design and experiment. Analytical Chemistry. 2009;**81**:130-138

[29] Douglass EF, Driscoll PF, Liu DL, Burnham NA, Lambert CR, McGimpsey WG. Effect of electrode roughness on the capacitive behavior of self-assembled monolayers. Analytical Chemistry. 2008;**80**:7670-7677

[30] Menshykau D, Compton RG. The influence of electrode porosity on diffusional cyclic voltammetry. Electroanalysis. 2008;**20**:2387-2394

[31] Bae JH, Lim YR, Jung W, Silbey RJ, Sung J. Practical model for imperfect conductometric molecular wire sensors. Analytical Chemistry. 2009;**81**:578-583

[32] Lee CY, Bond AM. Evaluation of levels of defect sites present in highly ordered pyrolytic graphite electrodes using capacitive and faradaic current components derived

simultaneously from large-amplitude Fourier transformed ac voltammetric experiments. Analytical Chemistry. 2009;**81**:584-594

[33] Sheth DB, Diefes R, Gratzl M. Spatially averaging electrodes. Analytical Chemistry. 2009;**81**:2129-2134

[34] Ogurtsov VI, Beni V, Strutwolf J, Arrigan DWM. Study of the effects of nonlinear potential sweeps on voltammetry. Electroanalysis. 2009;**21**:68-76

[35] Rajendra NG, Vinod KG, Neeta B, Ram AS. Electrochemical sensor for the determination of dopamine in presence of high concentration of ascorbic acid using a Fullerene-C60 coated gold electrode. Electroanalysis. 2008;**20**(7):757-764

[36] Meulenkamp EA. Synthesis and growth of ZnO nanoparticles. The Journal of Physical Chemistry. B. 1998;**102**:5566-5572

[37] Radzimska AK, Jesionowski T. Zinc oxide-from synthesis to application: A review. Materials. 2014;7:2833-2881

Chapter 2

Fabrication and Characterization of Nanoscale Shape Memory Alloy MEMS Actuators

Cory R. Knick

Abstract

The miniaturization of engineering devices has created interest in new actuation methods capable of large displacements and high frequency responses. Shape memory alloy (SMA) thin films have exhibited one of the highest power densities of any material used in these actuation schemes and can thermally recovery strains of up to 10%. Homogenous SMA films can experience reversible shape memory effect, but without some sort of physical biasing mechanism, the effect is only one-way. SMA films mated in a multi-layer stack have the appealing feature of an intrinsic two-way shape memory effect (SME). In this work, we developed a near-equiatomic NiTi magnetron co-sputtering process and characterized shape memory effects. We mated these SMA films in several "bimorph" configurations to induce out of plane curvature in the low-temperature Martensite phase. We quantify the curvature radius vs. temperature on MEMS device structures to elucidate a relationship between residual stress, recovery stress, radius of curvature, and degree of unfolding. We fabricated and tested laser-irradiated and joule heated SMA MEMS actuators to enable rapid actuation of NiTi MEMS devices, demonstrating some of the lowest powers (5–15 mW) and operating frequencies (1–3 kHz) ever reported for SMA or other thermal actuators.

Keywords: MEMS, phase-change, shape memory alloy, microactuators, optical actuation, electrical actuation, microrobotics, smart materials, nanoscale

1. Introduction

One of the earliest reported thin film version of Nickel-Titanium shape memory alloy was done in 1990 [1]. The first several accounts of SMA film characterization on Si wafers showed measureable shape memory effects, but all transformations happened below ambient conditions, in part due to the fact that the films tended to be Ni-rich in composition if starting from equiatomic NiTi sputter target, due to the different sputter yields of Ni and Ti. Ni has a higher sputter rate than Ti, and Ti has a tendency to react with any residual oxygen in the deposition chamber. In order to make high performance thermal actuators, it is thus necessary to undergo the necessary processing to ensure transformations are measured above ambient conditions [2]. This is no easy task, but can nonetheless be done by carefully controlling Ni/Ti ratio and thermal processing (i.e., annealing). One of the first SMA-based MEMS actuators was reported out of Case Western University (CWU) in 2001, based on a sputtered NiTi film capable of recovering about 250 MPa according to their stress-temperature loop on 4" Si wafer [2].

Of the many SMAs available, NiTi has become one of the most widely used due to its exceptional physical and mechanical properties (SME and SE), including large recoverable strains [3]. To understand the reason behind the SME/SE in NiTi, it is necessary to first understand the crystallography. The basis for SME/SE is the switching between two different crystallographic phases, namely the high temperature phase known as austenite (or) the parent phase, and the low temperature phase known as Martensite. The crystal structure of the austenite phase is a CsCl type B2 cubic structure and the low temperature Martensite phase is a complex monoclinic crystal structure (B19'). The martensitic transformation is a diffusionless solid-state phase transformation. During the martensitic transformation, the metal atoms move cooperatively in the matrix under shear stresses. As a result a new phase is formed from the parent phase. To accommodate the internal stresses caused by the transformation to the B19' phase, the formation of a combination of up to 24 multiple martensitic variants is possible, resulting in a twinned Martensite crystal form, also known as self-accommodated Martensite.

TiNi thin films are in demand for applications in actuators for micro-electromechanical systems (MEMS) [4–12], because these films exhibit large displacement, accompanied by the shape memory effect (SME) through the B2 austenite to B19' monoclinic Martensite transformation. The majority of TiNi films are fabricated by RF or DC magnetron sputtering methods [13–20], and these films are amorphous, unless the substrates are heated during deposition [16, 20]. Post deposition annealing at a temperature above 700 K (equivalent to 427°C) for crystallization is necessary for the films initially deposited in amorphous condition to show the shape memory effect [21]. It is noted that Ti–Ni thin films sputter-deposited at ambient temperature are often amorphous, thus require post-sputtering crystallizing at elevated temperature to obtain the desired shape memory property. It is also possible to crystallize TiNi films during deposition by utilizing a heated substrate above an ambient temperature [22]. The TiNi films deposited in this manner exhibit interesting behaviors such as lowered crystallization temperature and oriented crystallographic structure [18, 19]. For example, Ikuda observed that the NiTi film deposited onto a glass substrate at 673 K (or 400°C) produced crystallinity in the NiTi film. Other, more recent studies also looked into the SMA properties of in-situ annealed NiTi films.

Regarding the in-situ crystallization of NiTi, Gisser also observed that the films deposited on (100) silicon (Si) substrates at 733 K (equivalent to 500°C) showed a (110)-oriented crystalline structure of the austenite phase [23]. By incorporation of a Ru seed layer, epitaxial growth of the NiTi alloy can be achieved at some of the lowest deposition and crystallization temperatures, and thinnest films reported to date [24]. Hou also observed that the films deposited onto quartz and polyimide substrates above 623 K (equivalent to 350°C) showed a strong (110)-oriented crystalline structure [25–28]. This range of crystallization temperatures of the TiNi films suggests that the crystallization process is affected by the surface condition of the substrates. However, it is not clear why the heated substrates lowers the crystallization temperature and enhances a particular orientation for the TiNi thin films. This is due to lack of understanding the process of film growth during deposition. The composition and structure of sputtered Ni-Ti shape memory alloy (SMA) films are significantly affected by the sputtering conditions: target power, gas pressure, target to substrate distance, deposition temperature, substrate bias voltage, etc.

Some fundamental limitations for shape memory MEMS are related to (1) how thin could one go and still be able to measure reversible shape memory effects,

(2) how fast could one conceivably actuate the SMA MEMS device (limitations previously existed) that would not allow for heat transfer to happen much faster than several 10s, or at most, 100 Hz. Several accounts claim that NiTi films should be at least 100–400 nm thick [15] to help ensure shape memory effects which can be suppressed by film/substrate interfacial strains and small grain sizes [29–33].

Combined with SMA's natural advantages of large displacements, and high work densities [8], our efforts have demonstrated major breakthroughs in the bandwidth, or speed with which NiTi could be actuated, and thus enabled additional possibilities to use NiTi in microelectronics and MEMS. Shape memory MEMS can certainly now be used for higher frequency actuation applications such as mechanical logic, signal routing, and switching, and at relatively low power and energy consumption. In thin films the roll of texture is extremely important in improving shape memory properties like reversible strain [34].

In the preceding paragraphs we overviewed the important developments in NiTi thin films processing and characterization. As such, we aim to use the next paragraph to highlight some of the more relevant MEMS device implementations of NiTi SMA films.

In 2004, high frequency actuation based on SMA MEMS was only 20–40 Hz [35], and considerable improvements have been made since then. By shrinking the volume and heat capacity of the SMA MEMS actuators, we showed for the first time, reversible actuation beyond 1 kHz frequencies, verifying that the heat transfer (in other words, the heating and cooling of shape memory alloy), could happen more than 1000 times per second. For example, green laser actuation of shape memory MEMS bimorph actuators was characterized in [9], whereby actuation response happened in just a few milliseconds. Here, the authors showed that NiTi bimorph cantilevers with nanometer thickness NiTi, could be actuated in under 100 ms, with as little as 2 W/cm^2. In another paper, the high cycle frequency of actuation and electrical characterization of SMA MEMS device was characterized to include (resistance, current, and power requirements) [5]. Here, it was demonstrated that NiTi bimorph resistor actuators could be actuated with as little as 0.5 V, requiring just 5–15 mA of power, and at rates faster than 1000 times per second (up to 3 kHz) due to the small volume and rapid heat transfer facilitated by large surface to volume ratios. Expanding upon this work even further, these same NiTi films were integrated with nanoscale 3D printing to enable some impressive actuation metrics [36, 37]. Specifically, by 3D printing polymeric materials mated with NiTi films, the following metrics were achieved: >5000 reversible actuation cycles with very limited degradation, low voltage actuation of 3.7 V (which is compatible with common Li-ion batteries), large strokes (85 μm for 415 μm length cantilever), and large force-displacement product of 1.2×10^{-7} N-m, with an impressively small volume and weight (1.04×10^{-5} cm^3 and 1.27×10^{-5} g, respectively). By comparison in [38], 1.6 kHz actuation frequency was achieved using a pulsed laser to actuate a FIB cut NiTi SMA microactuator spring of 25 μm thickness. Most recently NiTi SMA has even recently been integrated with Si photonics to form a physically actuated optical coupler/de-coupler type device with excellent nano-positioning accuracy to within 4 nm and on/off ratio of 9 dB [7, 39].

2. Methods

2.1 Stress vs. temperature measurements

Stress versus temperature measurements were performed using a Toho FLX-2320-S wafer bow tool with controlled heating and cooling from 25 to 100°C with a heating and cooling rate of 1°C/min. For these experiments, we prepared films of NiTi by sputtering onto 4-inch silicon (Si) wafers and vacuum annealing at 600, 500, and 450°C to crystallize the material. Additionally, we measured several NiTi on Si wafers where the NiTi was sputtered under 600°C substrate conditions. In later efforts, we characterized these films with NiTi in-situ anneals of 325, 350, 375, 400, 425, and 500°C. Wafer bow was measured experimentally from 25 to 100°C at a 1°C/min heating and cooling rates, which allowed us to calculate and plot the temperature-dependent residual stress in the NiTi film for each wafer sample based on Stoney's equation (1).

$$\sigma = \frac{E}{6(1-\nu)}\frac{h_s^2}{h}\left(\frac{1}{R}-\frac{1}{R_o}\right) \tag{1}$$

Here, σ is the stress in the thin film, and E, ν, and hs are Young's modulus of Si, Poisson ratio of the Si substrate and the thickness of the Si substrate, respectively. h represents the NiTi thin film thickness and R and R_o represent the radii of curvature of the NiTi film- Si substrate composite and the curvature of the bare Si substrate. We used an extended version of Stoney's equation (2) in order to calculate the stress in the NiTi layer when deposited on a thin Pt film on Si wafer. Here, σ_{NiTi} is the stress in the NiTi layer, and σ_{Pt} represents the stress in the annealed Pt layer. The variable h_{NiTi} represents the NiTi thin film thickness, and R and R_o represent the radii of curvature of the NiTi film and the annealed Pt/Si substrate, respectively.

$$\sigma_{NiTi} = \frac{E\,h_s^2}{6(1-\nu)\,h_{NiTi}}\left(\frac{1}{R}-\frac{1}{R_o}\right) - \left(\frac{6(1-\nu)}{E}\right)\left(\frac{\sigma_{Pt}(h_{NiTi+Pt}-h_{NiTi})}{h_s^2}\right) \tag{2}$$

2.2 Laser actuation of SMA MEMS

We also used a 400 mW, 532 nm green laser exiting a 400 μm diameter optical fiber to irradiate and heat released cantilevers with a known optical intensity level. We used optical density filters (ThorLabs) to control the laser irradiance levels. The distance from the optical fiber exit and therefore laser spot size were fixed at 1 mm, which allowed calculation of the optical intensity. We used a Photron Fastcam camera connected to a microscope to record video at 2000 frames per second (fps) or more to measure temporal data on activation time.

2.3 Electrical actuation (joule heating) SMA MEMS

We build stressed bimorph actuators out of SU-8 and NiTi, whereby a pulsed current through the freestanding NiTi 'resistor' caused rapid heating and cooling through Joule Heating. Deflection was monitored using laser Doppler Vibrometry (LDV) experimental setup. A Keithly power meter was used to pulse current (square wave) at various frequencies (2–3000 Hz) through the NiTi resistively heated MEMS actuator.

3. Results and discussion

3.1 Stress vs. temperature measurements

Figure 1A shows a reversible phase change with onset at 60°C upon heating for two different NiTi sputter deposition pressures of $Ni_{50}Ti_{50}$ on 200 nm Pt. Since NiTi was sputtered onto a thin film of Pt on Si for this set, the modified Stoney's equation (2) was used to figure the NiTi film stress. Assuming the thickness of the two films to be similar, the film sputtered at 5 mTorr exhibited a higher value of recovery stress,

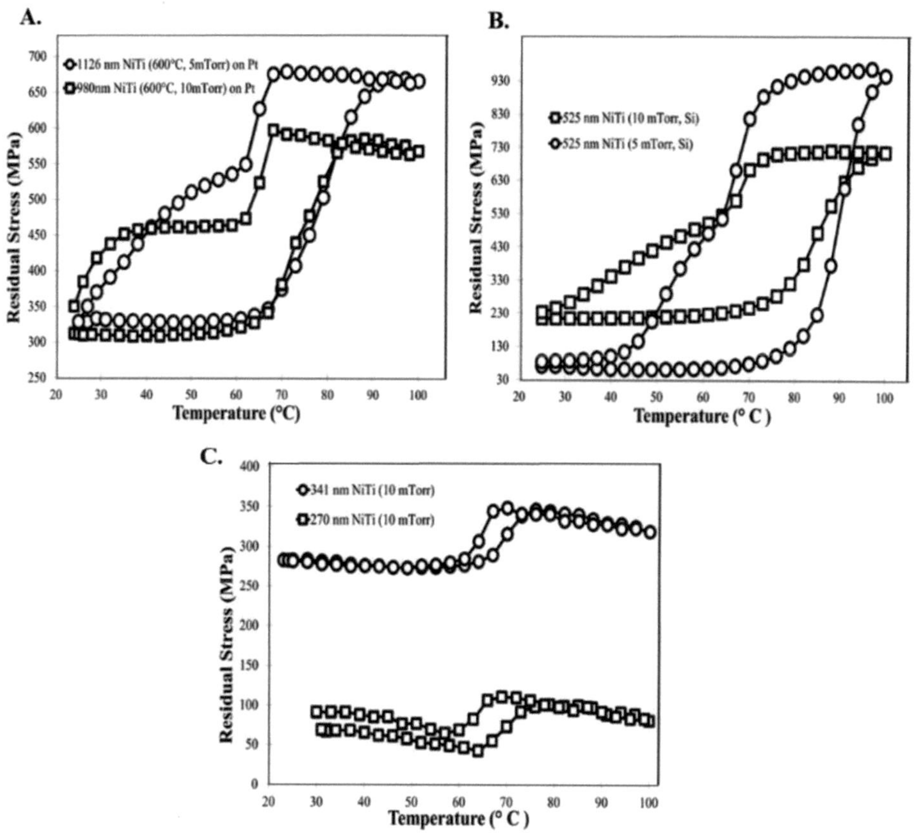

Figure 1.
(A) Stress vs. temperature plots for NiTi sputtered at 600°C under different pressures onto 200 nm Pt for NiTi (A) near 1 μm thickness, (B) near half micron thickness, and (C) approaching 200 nm minimum film thickness for SMA properties.

defined as the difference between initial stress and stress in the presumably austenitic phase at elevated temperature. Both films had a similar initial or residual stress of around 300 MPa. The maximum residual stress values peaked around 70°C for each wafer in this experiment, and the process was reversible when cooled back to RT.

Figure 1B shows that the trend of higher recovery stress (approximately 900 MPa) at lower deposition pressure was the same for two 525 μm thick NiTi films, which in this case were deposited onto Si and stress values were determined with standard Stoney equation (1). Residual stress was lower (70 MPa) for NiTi sputtered onto Si at 5 mTorr compared to the NiTi sputtered at 10 mTorr (230 MPa). Lower residual stress would generally be desired to reduce unwanted deformation of MEMS structures fabricated based on NiTi. These results are also useful, providing confirmation that the SME is similar when we deposited NiTi onto Pt or Si.

We also performed wafer bow stress measurements on thinner films of 341 and 270 nm NiTi which were sputtered onto Si at 600°C substrate temperature. **Figure 1C** shows a reversible SME in 341 and 270 nm films sputtered at 10 mTorr. Therefore, significant micro actuation should be achievable in even thinner films.

3.2 Laser actuation of SMA MEMS

We measured cantilever actuation under optical irradiation. Devices demonstrated rapid actuation ranging from 2 to 90 ms, depending on optical power

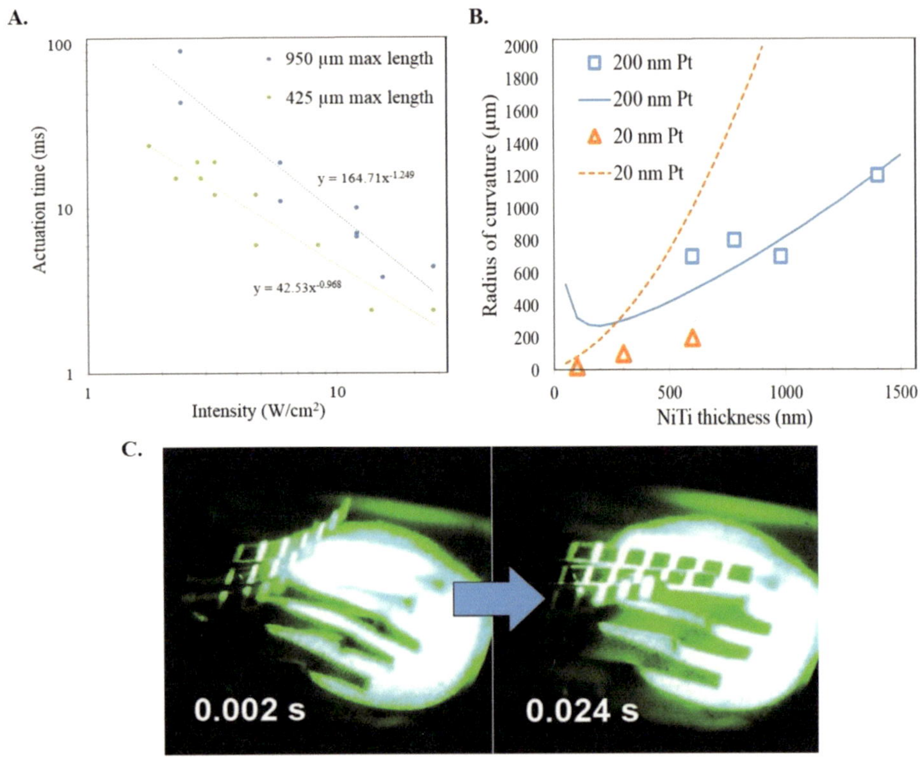

Figure 2.
(A) Actuation time vs. laser irradiance for 600 nm NiTi on 20 nm Pt bimorphs, (B) calculated and measured curvature radius for thermally activated NiTi on Pt bimorphs (1.4 µm on 200 nm Pt), and (C) demonstration of passive laser irradiated device-specific actuation with a 532 nm "green" laser (~7.2 W/cm²) in 24 ms for the 1.4 µm thick NiTi device stack.

density as shown in **Figure 2A**. As would be expected for a cantilever beam clamped on one end to a heat sink, the actuation time followed a 1/I2 (I, being intensity) power law. Overall, the devices could be fully actuated in under 20 ms with intensities as low as 2 W/cm^2. The response time decreased to 3 ms with intensities over 14 W/cm^2. As shown in **Figure 2C**, 1.4 µm thick NiTi on 200 nm Pt devices actuated into their downward state within 25 ms when irradiated at 7.2 W/cm^2. A slower actuation time of 230 ms was observed at 1.44 W/cm^2. The radius of curvature for the 600 nm NiTi/20 nm Pt stack was 5.4X tighter (200 µm), compared to the 1.2 mm curvature in the 1.4 µm thick NiTi stack. For the tightest curling (200 µm) 600 nm thick NiTi devices, we performed a dynamic optical actuation experiment where we measured actuation time at various laser intensities. These results are plotted in **Figure 2A**.

3.3 Electrical actuation (joule heating) SMA MEMS

Figure 3A shows schematic of the joule heated SEM MEMS resistor actuator including cross section. The bond pads for electrical probe pads or wire bonding are comprised on 200 micron squares of NiTi alloy. The cross section is comprised of 1 micron thick SU-8 epoxy on top of the 270 nm thick NiTi, with reversible SMA properties. We fabricated joule heaters with various widths and lengths (10, 15, 20) and (100, 150, 200, 300, and 400) microns, respectively. The large CTE mismatch between SU-8 and NiTi drives the upward curvature of the MEMS actuator post-release. **Figure 3B** shows SEM of the released actuator. **Figure 3C** depicts the fabrication process flow used to build the actuator, whereby NiTi is patterned with ion milling, and the release etch is done in XeF_2. **Figure 3D** shows the thermal actuation of the actuator which is characterized by large, non-linear changes in deflection upon subsequent heating and cooling cycles.

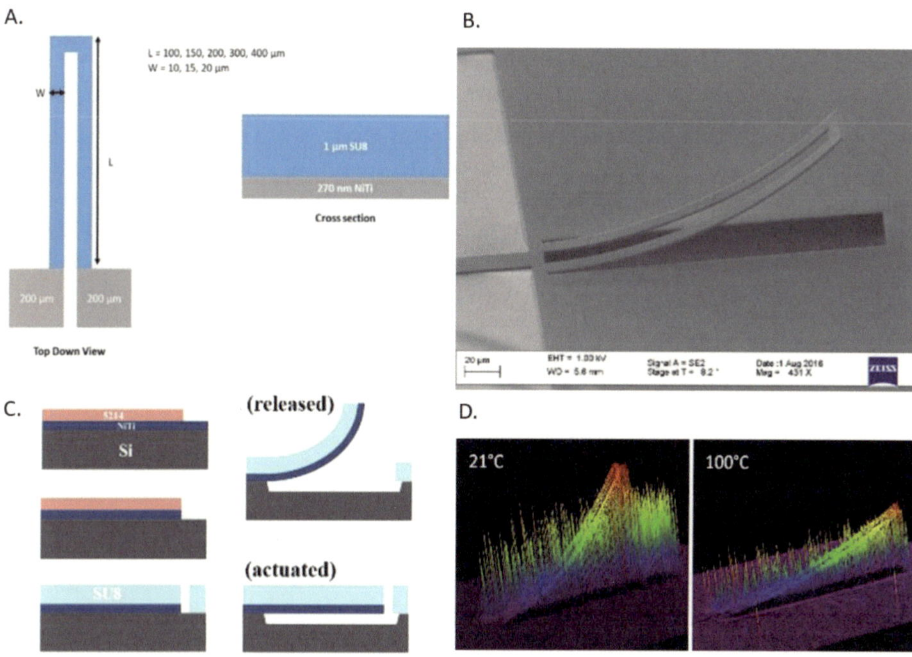

Figure 3.
(A) Schematic of the joule heated SEM MEMS resistor actuator including cross section, (B) SEM of released actuator, (C) fabrication process flow used to build the actuator, and (D) thermal actuation of the actuator.

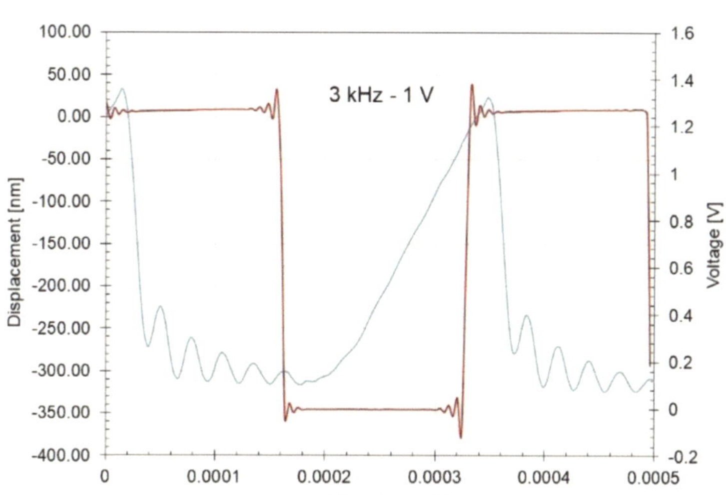

Figure 4.
Measured displacement of SMA MEMS resistively heated actuator vs. time for a 1 V pulsed square wave (50% duty cycle) at 3 kHz.

Figure 4 shows the measured deflection of the actuator using a 1 V pulsed current with 50% duty cycle at 3 kHz. For these devices the total power drawn was measured to be 5–15 mW.

4. Conclusion

In conclusion, we were able to demonstrate novel applications of the SMA MEMS actuator including both low power laser and low power and fast electrical joule heating. In other words, the SMA can be actuated by absorbing laser energy in the form of heat, or through pulsed electrical current. This was enabled by the development of deposition and characterizations of nanoscale thickness NiTi thin films. NiTi thin films with reversible changes in recovery stress were mated with other residual stressed thin films to enable reversible thermal actuation at MEMS scale. By incorporating nanoscale thin film SMA MEMS with good heat sinking design, we could achieve reversible actuations up to 3 kHz which is significantly greater than many previous reports for SMA actuators.

Acknowledgements

Special thanks to Mr. Brian Iassacson for assistance with processing SMA MEMS device wafers in the ARL cleanroom. Thanks also to Dr. Christopher Morris and Gabriel Smith of ARL, and Dr. Hugh Bruck of University of Maryland for their guidance and support.

Conflict of interest

No conflict of interests.

Notes/Thanks/Other declarations

Place any other declarations, such as “Notes”, “Thanks”, etc. in before the References section. Assign the appropriate heading.

Author details

Cory R. Knick
US Army Research Laboratory (ARL), US Army Night Vision and Electronic Sensors Directorate, Ft. Belvoir, VA, USA

*Address all correspondence to: cory.r.knick.civ@mail.mil

References

[1] Busch JD, Johnson AD, Lee CH, Stevenson DA. Shape-memory properties in Ni-Ti sputter-deposited film. Journal of Applied Physics. 1990;**68**(12):6224-6228. DOI: 10.1063/1.346914

[2] Chen-Luen Shih B-KL, Kahn H, Phillips SM, Heuer AH. A robust co-sputtering fabrication procedure for TiNi shape memory alloys for MEMS. JMEMS. 2001;**10**:69-79

[3] Otsuka K, Ren X. Physical metallurgy of Ti–Ni-based shape memory alloys. Progress in Materials Science. 2005;**50**(5):511-678. DOI: 10.1016/j.pmatsci.2004.10.001

[4] Choudhary N, Kaur D. Shape memory alloy thin films and heterostructures for MEMS applications: A review. Sensors and Actuators A: Physical. 2016;**242**:162-181. DOI: 10.1016/j.sna.2016.02.026

[5] Cory R Knick DJS, Wilson AA, Smith GL, Morris CJ, Bruck HA. High frequency, low power, electrically actuated shape memory alloy MEMS bimorph thermal actuators. Journal of Micromechanics and Microengineering. 2019;**29**(7):23

[6] Dahmardeh M et al. High-power MEMS switch enabled by carbon-nanotube contact and shape-memory-alloy actuator. Physica Status Solidi A. 2013;**210**(4):631-638. DOI: 10.1002/pssa.201228678

[7] Lambrecht IAF, Chernenko V, Kohl M. Integrated SMA-based NEMS actuator for optical switching. In: IEEE 29th International Conference on Micro Electro Mechanical Systems (MEMS); Shanghai, China; 2016. pp. 79-82. DOI: 10.1109/MEMSYS.2016.7421562

[8] Wolf RH, Heuer AH. TiNi (shape memory) films on Si for MEMS applications. Journal of Microelectromechanical Systems. 1995;**4**:4

[9] Knick CR, Smith GL, Morris CJ, Bruck HA. Rapid and low power laser actuation of sputter-deposited NiTi shape memory alloy (SMA) MEMS thermal bimorph actuators. Sensors and Actuators A: Physical. 2019;**291**:48-57. DOI: 10.1016/j.sna.2019.03.016

[10] Mohamed Ali MS, Bycraft B, Bsoul A, Takahata K. Radio-controlled microactuator based on shape-memory-alloy spiral-coil inductor. Journal of Microelectromechanical Systems. 2013;**22**(2):331-338. DOI: 10.1109/jmems.2012.2221161

[11] Namazu T, Tashiro Y, Inoue S. Ti–Ni shape memory alloy film-actuated microstructures for a MEMS probe card. Journal of Micromechanics and Microengineering. 2007;**17**(1):154-162. DOI: 10.1088/0960-1317/17/1/020

[12] Wibowo E et al. Two-way actuation of bilayer cantilever of nickel titanium and silicon nitride thin films by shape memory effect and stress relaxation. In: Presented at The Device and Process Technologies for MEMS, Microelectronics, and Photonics III; 2004

[13] Isalgue VTA, Seguin JL, Bendahan M, Esteve-Cano. Shape memory NiTi thin films deposited at low temperature. Materials Science and Engineering A. 1999;**273-275**:717-721

[14] Bechtold C, Chluba C, Lima de Miranda R, Quandt E. High cyclic stability of the elastocaloric effect in sputtered TiNiCu shape memory films. Applied Physics Letters. 2012;**101**(9):091903. DOI: 10.1063/1.4748307

[15] Fu YQ et al. On the lower thickness boundary of sputtered TiNi films

for shape memory application. Thin Solid Films. 2006;**515**(1):80-86. DOI: 10.1016/j.tsf.2005.12.039

[16] Huang X, Liu Y. Surface morphology of sputtered NiTi-based shape memory alloy thin films. Surface and Coatings Technology. 2005;**190**(2-3):400-405. DOI: 10.1016/j.surfcoat.2004.02.029

[17] Inoue S, Morino K, Yoshiki K, Namazu T. Effect of substrate temperature on the shape memory behavior of Ti-Ni-Cu ternary alloy sputtered films. Materials Science Forum. 2012;**706-709**:1903-1908. DOI: 10.4028/www.scientific.net/MSF.706-709.1903

[18] Martins RMS, Schell N, Mahesh KK, Pereira L, Silva RJC, Braz Fernandes FM. Texture development and phase transformation behavior of sputtered Ni-Ti films. Journal of Materials Engineering and Performance. 2009;**18**(5-6):543-547. DOI: 10.1007/s11665-009-9484-9

[19] Martins RMS et al. Texture development, microstructure and phase transformation characteristics of sputtered Ni–Ti shape memory alloy films grown on TiN<111>. Thin Solid Films. 2010;**519**(1):122-128. DOI: 10.1016/j.tsf.2010.07.078

[20] Wibowo E, Kwok CY. Fabrication and characterization of sputtered NiTi shape memory thin films. Journal of Micromechanics and Microengineering. 2006;**16**(1):101-108. DOI: 10.1088/0960-1317/16/1/014

[21] Satoh G, Birnbaum A, Yao YL. Annealing effect on the shape memory properties of amorphous NiTi thin films. Journal of Manufacturing Science and Engineering. 2010;**132**(5):051004-1-051004-9. DOI: 10.1115/1.4002189

[22] Lee H-J, Ramirez AG. Crystallization and phase transformations in amorphous NiTi thin films for microelectromechanical systems. Applied Physics Letters. 2004;**85**(7):1146-1148. DOI: 10.1063/1.1783011

[23] Gisser KRC, Busch JD, Johnson AD, Ellis AB. Oriented nickel-titanium shape memory alloy films prepared by annealing during deposition. Applied Physics Letters. 1992;**61**(14):1632-1634. DOI: 10.1063/1.108434

[24] Kenneth Ainslie CK, Smith G, Li J, Troxel C, Mehta A, Kukreja R. Controlling shape memory effects in NiTi thin films grown on Ru seed layer. Sensors and Actuators A: Physical. 2019;**294**:7. DOI: 10.1016/j.sna.2019.04.047

[25] Ishida A, Sato M. Development of polyimide/SMA thin-film actuator. Materials Science Forum. 2010;**654-656**:2075-2078. DOI: 10.4028/www.scientific.net/MSF.654-656.2075

[26] Ishida A, Sato M. Ti–Ni–Cu shape-memory alloy thin film formed on polyimide substrate. Thin Solid Films. 2008;**516**(21):7836-7839. DOI: 10.1016/j.tsf.2008.04.091

[27] Kishi Y, Ikenaga N, Sakudo N, Yajima Z. Shape memory behavior of TiNi alloy films sputter-deposited on polyimide substrate. Journal of Alloys and Compounds. 2013;**577**:S210-S214. DOI: 10.1016/j.jallcom.2012.02.020

[28] Kotnur VG, Tichelaar FD, Fu WT, De Hosson JTM, Janssen GCAM. Shape memory NiTi thin films deposited on polyimide at low temperature. Surface and Coatings Technology. 2014;**258**:1145-1151. DOI: 10.1016/j.surfcoat.2014.07.018

[29] Kabla M, Seiner H, Musilova M, Landa M, Shilo D. The relationships between sputter deposition conditions, grain size, and phase transformation temperatures in NiTi thin films.

Acta Materialia. 2014;**70**:79-91. DOI: 10.1016/j.actamat.2014.02.009

[30] Kumar A, Singh D, Kaur D. Grain size effect on structural, electrical and mechanical properties of NiTi thin films deposited by magnetron co-sputtering. Surface and Coatings Technology. 2009;**203**(12):1596-1603. DOI: 10.1016/j.surfcoat.2008.12.005

[31] Pan G, Cao Z, Wei M, Shi J, Xu L, Meng X. Thickness and grain size dependence of B2–R martensitic transformation behaviors in nanoscale TiNi films. Materials Letters. 2014;**130**:285-288. DOI: 10.1016/j.matlet.2014.05.129

[32] Ryklina EP, Polyakova KA, Tabachkova NY, Resnina NN, Prokoshkin SD. Effect of B2 austenite grain size and aging time on microstructure and transformation behavior of thermomechanically treated titanium nickelide. Journal of Alloys and Compounds. 2018;**764**:626-638. DOI: 10.1016/j.jallcom.2018.06.102

[33] Shi XB, Guo FM, Zhang JS, Ding HL, Cui LS. Grain size effect on stress hysteresis of nanocrystalline NiTi alloys. Journal of Alloys and Compounds. 2016;**688**:62-68. DOI: 10.1016/j.jallcom.2016.07.168

[34] Shu YC, Bhattacharya K. The influence of texture on the shape memory effect in polycrystals. Acta Metallurgica. 1998;**46**(15)

[35] Shin DD, Mohanchandra KP, Carman GP. High frequency actuation of thin film NiTi. Sensors and Actuators A: Physical. 2004;**111**(2-3):166-171. DOI: 10.1016/j.sna.2003.09.026

[36] Velez C, Patel D, Kim S, Babaei M, Knick C, Smith G, et al. Combining micro fabrication and additive manufacturing for microrobotic mechanisms. In: Special Proceedings of JMEMS Combining Micro Fabrication and Additive Manufacturing for Microrobotic Mechanisms; 2020

[37] Camilo Velez SK, Patel D, Babaei M, Knick CR, Smith G, Bergbreiter S. Rapid prototyping of microactuators by integrating 3D printed polymeric structures with NiTi thin film. In: Presented at the 33rd IEEE International Conference on Micro Electromechanical Systems (MEMS); 2020

[38] Hyun-Taek Lee MSK, Lee G-Y, Kim C-S, Ahn S-H. Shape memory alloy (SMA)-based microscale actuators with 60% deformation rate and 1.6 kHz actuation speed. Small. 2018;**14**(3)

[39] Rastjoo S, Fechner R, Bumke L, Kötz M, Quandt E, Kohl M. Development and co-integration of a SMA/Si bimorph nanoactuator for Si photonic circuits. Microelectronic Engineering. 2020;**225**:111257-11262. DOI: 10.1016/j.mee.2020.111257

Chapter 3

Optimization of MEMS Actuator Driven by Shape Memory Alloy Thin Film Phase Change

Cory R. Knick

Abstract

At the microscale, shape memory alloy (SMA) microelectromechanical system (MEMS) bimorph actuators offer great potential based on their inherently high work density. An optimization problem relating to the deflection and curvature based on shape memory MEMS bimorph was identified, formulated, and solved. Thicknesses of the SU-8 photoresist and nickel-titanium alloy (NiTi) was identified that yielded maximum deflections and curvature radius based on a relationship among individual layer thicknesses, elastic modulus, and cantilever length. This model should serve as a guideline for optimal NiTi and SU-8 thicknesses to drive large deflections and curvature radius that are most suitable for microrobotic actuation, micromirrors, micropumps, and microgrippers. This model would also be extensible to other phase-change-driven actuators where nonlinear and significant residual stress changes are used to drive actuation.

Keywords: shape memory alloy, thin film, microactuators, MEMS, optimization, radius of curvature, phase change

1. Introduction

In certain applications for MEMS microactuators, large deflections would be desired such as the case of micro-robotics [1–3], micromirrors [4–6], and microgrippers [3, 7]. Using a shape memory alloy (SMA), a material that undergoes large changes in stress during a temperature cycle due to a solid-solid phase change can be used to generate large, nonlinear deflections. We aim to find a relationship between deflections of a SMA MEMS actuator, and maximize the deflection of SMA MEMS bimorph. Shape memory alloy films based on sputtered NiTi have been exhaustively characterized in previous decades, leading to a wealth of information about the intricate interplay between Ni/Ti ratio, annealing temperatures and times, and thickness [8–31]. Bimorphic actuators can impart reversible deflection shape memory microactuators as previously demonstrated [32, 33].

To date, optimization of parameters for improving shape memory-induced actuation has not been explored. We chose for our candidate system an SU-8 patterned on top of NiTi SMA bimorph actuator. In this case, residual strains develop during the processing of MEMS actuators, and upon release from substrate, the device curls upward to relieve these strains. Thermal input converts the material into austenite, and shape memory effect drives the actuator into a more flat position, a process that is reversible upon subsequent thermal cycles. Thermal effects can be delivered to the SMA MEMS using laser irradiation [34], and joule-heating [35], at frequencies up to at least 1 kHz. SU-8 is an ideal material due to its relative ease of use in MEMS, low modulus of elasticity enables more flexible devices with large deflection, and good chemical stability.

Much literature exists for thin film development and characterization of nickel-titanium shape memory alloy [1–4]. Although many demonstrations of SMA MEMS actuators have been shown [9, 34–43], none of these citations perform design optimization studies to maximize deflection or curvature radius due to residual stress changes due to phase change. When the nitinol is thermally cycled between martensite and austenite phases, there is a corresponding change in residual stress, which is used to drive the nonlinear deflections. This nonlinear and large change in stress is defined as the recovery stress, and is a principal factor influencing the deflection and curvature radius. Our novel contributions take a realistic SMA MEMS bimorph design based on SU-8 on NiTi, and determine optimal thickness combinations to yield maximized deflections, which would be desirable in certain applications were large strokes are desired. We feed into the model the Young's modulus values for NiTi thin films that have been determined previously using nanoindentation techniques [44, 45].

2. Building and characterizing the SMA MEMS actuators

The nickel-titanium alloy (NiTi) would be co-sputtered onto a 4″ silicon wafer based on the methods reported in previous works [8, 34–36]. The substrate is rotated and heated during deposition to ensure crystallization of the film. The wafer stress vs. temperature measurements are performed, using Stoney's equation to determine recovery stress, hysteresis, and residual stress in the NiTi film. After verification of good shape memory alloy properties in the film at wafer level, a photomask is used to pattern bimorph actuator. Ion milling is used to remove portions of the NiTi film on the wafer. The SU-8 2000.5 is spin coated (where the RPM is used to control SU-8 thickness) and another mask plate is used to pattern SU-8 on top of the NiTi cantilever. Finally, the device is released by etching the Si substrate away in xenon difluoride (XeF2) gas. In practice, SU-8 thickness would be controlled by varying spin speed, and NiTi thickness based on sputtering time.

2.1 Problem identification

The design problem is to maximize the deflection of a MEMS bimorph cantilever beam based on the nonlinear shape memory alloy (SMA) as the actuating mechanism. The deflection is dependent in large part on the parameter called recovery stress. The larger the recovery stress, the larger the deflection. We may also wish to decrease the overall mass or volume of the actuator, of minimize the curvature radius. The objectives are competing in that reduction in the shape memory alloy thickness, generally leads to reduction of the recovery stress. The bimorph actuator could consist of SU-8 on top of NiTi thin film, but this optimization model would be easily extensible to other cases of interest.

We should consider that the equation describing the recovery stress-induced deflection in shape memory alloy MEMS actuator is (Eq. 1).

Initially, the contour plots of SU-8 and NiTi thickness showed that the optimization problem was not interesting for the simplest case of constant recovery stress over the range of NiTi thickness. To our advantage, the NiTi recovery stress is a parameter that depends on NiTi thickness, which makes the optimization problem more interesting. The equation describing the recovery stress-induced deflection in shape memory alloy (SMA) MEMS actuator is:

$$d = \frac{3E_{NiTi}\sigma_{rec}t_{NiTi}t_{SU\text{-}8}(t_{NiTi}+t_{SU\text{-}8})l^2}{E_{NiTi}^2 t_{NiTi}^4 + E_{SU\text{-}8}E_{NiTi}\left(4t_{NiTi}^3 t_{SU\text{-}8} + 6t_{NiTi}^2 t_{SU\text{-}8}^2 + 4t_{NiTi}t_{SU\text{-}8}^3\right) + E_{SU\text{-}8}^2 t_{SU\text{-}8}^4} \quad (1)$$

where σ_{rec} = recovery stress of the SMA MEMS actuator; d = deflection of the SMA MEMS actuator; l = total length of the SMA MEMS actuator; E_{NiTi} = elastic modulus of NiTi layer; $E_{SU\text{-}8}$ = elastic modulus of SU-8 layer; T_{NiTi} = thickness modulus of NiTi layer; $T_{SU\text{-}8}$ = thickness modulus of SU-8 layer.

Figure 1 shows stress vs. temperature curves for NiTi on Si wafer. These curves are experimentally generated, and indicate the recovery stress (difference between highest and lowest stress values), and the thermal hysteresis. Here, as an illustrative example, the NiTi thickness is 900 nm, and the temperature cycle is performed using a heating and cooling rate of 1°C/min.

Assumptions: we assume operating temperatures go between RT and 100°C to ensure full phase change. In all calculations, for simplicity we use Young's modulus of NiTi as a fixed value. In reality, the Young's modulus changes curing the phase change. Martensite (lower temperature phase usually has a lower elastic modulus compared to the higher temperature austenite phase).

Figure 2 shows the process used to build the SMA MEMS bimorph actuator comprised of the NiTi shape memory (SMA) layer underneath the SU-8 elastic layer. In step (a) deposition of SMA onto Si wafer and pattern using photolithography. In step (b) ion milling is performed to transfer the pattern into the SMA layer. In step (c) we spin on SU-8 and pattern it with mask plate and photolithography. In step (d) we release the MEMS bimorph by etching Si substrate with xenon difluoride (XeF_2) gas. In step (e) we thermally actuate the two-way shape memory MEMS device between curled and flat states.

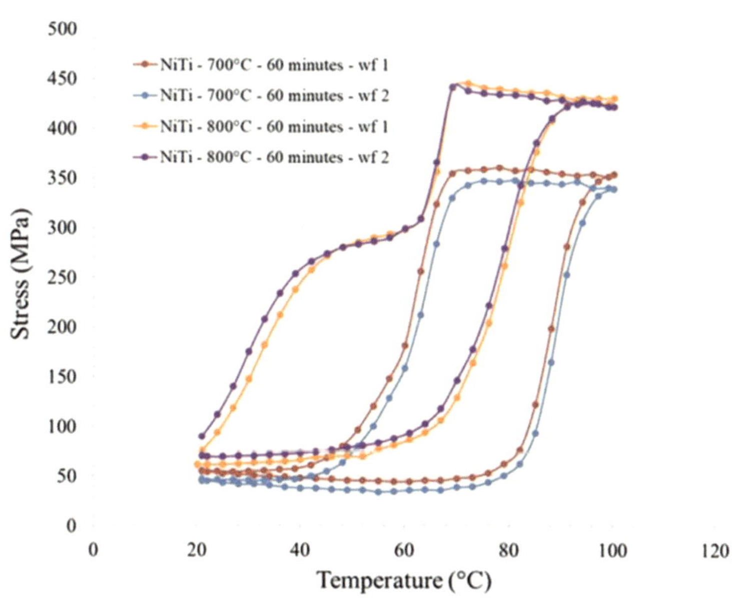

Figure 1.
Stress vs. temperature curves for NiTi on Si wafer. These curves are experimentally generated, and indicate the recovery stress (difference between highest and lowest stress values), and the thermal hysteresis.

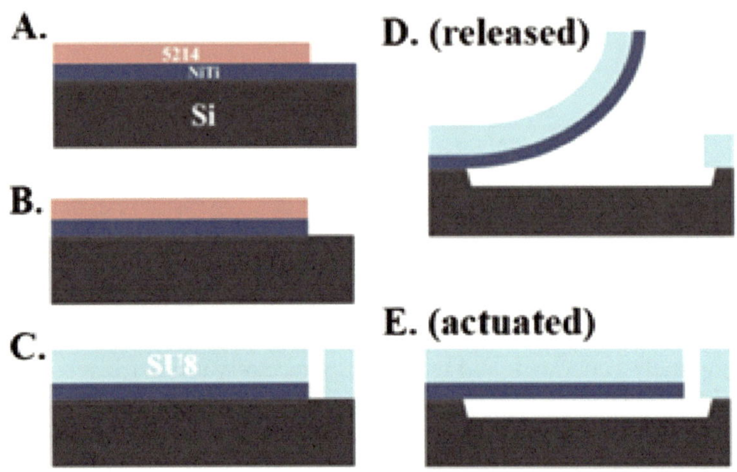

Figure 2.
SMA MEMS fabrication process for SU-8 on NiTi bimorph.

3. Methods, results, and discussions

3.1 Single-objective optimization

Regarding NiTi recovery stress, there would appear to be an optimal thickness range for which recovery stress reaches max values as depicted in **Figure 3**, below which there is a sharp drop off. Therefore the tendency for increased deflections for thinner materials reaches a point of diminishing returns due to the effect of decreasing recovery stress. Below 100–150 nm, shape memory properties have been

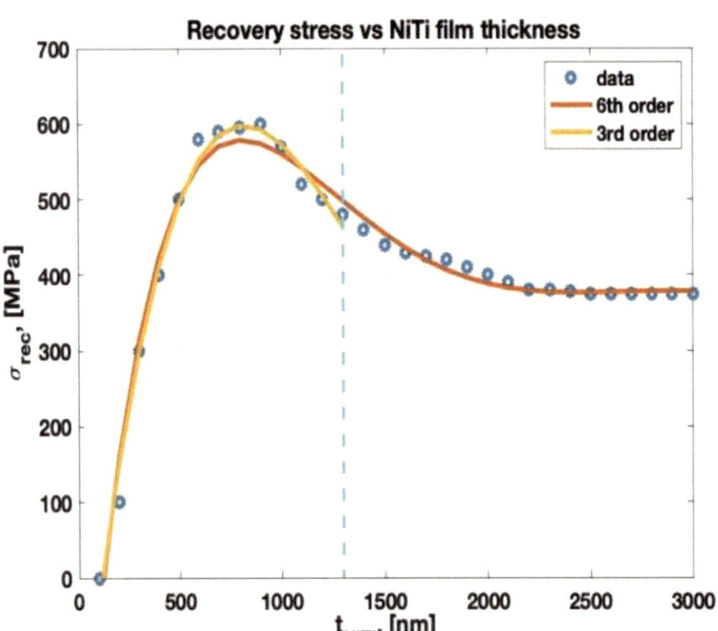

Figure 3.
Relation between recovery stress (MPa) and NiTi film thickness (nm).

shown to drop off completely, so we impose constraints for NiTi thickness to vary between 150 and 1300 nm.

Based on the curve fitting equation (*third-order polynomial*), the single-objective optimization problem can be written as following:

$$maximize:$$

$$d = \frac{3E_{NiTi}\sigma_{rec}t_{NiTi}t_{SU\text{-}8}(t_{NiTi} + t_{SU\text{-}8})l^2}{E_{NiTi}^2 t_{NiTi}^4 + E_{SU\text{-}8}E_{NiTi}\left(4t_{NiTi}^3 t_{SU\text{-}8} + 6t_{NiTi}^2 t_{SU\text{-}8}^2 + 4t_{NiTi}t_{SU\text{-}8}^3\right) + E_{SU\text{-}8}^2 t_{SU\text{-}8}^4} \quad (2)$$

subjected to::

$$100\ \mu m \le l \le 300\ \mu m \quad (3)$$

$$150\ nm \le t_{NiTi} \le 1000\ nm \quad (4)$$

$$200\ nm \le t_{SU\text{-}8} \le 2000\ nm \quad (5)$$

$$\sigma_{rec} \ge 0 \quad (6)$$

$$\sigma_{rec} = 5.36E26 t_{NiTi}{}^3 - 2.15E21 t_{NiTi}{}^2 + 2.45E15 t_{NiTi} - 2.58E08\ \text{(SI units)} \quad (7)$$

Covert to standard form in SI units:

$$minimize:$$

$$f = \frac{-3E_{NiTi}\sigma_{rec}t_{NiTi}t_{SU\text{-}8}(t_{NiTi} + t_{SU\text{-}8})l^2}{E_{NiTi}^2 t_{NiTi}^4 + E_{SU\text{-}8}E_{NiTi}\left(4t_{NiTi}^3 t_{SU\text{-}8} + 6t_{NiTi}^2 t_{SU\text{-}8}^2 + 4t_{NiTi}t_{SU\text{-}8}^3\right) + E_{SU\text{-}8}^2 t_{SU\text{-}8}^4} \quad (8)$$

subjected to::

$$g_1 : 100 \times 10^{-6} - l \le 0; \quad (9)$$

$$g_2 : l - 300 \times 10^{-6} \le 0; \quad (10)$$

$$g_3 : 150 \times 10^{-9} - t_{NiTi} \le 0; \quad (11)$$

$$g_4 : t_{NiTi} - 1300 \times 10^{-9} \le 0; \quad (12)$$

$$g_5 : 200 \times 10^{-9} - t_{SU\text{-}8} \le 0; \quad (13)$$

$$g_6 : t_{SU\text{-}8} - 2000 \times 10^{-9} \le 0; \quad (14)$$

$$g_7 : -\sigma_{rec} \le 0 \quad (15)$$

$$h_1 : \sigma_{rec} - 5.36 \times 10^{26} t_{NiTi}{}^3 + 2.15 \times 10^{21} t_{NiTi}{}^2 - 2.45 \times 10^{15} t_{NiTi} + 2.58 \times 10^8 = 0 \quad (16)$$

3.2 MATLAB optimization toolbox (fmincon)

According to the toolbox (and as shown in **Figure 4**), optimal solution is: $t_{NiTi} = 359\ nm, t_{SU\text{-}8} = 824\ nm, l = 300\ \mu m.$

3.3 Multi-objective optimization

The curvature of a bilayer elastic material [46] is given as

$$K = \frac{-E'_{SU\text{-}8}t_{SU\text{-}8}E'_{NiTi}t_{NiTi}(t_{NiTi} + t_{SU\text{-}8})}{G\left(E'_{SU\text{-}8}t_{SU\text{-}8} + E'_{NiTi}t_{NiTi}\right)}\Delta\varepsilon \quad (17)$$

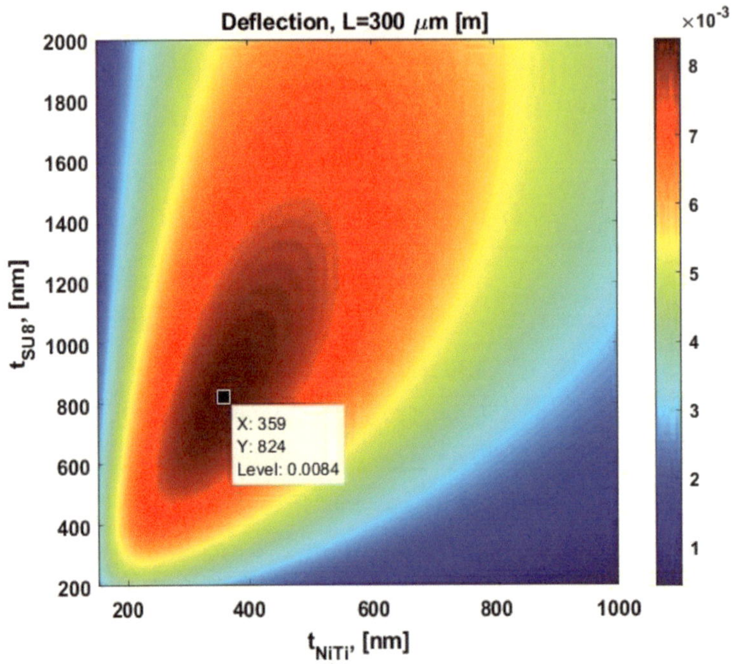

Figure 4.
Optimization contours for the case where SU-8 elastic modulus is 2 GPa. Variables considered are individual layer thicknesses: NiTi (x-axis) and SU-8 (y-axis).

$$G = E'_{SU\text{-}8}t^2_{SU\text{-}8}\left(\frac{t_{NiTi}}{2} - \frac{t_{SU\text{-}8}}{6} - \theta\right) - E'_{NiTi}t_{NiTi}\left[t_{SU\text{-}8}\left(t_{SU\text{-}8} + \frac{t_{NiTi}}{2}\right) + \frac{t^2_{NiTi}}{6} + \theta(2t_{SU\text{-}8} + t_{NiTi})\right] \quad (18)$$

$$\theta = \frac{t_{NiTi}t_{SU\text{-}8}\left(E'_{NiTi} - E'_{SU\text{-}8}\right)}{2\left(E'_{SU\text{-}8}t_{SU\text{-}8} + E'_{NiTi}t_{NiTi}\right)} \quad (19)$$

$$\Delta\varepsilon = (\alpha_{SU\text{-}8} - \alpha_{NiTi})\Delta T \quad (20)$$

ρ is the curvature radius generally expressed in units of μm. Δε is a strain differential term resulting from CTE mismatch and temperature difference experienced during the processing. θ is a correction factor used in the placement of neutral plane. E′ is the biaxial modulus defined as E/(1 − v) where v is Poisson ratio and E is Young's modulus. Poisson ratios are assumed to be 0.22 for SU-8 and 0.33 for NiTi. α_SU-8 is reported to be 52×10^{-6}/°C. α_NiTi (depending on austenite or martensite phase) is reported to be 6.6 or 11×10^{-6}/°C. For simplicity sake, we assume an intermediate value of α_NiTi = 9×10^{-6}/°C. Units for theta term is nm or m. Units for G term is Pa × nm^3 or Pa × m^3. Therefore units for curvature is in nm or m. Δε term is unit less.

The objective number 2 is to maximize curvature radius. We determine the pareto frontier and strong pareto points using the epsilon constrained method. In this epsilon constrained method, we minimize f1 while keeping f2 less than or equal to different values of epsilon. As a first step for objective function 2 (curvature of bimorph) we coded MATLAB script to generate contour plots as a function of the two main design variables (i.e., thickness of NiTi and SU-8). The problem formulation for objective function 2 is as follows (and contour plot is shown in **Figure 5**).

Curvature is:

$$K = -\frac{E'_{SU\text{-}8}t_{SU\text{-}8}E'_{NiTi}t_{NiTi}\left(t_{NiTi} + t_{SU\text{-}8}\right)}{G\left(E'_{SU\text{-}8}t_{SU\text{-}8} + E'_{NiTi}t_{NiTi}\right)}\Delta\varepsilon \quad (21)$$

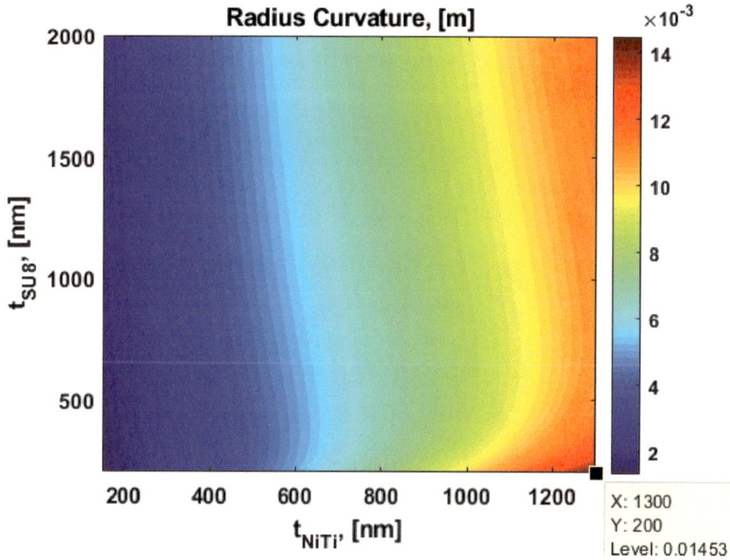

Figure 5.
MATLAB-generated contour plot of curvature radius (m), against the primary design variables (i.e., tNiTi and $t_{SU\text{-}8}$). Curvature radius is maximized for the thickest values of NiTi and thinnest values of SU-8. The result is intuitive because this is the stiffest beam (from the perspective of thickest NiTi with much larger Young's modulus compared to SU-8). Thinner SU-8 means the effect from strain differential and CTE mismatch is minimized and contributes less to curvature radius. Overall, this means that upper bound on NiTi thickness and lower bound on SU-8 thickness are active constraints for objective function 2.

Maximize:

$$\rho = \frac{1}{K} = -\frac{G\left(E'_{SU\text{-}8}t_{SU\text{-}8} + E'_{NiTi}t_{NiTi}\right)}{E'_{SU\text{-}8}t_{SU\text{-}8}E'_{NiTi}t_{NiTi}(t_{NiTi} + t_{SU\text{-}8})\Delta\varepsilon} \tag{22}$$

Subjected to:

$$g_1 : 150 \times 10^{-9} - t_{NiTi} \leq 0; \tag{23}$$

$$g_2 : t_{NiTi} - 1300 \times 10^{-9} \leq 0; \tag{24}$$

$$g_3 : 200 \times 10^{-9} - t_{SU\text{-}8} \leq 0; \tag{25}$$

$$g_4 : t_{SU\text{-}8} - 2000 \times 10^{-9} \leq 0; \tag{26}$$

$$\begin{aligned} h_1 : G - E'_{SU\text{-}8}t^2_{SU\text{-}8}\left(\frac{t_{NiTi}}{2} - \frac{t_{SU\text{-}8}}{6} - \theta\right) \\ - E'_{NiTi}t_{NiTi}\left[t_{SU\text{-}8}\left(t_{SU\text{-}8} + \frac{t_{NiTi}}{2}\right) + \frac{t^2_{NiTi}}{6} + \theta(2t_{SU\text{-}8} + t_{NiTi})\right] \\ = 0 \end{aligned} \tag{27}$$

$$h_2 : \theta - \frac{t_{NiTi}t_{SU\text{-}8}\left(E'_{NiTi} - E'_{SU\text{-}8}\right)}{2\left(E'_{SU\text{-}8}t_{SU\text{-}8} + E'_{NiTi}t_{NiTi}\right)} = 0 \tag{28}$$

For multi-objective optimization, the deflection and curvature radius of *SMA* bimorph actuator are maximized simultaneously. So, the multi-objective optimization problem can be stated as follows:

maximize:

$$f_1 : \rho = -\frac{G\left(E'_{SU\text{-}8}t_{SU\text{-}8} + E'_{NiTi}t_{NiTi}\right)}{E'_{SU\text{-}8}t_{SU\text{-}8}E'_{NiTi}t_{NiTi}(t_{NiTi} + t_{SU\text{-}8})\Delta\varepsilon} \tag{29}$$

$$f_2 : d = \frac{3E_{NiTi}\sigma_{rec}t_{NiTi}t_{SU\text{-}8}(t_{NiTi} + t_{SU\text{-}8})l^2}{E_{NiTi}^2 t_{NiTi}^4 + E_{SU\text{-}8}E_{NiTi}\left(4t_{NiTi}^3 t_{SU\text{-}8} + 6t_{NiTi}^2 t_{SU\text{-}8}^2 + 4t_{NiTi}t_{SU\text{-}8}^3\right) + E_{SU\text{-}8}^2 t_{SU\text{-}8}^4} \quad (30)$$

Subjected to:

$$g_1 : 100 \times 10^{-6} - l \leq 0; \quad (31)$$

$$g_2 : l - 300 \times 10^{-6} \leq 0; \quad (32)$$

$$g_3 : 150 \times 10^{-9} - t_{NiTi} \leq 0; \quad (33)$$

$$g_4 : t_{NiTi} - 1300 \times 10^{-9} \leq 0; \quad (34)$$

$$g_5 : 200 \times 10^{-9} - t_{SU\text{-}8} \leq 0; \quad (35)$$

$$g_6 : t_{SU\text{-}8} - 2000 \times 10^{-9} \leq 0; \quad (36)$$

$$g_7 : -\sigma_{rec} \leq 0; \quad (37)$$

$$h_1 : \sigma_{rec} - 5.36 \times 10^{26} t_{NiTi}{}^3 + 2.15 \times 10^{21} t_{NiTi}{}^2 - 2.45 \times 10^{15} t_{NiTi} + 2.58 \times 10^8 = 0; \quad (38)$$

$$h_2 : G - E'_{SU\text{-}8}t_{SU\text{-}8}^2\left(\frac{t_{NiTi}}{2} - \frac{t_{SU\text{-}8}}{6} - \theta\right) - E'_{NiTi}t_{NiTi}\left[t_{SU\text{-}8}\left(t_{SU\text{-}8} + \frac{t_{NiTi}}{2}\right) + \frac{t_{NiTi}^2}{6} + \theta(2t_{SU\text{-}8} + t_{NiTi})\right] = 0; \quad (39)$$

$$h_3 : \theta - \frac{t_{NiTi}t_{SU\text{-}8}\left(E'_{NiTi} - E'_{SU\text{-}8}\right)}{2\left(E'_{SU\text{-}8}t_{SU\text{-}8} + E'_{NiTi}t_{NiTi}\right)} = 0. \quad (40)$$

Due to the conflicting nature of the two objective functions, the contour plot for the multi-objective function has changed substantially. Maximizing the radius is favored by a larger t_{NiTi} as opposed to a smaller thickness required to maximize

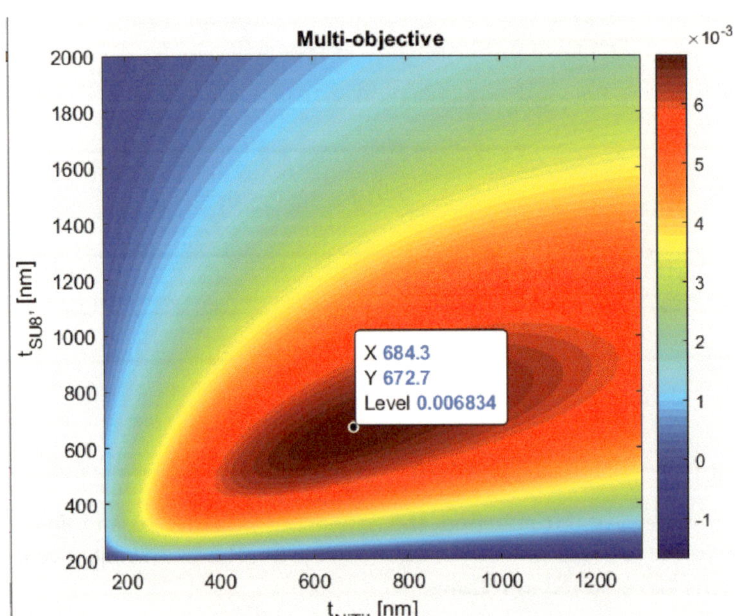

Figure 6.
Optimal solution for simultaneous multi-objective optimization of deflection and curvature radius.

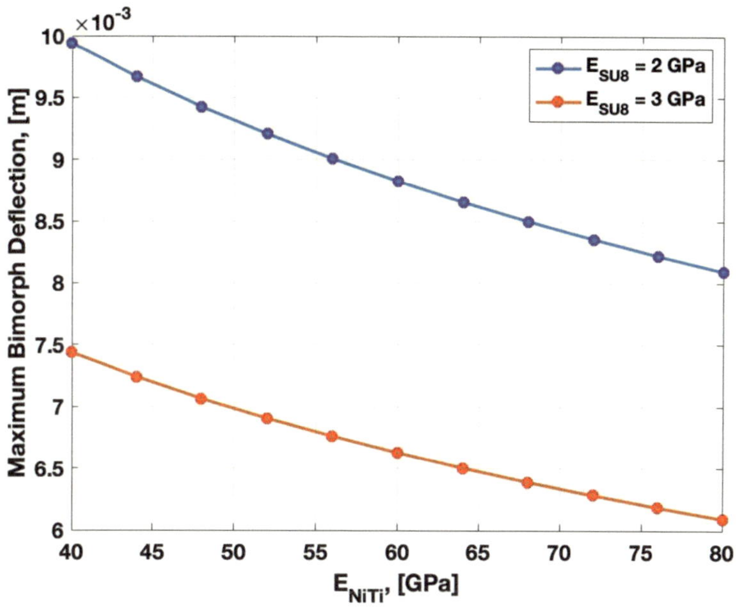

Figure 7.
Maximum bimorph deflection with a variation of Young's modulus of NiTi and SU-8 layer.

deflection. The optimal solution of multi-objective function as shown in **Figure 6** has a larger t_{NiTi}.

Once we have established the optimal objective values for deflection and curvature, we perform a sensitivity analysis regarding the following variables, for which experimentally could be varied with relative ease. These thickness values x1 and x2, corresponding the NiTi and SU-8 thicknesses, can be changed by varying the spin speed for SU-8 coating: faster spins corresponding to thinner films of SU-8 and vice versa. For NiTi, longer sputter time would be used for thicker films and vice versa. Young's modulus can be varied by deposition conditions for NiTi and curing/baking temperatures and conditions for SU-8. To perform the sensitivity analysis for Objective 1, we keep fixed the optimal thickness for SU-8 and vary the NiTi thickness to see how it changes, and plot a function (as shown in **Figure** 7) and generate a table of values. Similarly, we keep fixed the optimal value of NiTi thickness and recovery stress, and plot the deflection over a range of SU-8 thicknesses.

4. Conclusions

In conclusion, an interesting optimization problem was identified whereby the deflection of shape memory MEMS bimorph actuator was maximized. Original calculations showed that reductions in the thickness of the bimorph layers would yield maximized deflections (for the simplest case assuming constant values of recovery stress in NiTi layer). In the literature, a more complex relationship among recovery stress and the NiTi thickness was identified. A curve fit to this data yielded a much more interesting optimization problem, which was solved graphically (contour plots) and using the Optimization Toolbox in MATLAB. Optimal NiTi and SU-8 thickness were determined to be for the case where SU-8 modulus was 2 GPa to be t_{NiTi} = 359 nm, $t_{SU\text{-}8}$ = 824 nm. After solving the single-objective optimization problem using fmincon, Excel solver, and a hand-coded algorithm, we formulated a second objective function to maximize curvature radius (i.e., to maximize the flatness of the beam because larger curvature radius is a flatter beam). We used fmincon to solve for the optimal values of NiTi and SU-8 to maximize the curvature radius. We determined that the objective functions were conflicting (i.e., there was clearly a tradeoff in order to satisfy both conditions simultaneously), and therefore suitable for multi-objective optimization. We formulated a multi-objective optimization method and solved it using fmincon. Finally, a parametric study or sensitivity analysis was performed pertaining to NiTi and SU-8 Young's modulus.

Acronyms and abbreviations

α_{NiTi}	CTE of NiTi layer
$\alpha_{SU\text{-}8}$	CTE of SU-8 layer
Bimorph	composite cantilever beam consisting of two materials with different Young's modulus and thickness
CTE	coefficient of thermal expansion
d	deflection of the *SMA MEMS* actuator (micrometer)
E_{NiTi}	elastic modulus of *NiTi* layer (GPa)
$E_{SU\text{-}8}$	elastic modulus of *SU-8* layer (GPa)
E'_{NiTi}	biaxial elastic modulus of *NiTi* layer (GPa)
$E'_{SU\text{-}8}$	biaxial elastic modulus of *SU-8* layer (GPa)
$\Delta\varepsilon$	strain differential arising from thermal processing and CTE mismatch
l	total length of the *SMA MEMS* actuator (micrometer)
MEMS	microelectromechanical system
ρ	curvature radius
σ_{rec}	recovery stress of the *SMA MEMS* actuator (MPa)
SMA	shape memory alloy (i.e., NiTi)
T_{NiTi}	thickness modulus of *NiTi* layer (nm)
$T_{SU\text{-}8}$	thickness modulus of *SU-8* layer (nm)
θ	correction factor term for location of neutral axis

Author details

Cory R. Knick
US Army Night Vision and Electronic Sensors Directorate, Ft. Belvoir, VA, USA

*Address all correspondence to: cory.r.knick.civ@mail.mil

References

[1] Lee HT, Kim MS, Lee GY, Kim CS, Ahn SH. Shape memory alloy (SMA)-based microscale actuators with 60% deformation rate and 1.6 kHz actuation speed. Small. 2018;**14**:e1801023

[2] Liu K, Cheng C, Cheng Z, Wang K, Ramesh R, Wu J. Giant-amplitude, high-work density microactuators with phase transition activated nanolayer bimorphs. Nano Letters. 2012;**12**: 6302-6308

[3] Leong TG, Randall CL, Benson BR, Bassik N, Stern GM, Gracias DH. Tetherless thermobiochemically actuated microgrippers. Proceedings of the National Academy of Sciences of the United States of America. 2009;**106**: 703-708

[4] Ohnishi T, Sugimura A. Optical actuation of micromirrors fabricated by the micro-origami technique. Applied Physics Letters. 2009;**83**

[5] Ritt G, Eberle B. Automatic laser glare suppression in electro-optical sensors. Sensors (Basel). 2015;**15**: 792-802

[6] Murrer JRL, Goodwin SH, Dausch DE, Solomon SL, Lamvik MK. Electrostatic artificial eyelid actuator as an analog micromirror device. In: Proceedings SPIE 5785, Technologies for Synthetic Environments: Hardware-in-the-Loop Testing. Vol. 5785. 2005. p. 59

[7] Malachowski K, Jamal M, Jin Q, Polat B, Morris CJ, Gracias DH. Self-folding single cell grippers. Nano Letters. 2014;**14**:4164-4170

[8] Ainslie K, Knick C, Smith G, Li J, Troxel C, Mehta A, et al. Controlling shape memory effects in NiTi thin films grown on Ru seed layer. Sensors and Actuators A: Physical. 2019;**294**:133-139

[9] Choudhary N, Kaur D. Shape memory alloy thin films and heterostructures for MEMS applications: A review. Sensors and Actuators A: Physical. 2016;**242**:162-181

[10] Kotnur VG, Tichelaar FD, Fu WT, De Hosson JTM, Janssen GCAM. Shape memory NiTi thin films deposited on polyimide at low temperature. Surface and Coatings Technology. 2014;**258**: 1145-1151

[11] Sanjabi S, Sadrnezhaad SK, Barber ZH. Sputter alloying of Ni, Ti and Hf for fabrication of high temperature shape memory thin films. Materials Science and Technology. 2013; **23**:987-991

[12] Koker MKA, Schaab J, Zotov N, Mittemeijer EJ. X-ray diffraction study of the reverse martensitic transformation in NiTi shape memory thin films. Thin Solid Films. 2013;**545**: 71-80

[13] König D, Buenconsejo PJS, Grochla D, Hamann S, Pfetzing-Micklich J, Ludwig A. Thickness-dependence of the B2–B19 martensitic transformation in nanoscale shape memory alloy thin films: Zero-hysteresis in 75 nm thick Ti51Ni38Cu11 thin films. Acta Materialia. 2012;**60**:306-313

[14] Chung CY, Chan PM. NiTi shape memory alloy thin film micro-cantilevers array. Thin Solid Films. 2011;**519**:5307-5309

[15] Annadurai A, Manivel Raja M, Prabahar K, Kumar A, Kannan MD, Jayakumar S. Stress analysis, structure and magnetic properties of sputter deposited Ni–Mn–Ga ferromagnetic shape memory thin films. Journal of Magnetism and Magnetic Materials. 2011;**323**:2797-2801

[16] Satoh G, Birnbaum A, Yao YL. Annealing effect on the shape memory

properties of amorphous NiTi thin films. Journal of Manufacturing Science and Engineering. 2010;**132**:051004

[17] Rao J, Roberts T, Lawson K, Nicholls J. Nickel titanium and nickel titanium hafnium shape memory alloy thin films. Surface and Coatings Technology. 2010;**204**:2331-2336

[18] Ounaies Z, Mandepudi SK, Li J, Ackler HD. Processing and characterization of composite shape memory alloy (SMA) thin film structures for microactuators. Behavior and Mechanics of Multifunctional Materials and Composites. 2010;**7644**: 76440N

[19] Martins RMS, Schell N, Reuther H, Pereira L, Mahesh KK, Silva RJC, et al. Texture development, microstructure and phase transformation characteristics of sputtered Ni–Ti shape memory alloy films grown on TiN< 111>. Thin Solid Films. 2010;**519**: 122-128

[20] Sanjabi S, Barber ZH. The effect of film composition on the structure and mechanical properties of NiTi shape memory thin films. Surface and Coatings Technology. 2010;**204**: 1299-1304

[21] Wang X. Crystallization and martensitic transformation behavior of NiTi shape memory alloy thin films [PhD thesis]. Harvard: School of Engineering and Applied Sciences; 2007

[22] Wibowo E, Kwok CY. Fabrication and characterization of sputtered NiTi shape memory thin films. Journal of Micromechanics and Microengineering. 2006;**16**:101-108

[23] Fu YQ, Zhang S, Wu MJ, Huang WM, Du HJ, Luo JK, et al. On the lower thickness boundary of sputtered TiNi films for shape memory application. Thin Solid Films. 2006;**515**: 80-86

[24] Tomozawa M, Kim HY, Miyazaki S. Microactuators using R-phase transformation of sputter-deposited Ti-47.3Ni shape memory Alloy thin films. Journal of Intelligent Material Systems and Structures. 2006;**17**:1049-1058

[25] Getchel DJ, Savage RN. Fabrication and composition control of NiTi sape memory thin films for microactuators. Materials Research Society Symposium Proceedings. 2005;**875**:O6.4.1-O6.4.3

[26] Ishida A, Sato M, Tabata O, Yoshikawa W. Shape memory thin films formed with carrousel-type magnetron sputtering apparatus. Smart Materials and Structures. 2005;**14**:S216-S222

[27] Liu YS, Xu D, Jiang BH, Yuan ZY, Houtte PV. The effect of crystallizing procedure on microstructure and characteristics of sputter-deposited TiNi shape memory thin films. Journal of Micromechanics and Microengineering. 2005;**15**:575-579

[28] Ho KK, Carman GP, Jardine PA. Bimorphic, cmpositionally graded, sputter-deposited. Thin Film Shape Memory Alloys. 2004. US 6,689,586 B2

[29] Tingbin Wu BJ, Qi X, Liu Y, Xu D, Wang L. Residual stress of TiNi shape memory alloy thin films with (111) single-crystal silicon wafer. Materials Transactions. 2002;**43**:566-570

[30] Zohar Y, Wong M, Wang RX. Residual stress-loaded titanium–nickel shape memory alloy thin-film micro-actuators. Journal of Micromechanics and Microengineering. 2002;**12**:323-327

[31] Lehnert T, Crevoiserat S, Gotthardt R. Transformation properties and microstructure of sputter-deposited Ni-Ti shape memory alloy thin films. Journal of Materials Science. 2002;**37**

[32] Xiaobo Zhang YW, Miao X, Zhang C, Ding G. An electro-thermal SU-8 cantilever micro actuator based on

bimorph effect. In: Proceedings of the 2010 5th IEEE International Conference on Nano/Micro Engineered and Molecular Systems. Xiamen, China: IEEE; 2010. pp. 362-365

[33] Roch I, Bidaud P, Collard D, Buchaillot L. Fabrication and characterization of an SU-8 gripper actuated by a shape memory alloy thin film. Journal of Micromechanics and Microengineering. 2003;**13**:330-336

[34] Knick CR, Smith GL, Morris CJ, Bruck HA. Rapid and low power laser actuation of sputter-deposited NiTi shape memory alloy (SMA) MEMS thermal bimorph actuators. Sensors and Actuators A: Physical. 2019;**291**:48-57

[35] Knick CR, Sharar DJ, Wilson AA, Smith GL, Morris CJ, Bruck HA. High frequency, low power, electrically actuated shape memory alloy MEMS bimorph thermal actuators. Journal of Micromechanics and Microengineering. 2019;**29**:075005

[36] Knick CR, Morris CJ, Smith GL. Rapid and low power laser actuation of sputter-deposited NiTi shape memory alloy (SMA) MEMS thermal bimorph actuators. Sensors and Actuators A: Physical. 2019;**291**:48-57

[37] Mukesh Kumar S, Lakshmi MV. Applications of shape memory alloys in MEMS devices. International Journal of Advanced Research in Computer and Communication Engineering. 2013;**2**: 1122-1127

[38] Dahmardeh M, Mohamed Ali MS, Saleh T, Hian TM, Moghaddam MV, Nojeh A, et al. High-power MEMS switch enabled by carbon-nanotube contact and shape memory-alloy actuator. Physica Status Solidi. 2013; **210**:631-638

[39] Namazu T, Tashiro Y, Inoue S. Ti–Ni shape memory alloy film-actuated microstructures for a MEMS probe card. Journal of Micromechanics and Microengineering. 2007;**17**:154-162

[40] Grummon DS, Gotthardt R, LaGrange T. Planar extrinsic biasing of thin film shape memory MEMS actuators. MRS Online Proceedings Library Archive. 2003;**741**:287-292

[41] Bergmann NW, Xu D, Cai B, Ding G, Zhou Y, Yu A, et al. Novel micropump actuated by thin film shape memory alloy. In: Electronics and Structures for MEMS. Queensland, Austrailia: SPIE; 1999. pp. 369-375

[42] Wolf RH, Heuer AH. TiNi (shape memory) films on Si for MEMS applications. Journal of Microelectromechanical Systems. 1995; **4**:206

[43] Fallon P. Gerratt A. Kierstead B, Robert W. Shape memory alloy and elastomer composite MEMS actuators. Technical Proceedings of 2008 NSTI Nanotechnology Conference and Tradeshow. Vol. 3. 2008

[44] Cole DP, Jin H, Lu W-Y, Roytburd AL, Bruck HA. Reversible nanoscale deformation in compositionally graded shape memory alloy films. Applied Physics Letters. 2009;**94**:193114

[45] Cole DP, Bruck HA, Roytburd AL. Nanoindentation studies of graded shape memory alloy thin films processed using diffusion modification. Journal of Applied Physics. 2008;**103**: 064315

[46] Klein CA, Miller RP. Strains and stresses in multilayered elastic structures: The case of chemically vapor-deposited ZnS/ZnSe laminates. Journal of Applied Physics. 2000;**87**(5): 2265-2272

Chapter 4

Advanced Carbon Functional Materials for Superior Energy Storage

Mujtaba Ikram, Sana Arbab, Huma Anwar, Arsalan Nadeem, Sidra Baber, Abdullah Khan Durrani, Muhammad Ikram, Muhammad Aamir Iqbal, Muhammad Umer Farooq and Asghari Maqsood

Abstract

In the developing world, energy crisis is the main reason for less progress and development. Renewable and sustainable energy may be of bright future for scientific lagging and low-income countries; further, sustainability through smart materials got a huge potential; so, hereby keeping in view the energy crisis which the developing world is facing for many decades, we are proposing to write a chapter project for obtaining energy through cheap, sustainable, and functional advanced carbon materials. Carbon materials are the future of energy storage devices because of their ability to store energy in great capacity. The graphene is a material with amazing properties like no band gap, which turns graphene a wonderful candidate for use in the photovoltaic. Shortly, this chapter will discuss how superior energy storage may be obtained through various routes like using pyrrolic (N5) and pyridinic (N6) doping in advanced carbon functional materials, or superior energy by KOH activation in carbon materials, or through carbonization in organic matter, respectively. Further, for the advanced carbon functional materials, the superior energy storage using pyrrolic (N5) and pyridinic (N6) doping, or KOH activation, or through carbonization will be discussed one by one for lithium ion batteries, supercapacitors, and relevant energy devices, respectively.

Keywords: renewable energy, graphene, physical properties, reduced graphene oxide, superior energy storage, chemical activation, supercapacitors, lithium ion batteries

1. Introduction

In the developing world, energy crisis is the main reason for less progress and development. Renewable and sustainable energy may be of bright future for scientific lagging and low-income countries; further, sustainability through smart materials got a huge potential; so, hereby keeping in view the energy crisis which developing world is facing for many decades, there exists a possibility for obtaining energy through cheap, sustainable, and smart carbon materials. The nanostructures which are made up of sp2-hybridized carbon such as fullerene, carbon nanotubes, and graphene remain pivot point of advanced carbon functional material research [1]. As we look into the reversible electrochemical strength keeping in view of their great electric conductivity, chemical and mechanical durability, and the custom fitted structures that can be developed from advanced carbon functional materials, carbon materials have been the most significant anode material for the Li-ion batteries [2]. Advanced carbon functional materials may include graphene, carbon nanotubes, and fullerenes, respectively. Moreover, CNTs and graphene with high explicit surface territories can store higher charge capacity by the adsorption/desorption of particles at the anode/electrolyte interface. The higher charge capacity is hence incredible subject to the qualities of the carbon material, for example, its pore structure, doping, and defects. In this way, the advancement of electrochemical capacity in carbon needs a judiciously planned structure. One issue identified with the utilization of CNTs and graphene for electrochemical storage is the agglomeration of the nanostructures that prompts diminished efficiency [3, 4]. There has been progressive demand for the electrochemical energy storage with high energy density and remarkable rate performance. Electrical double layer capacitors (EDLCs), additionally known as supercapacitors (SCs), have attracted a worldwide attention because of their long cycle life and high power density but comparatively low energy density of commercially available carbon-based SCs. Graphene has attracted an in-depth attention in energy storage applications relating to its distinctive options of high surface space, flexibility, chemical stability, and remarkable electrical and thermal conduction [5]. Energy storage capacity defines advanced energy technologies. Further, superior energy storage may be obtained through various routes like using pyrrolic (N5) and pyridinic (N6) doping in carbon materials, or superior energy by KOH activation in carbon materials, or through carbonization in organic matter, respectively. Further, energy storage using pyrrolic (N5) and pyridinic (N6) doping, or KOH activation, or through carbonization in organic matter will be discussed one by one.

2. Energy storage using pyrrolic/pyridinic nitrogen in advanced carbon functional materials

For energy storage, ideal structures may become complex due to agglomeration at nanolevel during carbon reconstruction process. The gathering of sp2-hybridized carbon nanostructures to frame a 3D arrangement can adequately hinder any agglomeration and sustain electrical properties of the building structures. Moreover, the 3D arrangement gives channels for particle relocation. Required carbon structures may be obtained through the freeze drying or aqueous treatment of graphene or CNT suspensions with the help of templates. With increasingly proficient electron move and a progressively powerful structure, a customized 3D carbon will give possible higher electrochemical charge capacity by interfacing the structure squares covalently rather than by van der Waals communications. A graphene with a SSA of ≈850 $m^2.g^{-1}$ has indicated astounding electrical and mechanical attributes for Li-ion batteries. Nitrogen doping may be used to increase electrochemical storage in the advanced functional carbon materials and consequently to increase the capacitance at the anode/electrolyte interface [4, 6]. At the edges of graphene, it is experimentally proved, in the electrolyte, that the nitrogen ions have benefited the wettability of solo carbon layers. Consequently, it would increase the overall capacitance, which may occur due to ideal Faradaic redox responses. For Li-ion storage capacity of the carbon (N-doped), it may provide extra dynamic sites, which helps adsorption (Li-ion), resulting in an enhanced gravimetric limit [6, 7]. Among the N-dopants in graphitic carbon, pyrrolic (N5), pyridinic (N6), and quaternary N, pyrrolic N, and pyridinic N are viewed as additional dynamic for electrochemical storage, while the graphitic structure is the most fragile among them. Also, topological imperfections, for example, di-vacancy and Hurler Stone-Ribs deserts in the graphitic carbon may give additional dynamic destinations to particle adsorption or on the other hand for charge move in electrochemical storage. Furthermore, various works have demonstrated the significant job of full-scale macropores and mesopores in the particle transportation for high power thickness or brilliant rate execution in supercapacitors or Li-ion batteries. Along these lines, a carbon for unrivaled Li-ion capacity would preferably have the accompanying highlights, a 3D arrangement which contains the mesopores for quick particle relocation and a covalently associated structure made of sp2-hybridized carbon for high electrical conductivity [6, 8]. Further, a finely tuned arrangement of dopant molecules and deformities for increasingly dynamic locales is another requirement. To accomplish a carbon with these highlights, scientists picked a high-energy carbon nanostructure, C60, which is more receptive than CNTs and graphene, as the antecedent, and treated it with KOH at temperatures of 500–700°C [6]. The mentioned treatment helps changing the C60 molecules to carbon which contains enhanced nitrogen along with defects. Nitrogen which presents in the doped carbon consists of two types (pyrrolic and pyridinic). With 7.8% nitrogen-doped carbon, porous carbon (when used as anode) has shown 600 mA h g^{-1} (storage capacity), which occurs at 5 A g^{-1} for the Li-ion batteries [6]. A first-standard calculation has proposed that the unrivaled anode execution of the N-doped permeable carbon is intently identified with the bending of the carbon layers (graphenes) and the pyrrolic/pyridinic N-doping in the carbon. FCC structure has been obtained when agglomeration happens in the C60 molecules, which turns C60 into different carbons (permeable carbon) in KOH activation. While in path B, ammonia stream has been act to apply during the tempering, resulting in porous carbon which is highly N-doped. Meanwhile, scientists have proved that nanopores have been prepared in KOH activation, while quantum dots (of carbon) were fabricated without KOH activation [6, 8, 9]. It is believed that enhanced handling time (while keeping proportion of KOH and C60 lower) led to higher interfacial interactions and amends into a progressive structure at certain phase of activation. Likewise, it was discovered that the N-content expanded with activation temperature in the range 500–700°C; yet a further temperature increment to 800°C prompted an extremely low yield of N-aC60 tests. Further, systematic fabrication of nitrogen-doped carbon through activation of C60 molecules (route A indicates normal activation in argon flow, while route B indicates N doping in NH_3 flow) is as shown in **Figure 1** [6].

X-ray diffraction (XRD) analysis of C60, aC60, and N7.5%-aC60 revealed important information regarding structures, which further demonstrates that the C60 molecules have been totally rebuilt by KOH actuation which gave porous

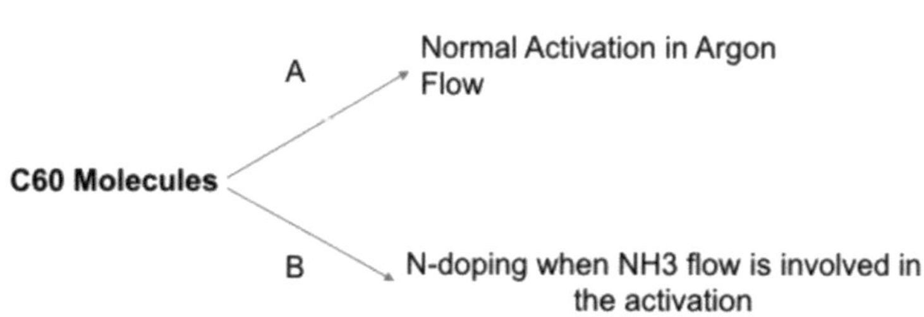

Figure 1.
Systematic fabrication of nitrogen-doped carbon through activation of C60 molecules (route A indicates normal activation in argon flow, while route B indicates N doping in NH_3 flow) [6].

carbon [6]. X-ray photoelectron spectroscopy (XPS) has been utilized to reveal significant constituents present in final product. FTIR results show that C–OH and C–O bonds are seen in both aC60 and N7.5%-aC60. Estimations utilizing the XPS information show that oxygen content increments from 1.3 at.% in C60 to 4.2 at.% in aC60 and to 9.5 at.% in N7.5%-aC60 samples, respectively. To additionally comprehend Li-ion battery's capacity, scientists explored the adsorption capacity of Li particles on graphene and C60 sections with and without nitrogen doping through atomic demonstrating. Two potential impacts, i.e., a curvature impact and an N-doping impact, have been thought to be important. For the former one, it has the adsorption capacity of a Li storage on a large portion of a C60 atom with edges immersed by H, i.e., $C_{30}H_{10}$, and a level graphene piece containing the equivalent number of carbon particles, i.e., $C_{30}H_{14}$, respectively [6, 10].

In synopsis, KOH activation has been utilized to totally convert C60 atoms to a 3D permeable carbon. In the porous carbon, the doping (of nitrogen) may additionally bring deformities and a large number of pores. The activation process may increase the doping level that depends upon activation conditions, resulting in a suitable storage capacity for Li-ion batteries [6]. KOH activation gave the bended layer structure also; further, N-doping, particularly pyrrolic nitrogen, has added to the high Li-ion stockpiling limit in the carbon.

3. Energy storage through sponge-templated activation of advanced carbon functional materials

There has been progressive demand for the electrochemical energy storage devices with high energy density and remarkable rate performance. Electrical double layer capacitors (EDLCs), additionally known as supercapacitors (SCs), have attracted a worldwide attention because of their long cycle lifespan and really high power density, but comparatively lower energy density has considerably limited the applications of the carbon-based supercapacitors [11, 12]. Graphene has enormous energy applications relating to distinctive physical properties of chemical stability, flexibility, and remarkable electrical conduction. Scientists have found that during wet chemical techniques, graphene platelets may tend to agglomerate, thus resulting in lower surface areas than the theoretical worth of 2630 m^2g^{-1}. Strategies are developed to assemble graphene-based platelets to 3D structures so as to stop the restacking of platelets for high surface areas while maintaining the intrinsic conduction of platelets [13]. For instance, scientists have demonstrated a graphene film which is 3D having macropores, within which PMMA functions as the main template. Scientists have prepared TiO_2-rGO sheets employing PU sponge (which is porous) as a model for the photoelectrochemical reaction of ethanol. A recent study reported that compressible all-solid-state SCs supported polyaniline-SWCNTs-sponge electrodes, in which sponge provides squeezability and polyaniline offers pseudo capacitance. As an economical model, sponge is ready to act as a wonderful support for the assembly of nanostructures for SC electrodes. It is found that graphene structures obtained using sponge templating may have lower surface area, which leads to lower interfacial interactions and limited SC performance. It is experimentally proved that graphene-based capacitors are of lower cathode thickness even if they possess higher surface areas, which led to lower performance of ultimate devices. Scientists have regenerated graphene colloidal gel films with a high packing density up to 1.33 gcm^{-3} what's more at the same time noticeable capacitive exhibitions (209 Fg^{-1}) and 261 Fcm^{-3} in organic electrolytes. The layer like stacking of graphene platelets may deteriorate the surface because of particle channels within the direction perpendicular to the layers [11, 14].

Figure 2.
Preparation of 3D carbon through sponge KOH activation via GO loading [11].

Chemical activation is an efficient methodology to make pores, e.g., within the preparation of activated carbons (ACs). Among various chemical activation strategies, KOH activation has been considered as an accustomed method. Recently, scientists have fabricated a porous carbon through chemical activation of GO; it is found that the selected capacitance of 166 Fg^{-1} has been demonstrated. Transforming low-thickness carbons to templates to get valuable thickness but yet with a moderately high porosity and high electrical conductivity is required for high-performance SCs [11, 13, 14].

Scientists have developed a carbon which was obtained using sponge templating followed by chemical activation (KOH activation) of GO. It is demonstrated that the GO platelets gather around the sponge's backbone. Meanwhile, KOH activation goes within the PU sponge which helps to create pores using temperature treatments, which may result in a conductive carbon. Using fabricated carbon for energy storage in two-electrode and three-electrode configurations, it has shown nearly perfect energy storage behavior, which may lead to acceptable superior power density [11, 15].

Preparation of the 3D carbon (aPG-10) is shortly illustrated within the **Figure 2** as shown [11].

After chemical activation of PU/GO mixture, the dried PU/GO/KOH mixture was toughened at 9000°C for 2 hours in inert gas flow, and hence subjected to drying, resulting in a final sample named as aPG-10 (final product) [11, 14].

The exhibition of the aPG-10 as an anode material for supercapacitors was evaluated using cycle voltammetry (CV) and galvanostatic charge-discharge (GCD) curves, respectively. In a three-electrode configuration, the execution of aPG-10 anode coated on a shiny carbon has been designed with 1.0 MH_2SO_4 as electrolyte. The particular capacitance determined from the charge/discharge curves at a current thickness of 5 Ag^{-1} is 401 Fg^{-1}, which stays 349 Fg^{-1} at a current density of 100 Ag^{-1}. Once aPG-10 was tried in two-electrode configuration, a particular capacitance of 227 Fg^{-1} at 5 Ag^{-1} was acquired [11].

In summary, a graded porous carbon was obtained through sponge KOH activation via GO loading [11, 14, 15]. The sponge has filled in as a proficient layout to pull KOH into its backbone, which helps to create large number of pores, in order to give highly porous carbon. Meanwhile, low sheet obstruction, high BET surface region, and much acceptable conductor density are accomplished based on the carbon activation through KOH which introduces sort of nitrogen which covers defects from graphitic lattice. Every one of these benefits leads to the great electrochemical execution of carbon terminals [11, 13, 15].

4. Energy storage through functional porous carbon obtained from frozen tofu (organic matter)

Frozen tofu is a source of carbon and nitrogen [16]. By using one-step carbonization activation method, it can be converted into oxygen-doped carbon and nitrogen-doped carbon, respectively. By one-step carbonization, sponge-like carbon (co-doped) has a maximum surface area of 3134 m^2g^{-1}. High volumes of mesopores (1.11 $cm^3 g^{-1}$) and micropores (0.71 $cm^3 g^{-1}$) are present in this hierarchical (graded) porous carbon [16, 17]. Scientists have discovered that this acquired carbon is used to make supercapacitors which acts as electrodes, in 1 M aqueous electrolyte sulfuric acid; it has a remarkable capacitance of 243 Fg^{-1} (evaluated at 0.1 Ag^{-1}); after 10,000 cycles, it has a capacitance retentiveness of 93% at 10 Ag^{-1}. In BMIMBF4 (1-butyl-3-methylimidazolium tetra fluoroborate) liquid ion electrolyte, the said carbon shows a precise capacity of 170 Fg^{-1} (assessed in 1 Ag^{-1}) along with a valuable effectiveness (135 F g^{-1} at 20 Ag^{-1}) ensuring the power density of 72 W h kg^{-1} (at 889 W kg^{-1}). A carbon supercapacitor (derived from frozen tofu) can comfortably drive 25 light-transmitting diodes in excess of 2 minutes [16]. The fabrication flow chart for porous carbon through one-step carbonization is as shown in **Figure 3** [16].

Further, systematic steps involved to get porous carbon through carbonization are freeze drying, KOH activation, temperature treatment, and carbonization, respectively, as are shown in flow chart diagram (**Figure 3**). Many energy storage devices are available worldwide but because of potential applications, electrochemical appliances such as lithium-ion batteries (LIBs) and supercapacitors have an appreciable fascination [16, 17]. On the one hand, supercapacitors have active charging/discharge performance and also a fine power density greater than 10 k W kg^{-1}; on the other hand, LIBs have a great energy density (usually 100–200 W h kg^{-1}), yet it has a longer charging time. Two mechanisms mainly used by supercapacitors in order to store energy are as follows [17, 18]:

- Pseudocapacitive electrodes store ions established at the electrode-electrolyte interface for immediate Faradaic reactions.
- Dual layer electrodes with capacitive electricity preserve energy through the desorption and adsorption of ions on a large field of spongy (porous) materials.

The lithium ions in LIBs are moved between anode and cathode, which results in lithium ions storage or discharge through distinct means according to the materials of the electrode. Because of many properties such as chemical stability, pattern porosity, and high electrical conductivity, for both commercial LIBs (e.g., graphite) and supercapacitors (e.g., activated carbon), carbon materials were selected as effective materials [16, 19]. It has been determined that macropores act as ion-buffering storage in porous carbon materials in supercapacitors and mesopores contribute channels for the transport of ions to micropores, where they were ultimately deposited. In the LIB (graphite) anodes, the intercalated LiC compound is the result of the complete intercalation of lithium ions that limits the reversible Li-ion storage potential for graphite (approximately 372 mA hg^{-1}) [16]. In addition,

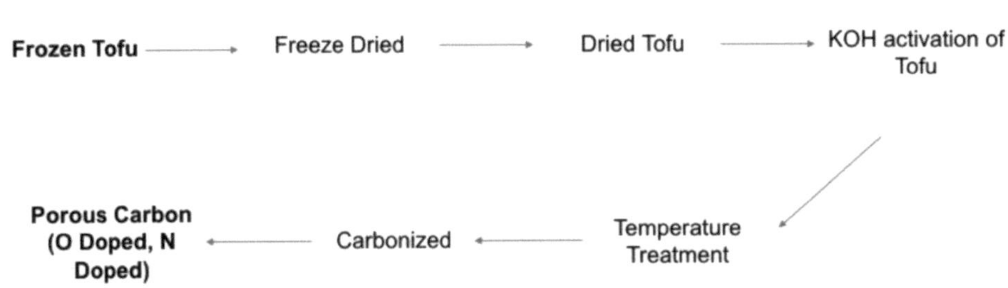

Figure 3.
The fabrication flow chart for porous carbon through one-step carbonization [16].

permeable carbon was too employed in the form of an electromagnetic scaffold or including the electrode in the LIBs, since the spongy (porous) arrangement is known to increase the contact areas of the electrode-electrolyte and decrease the length of the path for ions/electrons transport, which leads to an improvement in the transit kinetics and accordingly the energy density [16, 20].

Power density of industrial supercapacitors is usually lower than 6 W h kg^{-1} and that is beyond from long-term electronic equipment requirements [20]. However, the efficiency and performance of the LIB anodes must also be further enhanced. With regard to the two applications, spongy carbon of great electrical drivability and a customized 3D design is required. In the previous couple of centuries, by using different strategies like self-assembly, activation, and templating, several porous carbon materials have been researched [16, 18, 20]. Among available methods, an efficient method is activation, to increase the surface area of carbon by making several micropores. In supercapacitors, activated carbon is commercially employed due to its great specific surface area (greater than 2000 m^2g^{-1}). Activation will make effective manufacturing possible through treating of precursors of nanostructured carbon in order to get novel carbon compounds [21]. For instance, platelets of graphene may be entirely transformed to a three-dimensional porous (spongy) carbon with a SSA of up to 3100 m^2 g^{-1} and distribution of pore size is 0.6–5 nm by microwave-exfoliated graphite oxide activation of KOH [16, 21]. By KOH activation of C micro tubes, scientists have documented a porous (sponge-like) carbon made up of macropores (numerous microns in size) and micropores (0.47 nm in size). The conductivity and wettability of carbon compounds can be enhanced by adequate heteroatom doping. Additionally, nitrogen is a donor molecule of electrons, as well as in a graphite matrix, they have the ability to promote conductivity of the n-type. As the LIB anode, carbon N-doping has proved itself to support the Li injection because of the hybridization of the long pair of nitrogen electrons with carbon π electrons. In recent times, by activation of C with KOH in an ammonia environment, scientists have achieved porous N-doped carbon (7.5 wt %); the carbon was found to have a reversible power of 1900 mA h g^{-1} at 0.1 Ag^{-1}; furthermore, after 800 cycles (at 2 Ag^{-1}), the capacity is 600 mA hg^{-1}, respectively.

Porous carbon content obtained from renewables can be more environmentally sustainable relative to porous carbon products acquired from wood, polymer blends, tar, and other resources. Researchers have demonstrated that porous carbons can be obtained from various biomass sources such as rice husks, fungi, water bamboo, and rice straw for energy storage applications. For instance, researchers have informed that the willow catkin could be transformed into a cross-linked polymer carbon laminate co-doped with sulfur (S) and N by one-step pyrolysis-activation synthesis. In 1 M Na_2SO_4, the carbon demonstrated a remarkable electrochemical efficiency with a specific capacitance of 298 F g^{-1} at 0.5 A g^{-1} and magnificent cycling endurance at 5 A g^{-1} after 10,000 cycles with only 2% capacitance loss. Tofu consists of moisture, carbohydrates, proteins, and trace concentration of minerals; it is an available resource and is considered as a renewable fuel for nitrogen and carbon [16, 22]. Not long ago, scientists have stated the molten salt synthesis of strongly (N-doped) porous carbon which may be obtained from tofu, in which LiCl/KCl (45/55 in weight) is a eutectic mixture (which functioned as the activator) was used as the solvent to dilute $LiNO_3$. The collected carbon (N-content: 4.72 wt%, density: 0.84 g cm^{-3}, SSA: 1202 m^2 g^{-1}) in a 1 M Na_2SO_4 symmetric supercapacitor, showed 73.2 F g^{-1} capacitance (at 0.2 A g^{-1}). The porous carbon (obtained from frozen tofu) is known to be suitable anode specimens for LIBs, because of their bigger surface areas, porous hierarchical structures, and heteroatom doping. In short, we can easily obtain doped N porous hierarchical carbon from frozen tofu, using single-step carbonization activation. Frozen tofu is environment-friendly, cheap, and extendable biomaterial precursor. It has showed significant SSA of 3134 $m^2 g^{-1}$ and significant pore diameter of 1.82 $m^3 g^{-1}$ on the activation conditions, and this is better than traditional biomass-originated active carbon products. In 1 M H_2SO_4, supercapacitors based on porous carbon (from frozen tofu) revealed 243 F g^{-1} specific capacitance, and in BMIMBF4/AN, it showed an extraordinary power density of about 72 W h kg^{-1} at an ordinary 889 W kg^{-1} energy density. Such carbonization procedure offers a potentially helpful strategy from abundant supportable resources to design carbon electrode materials with supreme execution for supercapacitors and LIBs, respectively [6, 11, 16, 22].

5. Conclusions

In synopsis, this chapter has explained how KOH activation has been utilized to convert C60 molecules to a 3D carbon, while at the same time, doping with pyridinic and pyrrolic nitrogen has huge impact in energy storage capacity. Meanwhile, a graded porous carbon was obtained through sponge KOH activation via GO loading. Further, physical properties such as low sheet obstruction, high BET surface, and higher conductivities are accomplished based on the carbon activation through KOH, which introduces sort of nitrogen which covers defects from graphitic lattice eventually. Every one of these benefits leads to the great electrochemical execution of carbon terminals. Further, carbonization procedure offers a potentially helpful strategy from abundant supportable resources to design carbon electrode materials with supreme storage for supercapacitors and LIBs, respectively.

Conflict of interest

The authors have declared no “conflict of interest.”

DOI: http://dx.doi.org/10.5772/intechopen.93355

Author details

Mujtaba Ikram[1]*, Sana Arbab[1], Huma Anwar[1], Arsalan Nadeem[1], Sidra Baber[1], Abdullah Khan Durrani[1], Muhammad Ikram[2], Muhammad Aamir Iqbal[3], Muhammad Umer Farooq[4] and Asghari Maqsood[5]

1 Applied Physics Lab, Institute of Chemical Engineering and Technology (ICET), University of the Punjab (PU), Lahore, Pakistan

2 Department of Physics, Government College University (GCU), Lahore, Pakistan

3 Centre for Excellence in Solid State Physics, University of the Punjab, Lahore, Pakistan

4 Department of Physics, University of Education, Faisalabad Campus, Pakistan

5 Nanoscale Laboratory, Department of Physics, Air University, Islamabad, Pakistan

*Address all correspondence to: mujtaba.icet@pu.edu.pk

References

[1] Xia Y, Xiong Y, Lim B, Skrabalak SE. Shape-controlled synthesis of metal nanocrystals: Simple chemistry meets complex physics? Angewandte Chemie International Edition. 2009;**48**:60-103. DOI: 10.1002/ anie.200802248

[2] Liu ZB, Xu YF, Zhang XY, Zhang XL, Chen YS, Tian JG. Porphyrin and fullerene covalently functionalized graphene hybrid materials with large nonlinear optical properties. The Journal of Physical Chemistry B. 2009;**113**:9681-9686. DOI: 10.1021/ jp9004357

[3] Markandan K, Chin JK, Tan MT. Recent progress in graphene based ceramic composites: A review. Journal of Materials Research. 2017;**32**:84-106. DOI: 10.1557/ jmr.2016.390

[4] Neto AC, Guinea F, Peres NM, Novoselov KS, Geim AK. The electronic properties of graphene. Reviews of Modern Physics. 2009;**81**:109. DOI: 10.1103/RevModPhys.81.109

[5] Kim KS, Zhao Y, Jang H, Lee SY, Kim JM, Kim KS, et al. Large-scale pattern growth of graphene films for stretchable transparent electrodes. Nature. 2009;**457**:706-710. DOI: 10.1038/ nature07719

[6] Ziqi T, Kun N, Guanxiong C, Wencong Z, Zhuchen T, Mujtaba I, et al. Incorporating pyrrolic and pyridinic nitrogen into a porous carbon made from C60 molecules to obtain superior energy storage. Advanced Materials. 2017;**29**:1603414. DOI: 10.1002/ adma.201603414

[7] Novoselov KS, Geim AK, Morozov SV, Jiang D, Zhang Y, Dubonos SV, et al. Electric field effect in atomically thin carbon films. Science. 2004;**306**:666-669. DOI: 10.1126/ science.1102896

[8] Bhowmik K, Chakravarty A, Bysakh S, De G. γ-Alumina nanorod/ reduced graphene oxide as support for poly (ethylenimine) to capture carbon dioxide from flue gas. Energy Technology. 2016;**4**:1409-1419. DOI: 10.1002/ ente.201600186

[9] Huang D, Yang Z, Li X, Zhang L, Hu J, Su Y, et al. Three-dimensional conductive networks based on stacked SiO_2@graphene frameworks for enhanced gas sensing. Nanoscale. 2017;**9**:109-118. DOI: 10.1039/ C6NR06465E

[10] Lv L, Huang L, Zhu P, Li G, Zhao T, Long J, et al. SiO_2 particle-supported ultrathin graphene hybrids/polyvinylidene fluoride composites with excellent dielectric performance and energy storage density. Journal of Materials Science: Materials in Electronics. 2017;**28**: 13521-13531. DOI: 10.1007/ s10854-017-7191-0

[11] Jin X, Ziqi T, Wencong Z, Guanxiong C, Shuilin W, Yuan Z, et al. A hierarchical carbon derived from sponge-templated activation of graphene oxide for high-performance supercapacitor electrodes. Advanced Materials. 2016;**28**(26):5222-5228. DOI: 10.1002/adma.201600586

[12] Qian R, Yu J, Wu C, Zhai X, Jiang P. Alumina-coated graphene sheet hybrids for electrically insulating polymer composites with high thermal conductivity. RSC Advances. 2013;**3**:17373-17379. DOI: 10.1039/ c3ra42104j

[13] Akhtar F, Rehman Y, Bergström L. A study of the sintering of diatomaceous earth to produce porous ceramic monoliths with bimodal porosity and high strength. Powder Technology. 2010;**201**:253-257. DOI: 10.1016/j. powtec.2010.04.004

[14] Dervin S, Lang Y, Perova T, Hinder SH, Pillai SC. Graphene oxide reinforced high surface area silica aerogels. Journal of Non-Crystalline Solids. 2017;**465**:31-38. DOI: 10.1016/j.jnoncrysol.2017.03.030

[15] Lwin S, Li Y, Frenkel AI, Wachs IE. Nature of WOx sites on SiO_2 and their molecular structure–reactivity/selectivity relationships for propylene metathesis. ACS Catalysis. 2016;**6**:3061-3071. DOI: 10.1021/acscatal.6b00389

[16] Xumei S, Jianglin Y, Fei P, Jin X, Tao C, Xiangyang W, et al. Hierarchical porous carbon obtained from frozen tofu for efficient energy storage. New Journal of Chemistry. 2018;**42**:12421-12428. DOI: 10.1039/C8NJ01788C

[17] Ciriminna R, Fidalgo A, Pandarus V, Béland F, Ilharco LM, Pagliaro M. The sol–gel route to advanced silica-based materials and recent applications. Chemical Reviews. 2013;**113**:6592-6620. DOI: 10.1021/cr300399c

[18] Ramezanzadeh B, Haeri Z, Ramezanzadeh M. A facile route of making silica nanoparticles-covered graphene oxide nanohybrids (SiO_2-GO); fabrication of SiO_2-GO/epoxy composite coating with superior barrier and corrosion protection performance. Chemical Engineering Journal. 2016;**303**:511-528. DOI: 10.1016/j.cej.2016.06.028

[19] Zhibin L, Xue-Feng Y, Zhu P-K. Recent advances in cell-mediated nanomaterial delivery systems for photothermal therapy. Journal of Materials Chemistry B. 2018;**6**:1296-1311. DOI: 10.1039/C7TB03166A

[20] Islam A, Mukherjee B, Sribalaji M, Rahman OA, Arunkumar P, Babu KS, et al. Role of hybrid reinforcement of carbon nanotubes and graphene nanoplatelets on the electrical conductivity of plasma sprayed alumina coating. Ceramics International. 2017;**44**:4508-4511. DOI: 10.1016/j.ceramint.2017.12.021

[21] Huang L, Zhu P, Li G, Lu DD, Sun R, Wong C. Core–shell SiO_2@RGO hybrids for epoxy composites with low percolation threshold and enhanced thermo-mechanical properties. Journal of Materials Chemistry A. 2014;**2**:18246-18255. DOI: 10.1039/C4TA03702B

[22] Huang D, Li X, Wang S, He G, Jiang W, Hu J, et al. Three-dimensional chemically reduced graphene oxide templated by silica spheres for ammonia sensing. Sensors and Actuators B: Chemical. 2017;**252**:956-964. DOI: 10.1016/j.snb.2017.05.117

Chapter 5

Progress, Challenges and Opportunities in Divalent Transition Metal-Doped Cobalt Ferrites Nanoparticles Applications

Oana Cadar, Thomas Dippong, Marin Senila and Erika-Andrea Levei

Abstract

Engineered nanomaterials with tailored properties are highly required in a wide range of industrial fields. Consequently, the researches dedicated to the identification of new applications for existing materials and to the development of novel promising materials and cost effective, eco-friendly synthesis methods gained considerable attention in the last years. Cobalt ferrite is one of the nanomaterials with a wide application range due to its unique properties such as high electrical resistivity, negligible eddy current loss, moderate saturation magnetization, chemical and thermal stability, high Curie temperature and high mechanical hardness. Moreover, its structural, magnetic and electrical properties can be tailored by the selection of preparation route, chemical composition, dopant ions and thermal treatment. This chapter presents the recent applications of nanosized cobalt ferrites doped or co-doped with divalent transition ions such as Zn^{2+}, Cu^{2+}, Mn^{2+}, Ni^{2+}, Cd^{2+} obtained by various synthesis methods in ceramics, medicine, catalysis, electronics and communications.

Keywords: cobalt ferrite, doping, transition metal, nanoparticles, applications

1. Introduction

In the last decades, a wide range of nanomaterials were developed for applications in the field of magnetic recording and imaging, data and energy storage, refrigeration, electrical and communication devices, environmental depollution, catalysis, ceramics and pigments, sensors, medicine, etc. [1–8].

Spinel ferrite nanoparticles (NPs) with a general formula of MFe_2O_4 (M = divalent metal ion such as Mn, Cu, Co, Ni and Zn) have open a new and exciting research field because of their unique structural, magnetic, optical and electric properties. Among them, nanosized Co ferrite received a lot of attention due to its remarkable mechanical and chemical stability, wear resistance, dielectric character, electrical conductivity and excellent magnetic properties such as high coercivity (H_C) and moderate saturation magnetization (M_S), high Curie temperature (T_C) and large magnetocrystalline anisotropy [8, 9]. It also possesses some other unique characteristics such as good catalytic performance, small particle size, large surface area, narrow optical bandgap, non-toxicity and low production costs [10]. Depending on the composition, synthesis method and thermal treatment, Co ferrites have different structural, optical, electric, magnetic and biomedical properties [4, 11].

The doping with different elements adjusts the properties of ferrites by changing the structure, crystallinity and elements distribution among tetrahedral (A) and octahedral (B) sites [10]. The dopant amount, valency, size and site preferences define the structural, electrical and magnetic properties of doped Co ferrites [12]. The Co ferrite has an inverse spinel structure, but the doping with divalent metal ions could changes its structure into normal spinel [13]. The change from normal to inverse spinel depends also on the ratio between Co and the dopant ion. Co ferrite is a hard-magnetic material, but it may be softened by doping with non-magnetic ions [14].

The NPs embedding into or coating with insulating matrix such is silica could also considerably change the properties of the obtained nanocomposites (NCs), as the silica network can limit the particles growth, act as a buffer to protect the nanoparticles from mechanical stress and minimize the surface roughness and spin disorder [1]. Thus, by selecting the dopant, synthesis route and parameters, nano-sized doped Co ferrites with tailored properties were produced for a broad range of applications [9, 12].

Various methods for the preparation of undoped and doped Co ferrite NPs have been reported such as: sol-gel, co-precipitation, polymerized complex, hydrothermal, thermal plasma methods, sol-gel, solvothermal, thermal decomposition, ultrasonic cavitation, mechanical alloying, ball milling, pulsed laser deposition, reverse micelle, micro-emulsion, microwave assisted synthesis, thermal decomposition, electrochemical and auto-combustion [6–8, 15, 16]. Although by using these methods, the required sizes and microstructures can be achieved, they are difficult to apply on large scale due to their complex and expensive procedures, long reaction times, high reaction temperatures, hazardous reagents and by-products and potential harm to the environment [16]. Among different synthesis procedures, the sol-gel method and post-annealing treatment is one of the simplest, feasible and most effective routes that produces high purity NCs at low temperatures and permits a good control over the particle size, morphology and chemical composition [1].

In this review, we summarize the recent, significant developments related to applications of Co ferrite NPs doped with divalent transitional metals in different fields based on their coloristic, magnetic, antimicrobial, biological, catalytic and dielectric properties.

2. Applications of coloristic properties

The conventional coloring method of ceramics is based on the addition of pigments or dyes. Generally, the ceramic pigments are crystalline inorganic transition metal oxides powders with high chemical and thermal stability. They are soluble in glasses and glazes at high temperatures, have high tinting strength, high refractive index, low abrasive strength, and acid and alkali resistance. The color of pigments is determined by the presence of chromophore ions (usually transition metals) in an inert matrix (oxidic systems) or these ions may be part of their own matrix, as in the case of ferrites. The nano-pigments (nanoparticles dispersed in an organic vehicle) have a wide range of applications due to their high surface coverage, sharp spectral features, high scattering and uniform dispersion [17, 18]. The color performance of conventional ceramic pigments depends on the coloring efficiency and dissolution kinetics in the ceramic matrix, that are expected to be improved by small particle sizes. Magnetic inorganic pigments are also used in high-tech applications such as radar absorbing materials in military applications [2, 3].

Co ferrite is a black pigment widely used in the ceramic industry due to its excellent properties such as chemical and thermal stability [17]. There are only few studies reporting the use of divalent transition metal doped Co ferrites NPs as pigments. Sol-gel synthesis followed by post annealing pathway was used to obtain Zn doped Co ferrites ($Co_{0.3}Zn_{0.7}Fe_2O_4$ and $Co_{0.7}Zn_{0.3}Fe_2O_4$) embedded in SiO_2 matrix in order to be used as dark gray to black color ceramic pigments [19]. The coloring properties of the $Zn_{0.6}Co_{0.4}Fe_2O_4$ NPs were tested by embedding in opaque and transparent tile glazes, and their application on ceramic tile. For pigments, the cartesian coordinates confirmed the dark gray color, that becomes almost black in bulk, while by dispersion in glazes the dark pigment present a bright gray color [20].

3. Applications of magnetic properties

The magnetic properties of nanomaterials, associated with the spin of electrons, make them suitable for various applications in biotechnology, telecommunications and electronic industries. The magnetic properties of cubic spinel ferrites depend upon their metallic composition, particle size and cationic distribution between tetrahedral (A) and octahedral (B) sites [7, 21]. In case of magnetic NPs, the presence of large number of atoms at the surface due to high-surface-to-volume ratio and finite size effects result in several interesting and superior properties compared to bulk materials [22].

Co ferrite is well-known magnetic nanomaterial with high H_C and M_S. The M_S, H_C, T_C and anisotropy constant (K) of Co ferrite decrease by doping with non-magnetic ions decrease the hard-magnetic behavior and change the ferromagnetic to superparamagnetic behavior, leading to various applications [14, 23, 24].

Zn doped Co ferrites are soft magnetic materials with good chemical stability and high H_C [25]. The H_C, remanence magnetization (M_R) and squareness ratio (M_R/M_S) decrease by doping, as a result of the anisotropic nature of spinel Co-Zn ferrites and the non-magnetic moment of Zn^{2+} ions. The M_S values increase with the increasing content of dopant ions and their preference for tetrahedral (A) site. The dopant ions displace Fe^{3+} from tetrahedral (A) to octahedral (B) sites, resulting in weak magnetic interactions and low Neel temperature. High content of Fe^{3+} and Co^{2+} magnetic ions at the octahedral (B) sites leads to enhancement of B–B exchange interaction and weakening of A–B interaction. Nanosized magnetic zinc-cobalt ferrites with different Co to Zn ratio were obtained by co-precipitation [24], auto-combustion [13] and sol-gel [14] methods. The high M_S values make Zn-Co ferrites potential candidate for high-frequency inductors, information technology and communication [25].

In case of $Co_{1-x}Cd_xFe_2O_4$ (x = 0, 0.1) obtained by auto-combustion, the co-doping with diamagnetic (Zn^{2+} and Cd^{2+}) ions brings interesting change in the magnetic properties of Co ferrite. The Zn-Cd co-doped Co ferrites ($Zn_xCd_{0.375-x}Co_{0.625}Fe_2O_4$, x = 0.0, 0.075, 0.125, 0.25) synthesized via chemical co-precipitation route are recommended as soft magnets. The variation of M_R, M_S and H_C is due to the different chemical composition, crystal structure, particles size and arrangements at the lattice sites. The H_C also decreases by increasing Zn content, due to the lower magneto-crystalline anisotropy of Zn compared to Co and Cd [21].

The magnetic properties of $Co_{1-x-y}Sr_xZn_yFe_2O_4$ (x = 0.0, 0.01, 0.05, 0.3 and y = 0.0, 0.05, 0.1, 0.4, 0.5, 0.7) NPs synthetized by spontaneous gel auto-combustion (Pechini) technique were strongly influenced by the presence of both dopant ions, resulting in a superparamagnetic behavior [23]. The decrease of M_S values with increasing dopant ions content and decreasing particle size is due to the surface anisotropy of nanoferrites, while the decrease of H_C values is the result of some structural defects, such as dislocations, grain boundaries and anisotropy. The obtained results recommended the Zn-Sr co-doped Co ferrite as excellent candidate for various applications such as information storage devices, contrast agents in magnetic resonance imaging and gas sensors [23].

The addition of surfactants assures the control of the crystal nucleation and growth, due to their capability to act as a protective coating for NPs, reduces coalescence and enhances the crystallite size, porosity and specific surface. All these parameters further allow the control of the magnetic properties. In this regard, $Co_{0.5}Zn_{0.5}Fe_2O_4$ NPs prepared by co-precipitation method with ethanol as a surfactant show good M_S and large H_C [13]. When Co^{2+} ion with higher magnetic moment replaced Ni^{2+} ion with lower magnetic moment at B-sites, the H_C and M_R of $Co_xNi_{1-x}Fe_2O_4$, (x = 0.0–0.4) [26] and $Ni_xCo_{1-x}Fe_2O_4$ (x = 0, 0.25, 0.5, 0.75, 1.0) [27] increased, while M_S changed randomly. This increase is the result of cations distribution at the octahedral (B) and tetrahedral (A) sites in lattice structure, in spin canting and spin disorder [26, 27]. The $Ni_{1-x}Co_xFe_2O_4$ (x = 0.0, 0.15, 0.3, 0.45, 0.6, 0.75, 0.9, 1.0) synthesized by Pechini's sol-gel method showed an increase of M_S, H_C and T_C by Co^{2+} doping. Also, the number of magnetic domains increases and domain wall movement is facilitated by increasing particle size [28].

The magnetic properties of Co ferrite are also modified by incorporating Mn^{2+} ions. In case of $Mn_xCo_{1-x}Fe_2O_4$ (x = 0.2, 0.4, 0.6, 0.8) synthesized by sol-gel precipitation method, the M_S increases (up to x = 0.4) and then decreases (up to x = 0.8) with increasing Mn^{2+} content, due to the surface disorders resulted from the distortion of the magnetic moments at the surface and to the antiferromagnetic nature of the Mn^{2+} ions. The K decreased with increasing Mn^{2+} content, indicating the interaction between grains [29]. The M_S and magnetic moment increase with increasing Co^{2+} content in $Mn_{1-x}Co_xFe_2O_4$ (x = 0.2, 0.4, 0.6, 0.8) obtained by auto-combustion the ferrite structure [7].

4. Applications of antimicrobial activity

Nowadays, the alternative antimicrobials are highly considered due to the intense growing bacterial resistance towards conventional drugs [30]. In this regard, the development of novel multifunctional materials with antimicrobial properties that meet the requirements of a drug delivery system allowing the minimization of antibiotic concentration is of great interest. The essential characteristics of ferrite nanoparticles such as the high surface-to-volume ratio and nanoscale particle size, improve their reaction with pathogenic microbes. Also, the high surface area, low crystallite size and porosity have a significant role in improving the efficiency of NPs even at low (20 ppm) concentrations [31]. The main drawbacks in the use of these materials are that their antimicrobial properties easily change by varying their size, shape and crystallinity [32].

There are only few studies that investigate the antimicrobial effect of transition metal substituted Co ferrite nanopowders. Zhang et al. reported that the bactericidal effectiveness against gram-negative *E. coli* bacteria of $Cu_xCo_{1-x}Fe_2O_4$ (x = 0.0, 0.3, 0.5, 0.7, 1.0) NPs prepared by wet chemical co-precipitation method was enhanced by increasing Cu content [33]. The mechanisms involved in the antibacterial activity of NPs are: (*i*) decomposition of ferrite and formation of reactive oxygen species, (*ii*) electrostatic interaction of nanomaterials with cell membrane and (*iii*) photocatalytic light activation of nanoparticles [34–37]. The particle size, morphology, surface area, increase in oxygen vacancies, chemical molecule diffusion ability and discharge of metal ions also play important roles in the bactericidal activity [38]. Good antibacterial activity against *E. coli* and gram-positive *S. aureus* of $Cu_{0.5}Co_{0.5}Fe_{1.9}Bi_{0.1}O_4$ NPs synthetized by combustion technique was obtained, due to the co-doping of Cu and Bi in Co ferrite [39].

The bacterial growth rate inhibition of Zn-substituted Co ferrite ($Zn_xCo_{1-x}Fe_2O_4$, x = 0.0, 0.5, 1.0) nanoparticles (NPs) obtained via sol-gel route was found to be higher for the methicillin-resistant *S. aureus* (MRSA) than for *E. coli* strains [40]. Oppositely, the antibacterial activity of the $Zn_xCo_{1-x}Fe_2O_4$ (x = 0, 0.3, 0.5, 0.7, 1.0) NPs obtained by sol-gel process using citric acid as chelating agent was higher against gram negative bacteria (*E. coli*) than against gram-positive bacteria (*S. aureus*). Generally, the antibacterial capacities increased with increasing Zn content [41]. The *in vitro* antimicrobial activity of $Co_{0.6}Zn_{0.4}Fe_2O_4$ prepared by citrate-gel method tested against a wide range of gram-positive (*B. subtilis, S. aureus, M. luteus*) and gram-negative (*E. coli, P. aeruginosa, K. planticola*) bacteria revealed its efficiency in treatment of plants and trees affected by large microbial cells [42]. Good antibacterial effects of Zn-doped Co ferrite NPs prepared using curd as fuel via combustion method against gram-negative *S. typhi* and gram-positive *S. aureus* was also reported [43]. The obtained result indicated that Zn doped Co ferrite may be used as component in cosmetics, emulsions, creams, powders and lotions for dermatological and biomedical treatments (drug carriers, magnetically directed drug delivery, imaging factors and cancer therapy) [35].

The bactericidal activity of $Co_{0.5}Fe_{0.5}Fe_2O_4$ and $Co_{0.2}Fe_{0.8}Fe_2O_4$ NPs with average particle size of 5.0–6.4 nm, has been studied against gram-negative (*E. coli*), gram-positive (*S. aureus*), bacteria and fungi (*C. parapsilosis* and *C. albicans*), pathogens known to increasing mortality associated with multidrug resistance [5, 44]. $Co_{0.2}Fe_{0.8}Fe_2O_4$ NPs exhibited good antibacterial efficiency (21–70%) against all tested microorganisms. The number of colonies decreased considerable with increasing Co content in the investigated NPs [5].

$Mn_{1-x}Co_xFe_2O_4$ (x = 0.2, 0.4, 0.6, 0.8) prepared using open-air auto combustion was found to have excellent antifungal activity against *Rhizopus fungi* and its efficiency increase with increasing Co content [7]. Ashour et al. demonstrated the antimicrobial activity of metal (Zn, Mn, Cu) doped Co ferrite nanoparticles against *B. subtilis, S. aureus, E. coli, P. aeruginosa and C. albicans*. The Zn-substituted Co ferrite NPs, were more active against gram-positive than gram-negative bacteria and had strong antifungal activity against *C. albicans*. Gamma-irradiated Zn-substituted Co ferrite (150 kGy) was more active against *S. aureus* and *P. aeruginosa*, as a result of the decreasing crystallite size [35].

The $M_xCo_{1-x}Fe_2O_4$ (M = Zn, Cu, Mn; x = 0.00, 0.25, 0.50, 0.75) NPs synthetized via sol-gel method were investigated as antibacterial agents towards bacteria that commonly diffused on the surfaces of the medical operating room walls (*S. lentus, S. sciuri, S. vitulinus, S. aureus, A. viridians* and *E. columbae*). The antibacterial activity is enhanced in the following order: $Mn_xCo_{1-x}Fe_2O_4 > Cu_xCo_{1-x}Fe_2O_4 > Zn_xCo_{1-x}Fe_2O_4$. The most effective ferrite was $Zn_{0.75}Co_{0.25}Fe_2O_4$ NPs which exhibited the highest activity towards all investigated pathogenic bacteria. The highest activity of $Mn_{0.75}Co_{0.25}Fe_2O_4$ was against *S. vitulinus*, while $Cu_{0.75}Co_{0.25}Fe_2O_4$ NPs were effective against *S. aureus* [31].

The antimicrobial performance of $M_xCo_{1-x}Fe_2O_4$; (M = Zn, Cu, Mn; x = 0, 0.5) NPs prepared using a sol-gel method in the presence of citric acid and ethylene glycol upon pathogenic microorganisms infected urinary tract and blood samples was investigated by Maksoud et al. The tested pathogens were gram-positive bacteria (*S. epidermidis, S. aureus, MRSA and E. faecalis, B. subtilis*), gram-negative bacteria (*A. baumannii, E. cloacae, E. coli, K. pneumoniae, P. aeruginosa*) and uni-cellular fungi (*C. albicans*). Zn-Co ferrite NPs displayed a maximum growth inhibition against *K. pneumoniae*, *P. aeruginosa* and *C. albicans* [30].

The antimicrobial activity of co-doped $Co_{0.5}M_{0.5}Fe_2O_4$ NPs (M = Cu, Zn, Mn, Ni) obtained by the sol-gel process using citric acid as the chelating agent tested against *E. coli* and *S. aureus* revealed that substituted Co ferrite NPs exhibited the most effective biocidal property, while the substitution of Zn and Cu in Co ferrite NPs considerably enhanced the antibacterial activity [45].

5. Applications of biological properties

The applications of nanotechnology in various medical areas, especially in drug delivery have been extensively explored lately. Considering the ultra-small (1–100 nm) and controllable size, high surface-to-mass ratio and high reactivity of NPs, they easily interact with biological systems [46].

The nanosized spinel ferrites and transition metal-substituted ferrites could successfully substitute some antibiotics that are currently used to combat pathogenic bacteria in the gastrointestinal tract of animals, as well as other biomedical applications. Many studies reported the synthesis and characterization of transition metal substituted Co ferrite NPs, but the attention dedicated to their biocompatibility in view of *in vivo* biomedical applications to assure their safe clinical use is still limited. The key criteria for their clinical applications are good biocompatibility and safety [47].

The use of magnetic nanoparticles in biomedical applications demands appropriate shape and size, high magnetization, good ability to deliver the pharmacologically active compounds, non-toxicity and biodegradability. The overall biocompatibility of $Co_{0.5}M_{0.5}Fe_2O_4$ (M = Cu, Zn, Mn, Ni) NPs synthetized by the sol-gel process using citric acid as chelating agent decreased in the following order: $Co_{0.5}Mn_{0.5}Fe_2O_4 < Co_{0.5}Cu_{0.5}Fe_2O_4 < Co_{0.5}Zn_{0.5}Fe_2O_4 < Co_{0.5}Ni_{0.5}Fe_2O_4$. The biocompatibility of NPs depended on the toxicity of transition metal and the releasing rate of transition metal ions into the cell culture medium [48]. Some possible mechanisms of magnetic NP-based antimicrobial drug delivery to microorganisms could be: (*i*) the NPs fuse with microbial cell wall or membrane and release the carried drugs into the bacteria cell; (*ii*) the NPs bind to cell wall and continuously release the drug, which diffuses into the interior of the microorganisms [49]. Iqbal et al. reported the development of $Zn_{0.5}Co_{0.5}Fe_2O_4$ NPs, with the required shape and size, as anti-cancer drug with passive targeting NPs delivering system into cancerous cells by applying photodynamic therapy through controlling the particle size according to the human body (HepG2) cells [40].

The applications of ferrites in tissue engineering are limited due to their inertness towards bioactivity and release of some toxic elements into the human body fluid. However, the migration can be controlled by encapsulation of ferrite NPs by glass matrix. The addition of bioglass in the ferrite displays some biodegradability and supports better osteoblasts growth *in vitro*. In this regard, the bioactive glass containing $Co_{0.2}Cu_{0.8}Fe_2O_4$ prepared using self-propagating high-temperature synthesis, showed good potential in bone hyperthermia application [50]. In magnetic hyperthermia, the ferrite NPs are used as local heat dissolving agents in external magnetic field. After their introduction into the body through blood, the body's immune system identifies them as foreign substances and the body rejects the material. To overcome this problem, the biocompatible surface-coating (i.e. chitosan) helps to stabilize the ferrite NPs and provides an available surface area for the biomolecular conjugation for biomedical applications [51]. In this regard, the study of the effect of chitosan-coated $Co_{1-x}Mn_xFe_2O_4$ (x = 0.0, 0.2, 0.4, 0.6, 0.8, 1.0) NPs obtained via wet chemical co-precipitation on the hyperthermia temperature (directly related to specific loss power for cancer treatment) revealed that $Co_{0.2}Mn_{0.8}Fe_2O_4$ exhibited hyperthermia range [52].

6. Applications of catalytic activity

The catalysts are important players in numerous chemical processes, especially in organic synthesis or decomposition of persistent pollutants. In the last decade, the use of magnetic NPs as catalysts attracted considerable interest due to the enhancement of the reaction speed, possibility of catalyst separation from the reaction medium by using an external magnet, without any filtration or centrifugation, and its reuse or recycling [53, 54]. In this regard, several conventional synthesis methods were replaced with more eco-friendly options that use magnetic nano-sized catalysts. The catalytic processes that use magnetic NPs as catalysts include decomposition of recalcitrant organics, dehydrogenation, oxidation, alkylation and coupling reactions [9, 54].

The catalytic activity is influenced by the particle size, surface area, morphology, nature and concentration of the catalyst [10, 53]. In spinel ferrites, the presence of cations with different charges determines its catalytic properties as it allow internal redox reactions [55]. The distribution of metal ions between tetrahedral (A) and octahedral (B) sites also influences the catalytic activity. Thus, by doping transition metal ions in the ferrite structure, the cationic distribution is changed resulting in modified catalytic activity [10].

One of the main applications of magnetic NPs as catalysts is in the photocatalytic degradation or organics in the presence of visible or UV light. The photocatalytic activity of NPs is based on their capacity to efficiently absorb photons, that excite electrons from the valence band into the conduction band, leaving positively charged vacancy to react with the water molecules and to generate active radicals such are hydroxyl (·OH) or superoxide ($\cdot O_2$) that further react with the pollutants [56]. Beside the ability to absorb photons, the reusability, recyclability, low cost, chemical stability and high corrosion resistance are important factors in the selection of photocatalysts [57]. The crystallite size, surface area, band gap, cations distribution among tetrahedral (A) and octahedral (B) sites and magnetic properties are influenced by the dopant type and amount [10, 43].

In the last decades, a wide range of non-biodegradable organic dyes, inks and pigments were identified in wastewaters from the leather, textile, printing, paper, food and cosmetics industries. These dyes may pose carcinogenic and mutagenic risks and are difficult to treat using conventional water treatment methods. Nanosized Co ferrite is a magnetic material with high H_C and moderate M_S, narrow band gap, low toxicity, low price and good catalytic activity [10, 58]. By doping, its structural and catalytic properties may be further enhanced. The doped and co-doped Co ferrites are promising catalysts that may decompose recalcitrant organic chemicals from wastewaters or enhance the synthesis of organics [9, 52, 56, 59–61]. The doping of transition metal ions (Cr, Mn, Co, Zn) into the spinel lattice of Co ferrite influences the physicochemical properties and improves their stability [11]. Moreover, the doping favors the formation of mixed or inverse spinel structures and introduces new donor or acceptor levels, which boosts the visible light activated photocatalyst activity [62].

The photocatalytic activity of a wide range of doped Co ferrites were tested on rhodamine B (RhB), methyl orange (MO), methylene blue (MB) and congo red (CR), synthetic dyes known to be highly toxic and carcinogenic. Nanocrystalline magnetic $Zn_xCo_{1-x}Fe_2O_4$ (x = 0.0, 0.2, 0.4, 0.6, 0.8, 1.0) with good photocatalytic activity was obtained by reverse micelle technique [63]. The doping with Zn increased the RhB degradation rate and reduced the degradation time, while, the band gap increased with increasing Zn content [63]. The photocatalytic degradation of CR and Evans blue by $Zn_xCo_{1-x}Fe_2O_4$ (x = 0.0, 0.2, 0.4, 0.6) prepared using curd as a fuel through the combustion method was found also to increase with the increase of Zn doping up to x = 0.4, suggesting that Zn doped Co ferrite are better photocatalysts than Co ferrite [43]. The photocatalytic activity of $M_xCo_{1-x}Fe_2O_4$ (M = Zn, Cu, Mn; x = 0.0, 0.25, 0.50, 0.75) NPs synthesized by citrate sol-gel method enhanced with increasing M content, but were lower than that of undoped Co ferrite in case of M = Cu and Zn and higher in case of M = Mn used for MB degradation [31].

The photocatalytic performance of $Co_{0.6}Zn_{0.4}Cu_xFe_{2-x}O_4$ (x = 0.2, 0.4, 0.6, 0.8 and 1.0) obtained by sol-gel auto combustion method was evaluated by MO dye degradation under visible light and presence of hydrogen peroxide. The results showed that the degradation of MO enhances as the content of Cu in Co-Zn ferrites increases, due to the strong preference of Cu^{2+} ions for the octahedral (B) sites [61]. The photocatalytic degradation of CR by $Cu_{0.5}Co_{0.5}Fe_{1.9}Bi_{0.1}O_4$ NPs obtained by solution combustion technique was found to have around 90% efficiency, the photocatalyst being stable and reusable [39]. High removal percentage of CR and bisphenol A was reported for $Co_{0.5}Cu_{0.5}Fe_{1.95}Ce_{0.05}O_4$ after exposure to both visible and UV-light [64].

The $Zn_{1-x}Co_xFe_2O_4$ (x = 0.03, 0.1, 0.2) and $Cu_xCo_{0.5-x}Ni_{0.5}Fe_2O_4$ (x = 0.1, 0.2, 0.3, 0.4) NPs obtained by facile reduction-oxidation route and respectively precipitation method in the presence of oleic acid as a surfactant, were found to be able to photodegrade MB, the degradation efficiency decreasing with the increase of Zn and Cu content, respectively [57, 65].

Despite the high number of applications of transitional metal doped Co ferrites in the photocatalytic decomposition of various organic pollutants, there are only few studies on their use in organic synthesis. The Ni-substituted Co ferrite NPs supported on arginine-modified graphene oxide nanosheets ($Ni_{0.5}Co_{0.5}Fe_2O_4$@Arg–GO) were proven to be effective for the one-pot tandem oxidative synthesis of 2-phenylbenzimidazole derivatives [66]. $Ni_xCo_{1-x}Fe_2O_4$ (x = 0.0, 0.2, 0.4, 0.6, 0.8, 1.0) ferrite NPs obtained by microemulsion method were found to effectively reduce 4-nitrophenol to 4-aminophenol in the presence of $NaBH_4$ as reducing agent [67].

Another important application of magnetic spinel NPs is in the field of renewable energy production and storage as catalysts for driving the water electrolysis by enhancing the hydrogen and oxygen evolution reactions (HER, OER). Ni-Co ferrite ($Co_{0.5}Ni_{0.5}Fe_2O_4$) anchored on ultrathin conductive graphene oxide nanosheets acts as a highly active, stable and low-cost electrocatalyst in the water splitting processes, being a low-cost alternative to noble metal-oxides catalyst [68]. The OER and HER catalytic activity of $Co_xNi_{1-x}Fe_2O_4$ (x = 0.0, 0.25, 0.5, 0.75, 1.0) NPs prepared by citric acid assisted sol-gel combustion method was found to be lower than bulk Ni ferrite, the Ni content increase improving the catalytic activity and the electron transfer rate [69].

7. Applications of dielectric properties

The significant progress in information technology, electronics and wireless communication devices together with a new trend of miniaturization and multifunctionality led to the necessity of new materials with special characteristics. Considering the structural, electric and magnetic properties, the nanosized ferrites may become important candidates for applications in microwave communication systems, electromagnetic devices, resonators and filters for satellites, broadcasting equipment, batteries, supercapacitors and many other microwave devices [70].

The dielectric structure of ferrites consists in well conducting grains separated by highly resistive grain boundaries. The resistive grain boundaries are more effective at low frequencies, while the highly conducting grains act at high frequencies [29]. The dielectric constant at low applied frequencies is determined by the space charge polarization favored by the electrons grouped along the resistive grain boundary [6, 29]. In case of doped Co ferrites, the dielectric constant depends on the content of Fe^{3+}, Co^{2+} and other divalent transition ions present in the spinel structure [6]. Generally, the dielectric properties are strongly influenced by the grain size, porosity, synthesis method and annealing temperature [6, 29].

Considering its good chemical and thermal stability, high electrical resistivity, magnetic anisotropy, high H_C, moderate M_S, superparamagnetism, ferrimagnetism and dielectric structure, the Ni doped Co ferrite is a good candidate for microwave devices and data storage [71]. The $Co_{1-x}Ni_xFe_2O_4$ ($0.0 \le x \le 1.0$) synthesized using simple, low temperature auto-combustion method showed high resistance. As the Ni content increases, the dielectric constant and loss tangent decrease and remain constant at higher frequencies, while conductivity increases with increasing frequency [71]. The use as supercapacitors of mixed ternary Cu-Co-Ni ferrites obtained by sol-gel synthesis and citric acid as chelating agent was investigated by Bhujun et al. [72]. The cyclic voltammogram profiles showed that the capacitive behavior is close to ideal rectangular shape, confirming the reversibility of the system and the decrease of specific capacitance value with the increase of the cycle numbers. The specific capacitance of $Cu_{0.5}Co_{0.5}Fe_2O_4$ (221 Fg^{-1}) was higher than of $Ni_{0.5}Co_{0.5}Fe_2O_4$ (60 Fg^{-1}) and showed excellent cycling stability [72].

$Mn_xCo_{1-x}Fe_2O_4$ (x = 0.2, 0.4, 0.6, 0.8) NPs synthesized by sol-gel precipitation method have dielectric properties that decrease with the increase of the doping ratio for x = 0.2–0.6 [29]. However, the $Mn_{0.8}Co_{0.2}Fe_2O_4$ was found to have the highest dielectric constant (8.38) at 100 Hz, due to the increasing porosity and grain boundaries between the small sized grains. The low H_C and low dielectric loss between 100 and 100 kHz indicate its potential use as inductor and transformer for switch-mode power supplies [29]. $MnCoFeO_4$ NPs with the average particle size in the range of 30–40 nm synthesized via a simple one-pot co-precipitation method were also proven to be suitable as high-performance capacitors for electrical energy storage [73].

8. Conclusions and future perspectives

The Co ferrite continues to attract considerable attention due to its unique and exciting properties and opens new doors towards many potential applications. The properties of Co ferrite can be easily controlled by preparation technique, morphology, dopants type/content and cation distribution between tetrahedral (A) and octahedral (B) sites. There is a high number of studies that reported the physical, chemical, magnetic, electrical and optical properties of undoped and doped Co ferrites. Also, an increasing interest towards the incorporation of newer ions into the Co ferrite lattice in order to tailor its properties was noticed. The excellent properties of divalent transition metal doped Co ferrites, together with the possibility to tailor their particle size, shape, purity and chemical composition became a promising alternative for future generation nanomaterials designed for various industrial, environmental and medical applications.

Acknowledgements

This research was funded by the Romanian Research and Innovation Ministry, grant number 19PFE/2018. The APC was funded by the Romanian Research and Innovation Ministry, grant number 19PFE/2018 PROINSTITUTIO.

Author details

Oana Cadar[1], Thomas Dippong[2], Marin Senila[1] and Erika-Andrea Levei[1*]

1 INCDO-INOE 2000, Research Institute for Analytical Instrumentation, Cluj-Napoca, Romania

2 Department of Chemistry and Biology, North University Center of Baia Mare, Technical University of Cluj-Napoca, Baia Mare, Romania

*Address all correspondence to: erika.levei@icia.ro

References

[1] Wang L, Lu M, Liu Y, Li J, Liu M, Lin H. The structure, magnetic properties and cation distribution of $Co_{1-x}Mg_xFe_2O_4/SiO_2$ nanocomposites synthesized by sol–gel method. Ceramics International. 2015;**41**:4176-4181. DOI: 10.1016/j.ceramint.2014.12.099

[2] Jebeli Moeen S, Vaezi MR, Yousefi AA. Chemical synthesis of nano-crystalline nickel-zinc ferrite as a magnetic pigment. Progress in Color, Colorants and Coatings. 2010;**3**:9-17

[3] Debnath S, Deb K, Saha B, Das R. X-ray diffraction analysis for the determination of elastic properties of zinc doped manganese spinel ferrite nanocrystals ($Mn_{0.75}Zn_{0.25}Fe_2O_4$), along with the determination of ionic radii, bond lengths, and hopping lengths. Journal of Physics and Chemistry of Solids. 2019;**134**:105-114. DOI: 10.1016/j.jpcs.2019.05.047

[4] Priya AS, Geetha D, Kavitha N. Effect of Al substitution on the structural, electric and impedance behavior of cobalt ferrite. Vacuum. 2019;**160**:453-460. DOI: 10.1016/j.vacuum.2018.12.004

[5] Zalneravicius R, Paskevicius A, Mazeika K, Jagminas A. Fe(II)-substituted cobalt ferrite nanoparticles against multidrug resistant microorganisms. Applied Surface Science. 2018;**435**:141-148. DOI: 10.1016/j.apsusc.2017.11.028

[6] Kershi RM, Aldirham SH. Transport and dielectric properties of nanocrystallite cobalt ferrites: Correlation with cations distribution and crystallite size. Materials Chemistry and Physics. 2019;**238**:121902. DOI: 10.1016/j.matchemphys.2019.121902

[7] Naik AB, Naik PP, Hasolkar SS, Naik D. Structural, magnetic and electrical properties along with antifungal activity & adsorption ability of cobalt doped manganese ferrite nanoparticles synthesized using combustion route. Ceramics International. 2020. DOI: 10.1016/j.ceramint.2020.05.177 [in press]

[8] Dippong T, Levei EA, Cadar O, Goga F, Toloman D, Borodi G. Thermal behavior of Ni, Co and Fe succinates embedded in silica matrix. Journal of Thermal Analysis and Calorimetry. 2019;**136**:1587-1596. DOI: 10.1007/s10973-019-08117-8

[9] Kharisov BI, Rasika Dias HV, Kharissova OV. Mini-review: Ferrite nanoparticles in the catalysis. Arabian Journal of Chemistry. 2019;**12**:1234-1246. DOI: 10.1016/j.arabjc.2014.10.049

[10] Dou R, Cheng H, Ma J, Komarneni S. Manganese doped magnetic cobalt ferrite nanoparticles for dye degradation via a novel heterogeneous chemical catalysis. Materials Chemistry and Physics. 2020;**240**:122181. DOI: 10.1016/j.matchemphys.2019.122181

[11] Casbeer E, Sharma VK, Li XZ. Synthesis and photocatalytic activity of ferrites under visible light: A review. Separation and Purification Technology. 2012;**87**:1-14. DOI: 10.1016/j.seppur.2011.11.034

[12] Kavitha S, Kurian M. Effect of zirconium doping in the microstructure, magnetic and dielectric properties of cobalt ferrite nanoparticles. Journal of Alloys and Compounds. 2019;**799**:147-159. DOI: 10.1016/j.jallcom.2019.05.183

[13] El Foulani AH, Aamouche A, Mohseni F, Amaral JS, Tobaldi DM, Pullar RC. Effect of surfactants on the optical and magnetic properties of cobalt-zinc ferrite $Co_{0.5}Zn_{0.5}Fe_2O_4$. J. Alloys and Compounds. 2019;**774**:1250-1259. DOI: 10.1016/j.jallcom.2018.09.393

[14] Andhare DD, Patade SR, Kounsalye JS, Jadhav KM. Effect of Zn doping on structural, magnetic and optical properties of cobalt ferrite nanoparticles synthesized via co-precipitation method. Physica B. 2020;**583**:412051. DOI: 10.1016/j.physb.2020.412051

[15] Perales-Perez O, Cedeno-Mattei Y. Optimizing processing conditions to produce cobalt ferrite nanoparticles of desired size and magnetic properties. In: Seehra MS, editor. Magnetic Spinels. London: IntechOpen; 2016. p. 51. DOI: 10.5772/66842

[16] Jauhar S, Kaur J, Goyal A, Singhal S. Tuning the properties of cobalt ferrite: A road towards diverse applications. RSC Advances. 2016;**6**:100. DOI: 10.1039/c6ra21224g

[17] Medeiros PN, Gomes YF, Bomio MRD, Santos IMG, Silva MRS, Paskocimas CA, et al. Influence of variables on the synthesis of CoFe2O4 pigment by the complex polymerization method. ournal of Advanced Ceramics. 2015;**4**:135-141. DOI: 10.1007/s40145-015-0145-1

[18] Cavalcante PMT, Dondi M, Guarini G, Raimondo M, Baldi G. Colour performance of ceramic nano-pigments. Dyes and Pigments. 2009;**80**:226-232. DOI: 10.1016/j.dyepig.2008.07.004

[19] Dippong T, Goga F, Levei EA, Cadar O. Influence of zinc substitution with cobalt on thermal behavior, structure and morphology of zinc ferrite embedded in silica matrix. Journal of Solid State Chemistry. 2019;**275**:159-166. DOI: 10.1016/j.jssc.2019.04.011

[20] Dippong T, Levei EA, Goga F, Petean I, Avram A, Cadar O. The impact of polyol structure on the formation of $Zn_{0.6}Co_{0.4}Fe_2O_4$ spinel-based pigments. Journal of Sol-Gel Science and Technology. 2019;**92**:736-744. DOI: 10.1007/s10971-019-05140-x

[21] Shakil IU, Arshad MI, Nabi G, Khalid NR, Tariq NH, Shahd A, et al. Influence of zinc and cadmium co-doping on optical and magnetic properties of cobalt ferrites. Ceramics International. 2020;**46**:7767-7773. DOI: 10.1016/j.ceramint.2019.11.280

[22] Margabandhu M, Sendhilnathan S, Senthikumar S, Gajalakshmi D. Investigationofstructural,morphological, magnetic properties and biomedical applications of Cu^{2+} substituted uncoated cobalt ferrite nanoparticles. Brazilian Archives of Biology and Technology. 2016;**52**:1-10. DOI: 10.1590/1678-4324-2016161046

[23] Imanipour P, Hasani S, Afsharia M, Sheykha S, Seifoddinia A, Jahanbani-Ardakani K. The effect of divalent ions of zinc and strontium substitution on the structural and magnetic properties on the cobalt site in cobalt ferrite. Journal of Magnetism and Magnetic Materials. 2020;**510**:166941. DOI: 10.1016/j.jmmm.2020.166941

[24] Singh A, Pathak S, Kumar P, Sharma P, Rathi A, Basheed GA, et al. Tuning the magnetocrystalline anisotropy and spin dynamics in $Co_xZn_{1-x}Fe_2O_4$ ($0 \le x \le 1$) nanoferrites. Journal of Magnetism and Magnetic Materials. 2020;**493**:165737. DOI: 10.1016/j.jmmm.2019.165737

[25] Kaur H, Singh A, Kumar V, Ahlawat DS. Structural, thermal and magnetic investigations of cobalt ferrite doped with Zn^{2+} and Cd^{2+} synthesized by auto combustion method. Journal of Magnetism and Magnetic Materials. 2019;**474**:505-511. DOI: 10.1016/j.jmmm.2018.11.010

[26] Chakradhary VK, Ansaria A, Akhtar MJ. Design, synthesis, and testing of high coercivity cobalt doped

nickel ferrite nanoparticles for magnetic applications. Journal of Magnetism and Magnetic Materials. 2019;**469**:674-680. DOI: 10.1016/j.jmmm.2018.09.021

[27] Dippong T, Levei EA, Cadar O, Deac IG, Diamandescu L, Barbu-Tudoran L. Effect of nickel content on structural, morphological and magnetic properties of $Ni_xCo_{1-x}Fe_2O_4/SiO_2$ nanocomposites. Journal of Alloys and Compounds. 2019;**786**:330-340. DOI: 10.1016/j.jallcom.2019.01.363

[28] Pubby K, Babu KV, Narang SB. Magnetic, elastic, dielectric, microwave absorption and optical characterization of cobalt-substituted nickel spinel ferrites. Materials Science and Engineering B-Advanced. 2020;**255**:114513. DOI: 10.1016/j.mseb.2020.114513

[29] Jabbar R, Sabeeh SH, Hameed AM. Structural, dielectric and magnetic properties of Mn^{+2} doped cobalt ferrite nanoparticles. Journal of Magnetism and Magnetic Materials. 2020;**494**:165726. DOI: 10.1016/j.jmmm.2019.165726

[30] Maksoud MIAA, El-Sayyas GS, Ashour AH, El-Batal AI, Abd-Elmonem MS, Hendawy HAM, et al. Synthesis and characterization of metals-substituted cobalt ferrite $[M_xCo_{(1-x)}Fe_2O_4$; (M = Zn, Cu and Mn; x=0 and 0.5)] nanoparticles as antimicrobial agents and sensors for Anagrelide determination in biological samples. Materials Science & Engineering C-Materials. 2018;**92**:644-656. DOI: 10.1016/j.msec.2018.07.007

[31] Maksoud MIAA, El-Sayyad GS, Ashour AH, El-Batal AI, Elsayed MA, Gobara M, et al. Antibacterial, antibiofilm, and photocatalytic activities of metals-substituted spinel cobalt ferrite nanoparticles. Microbial Pathogenesis. 2019;**127**:144-158. DOI: 10.1016/j.micpath.2018.11.045

[32] Seil JT, Webster TJ. Antimicrobial applications of nanotechnology: Methods and literature. International Journal of Nanomedicine - UK. 2012;**7**:2767-2781. DOI: 10.2147/IJN.S24805

[33] Zhang L, Jiang Y, Ding Y, Povey M, York D. Investigation into the antibacterial behaviour of suspensions of ZnO nanoparticles (ZnO nanofluids). Journal of Nanoparticle Research. 2007;**9**:479-489. DOI: 10.1007/s11051-006-9150-1

[34] Schwartz VB, Thetiot F, Ritz S, Putz S, Choritz L, Lappas A, et al. Antibacterial surface coatings from zinc oxide nanoparticles embedded in poly(n-isopropylacrylamide) hydrogel surface layers. Advanced Functional Materials. 2012;**22**:2376-2386. DOI: 10.1002/adfm.201102980

[35] Ashour AH, El-Batal AI, Maksouda MIA, El-Sayyad GS, Labib S, Abdeltwab E, et al. Antimicrobial activity of metal-substituted cobalt ferrite nanoparticles synthesized by sol-gel technique. Particuology. 2018;**40**:141-151. DOI: 10.1016/j.partic.2017.12.001

[36] Yamamoto O, Sawai J. Investigation into the antibacterial behaviour of suspensions of ZnO nanoparticles (ZnO nanofluids). Bulletin of the Chemical Society of Japan. 2001;**74**:1761-1765. DOI: 10.1007/s11051-006-9150-1

[37] Sawai J, Shoji S, Igarashi H, Hashimoto A, Kokugan T, Shimizu M, et al. Hydrogen peroxide as an antibacterial factor in zinc oxide powder slurry. Journal of Fermentation and Bioengineering. 1998;**86**:521-522. DOI: 10.1016/S0922-338X(98)80165-7

[38] Samavati A, Ismail AF. Antibacterial properties of copper-substituted cobalt ferrite nanoparticles synthesized by co-precipitation method. Particuology. 2017;**30**:158-163. DOI: 10.1016/j.partic.2016.06.003

[39] Kirankumar VS, Sumathi S. Photocatalytic and antibacterial activity of bismuth and copper co-doped cobalt ferrite nanoparticles. Journal of Materials Science: Materials in Electronics. 2018;**29**:8738-8746. DOI: 10.1007/s10854-018-8890-x

[40] Iqbal A, Fakhar-e-Alam M, Atif M, Amin N, Alimgeer KS, Ali A, et al. Structural, morphological, antimicrobial, and in vitro photodynamic therapeutic assessments of novel Zn^{+2}-substituted cobalt ferrite nanoparticles. Results in Physics. 2019;**15**:102529. DOI: 10.1016/j.rinp.2019.102529

[41] Sanpo N, Berndt CC, Wang J. Microstructural and antibacterial properties of zinc-substituted cobalt ferrite nanopowders synthesized by sol-gel methods. Journal of Applied Physics. 2012;**112**:084333. DOI: 10.1063/1.4761987

[42] Vinutha CH, Naidu KCB, Sekhar CC, Ravinder D. Magnetic and antimicrobial properties of cobalt-zinc ferrite nanoparticles synthesized by citrate-gel method. International Journal of Applied Ceramic Technology. 2019;**16**:1944-1953. DOI: 10.1111/ijac.13276

[43] Naik MM, Naik HSB, Nagaraju G, Vinuth M, Vinu K, Viswanath R. Green synthesis of zinc doped cobalt ferrite nanoparticles: Structural, optical, photocatalytic and antibacterial studies. Nano-Structures & Nano-Objects. 2019;**19**:100322. DOI: 10.1016/j.nanoso.2019.100322

[44] Park JY, Choi ES, Baek MJ, Lee GH. Colloidal stability of amino acid coated magnetite nanoparticles in physiological fluid. Materials Letters. 2009;**63**:379-381. DOI: 10.1016/j.matlet.2008.10.057

[45] Sanpo N, Berndt CC, Wen C, Wang J. Transition metal-substituted cobalt ferrite nanoparticles for biomedical applications. Acta Biomaterialia. 2013;**9**:5830-5837. DOI: 10.1016/j.actbio.2012.10.037

[46] Zhang L, Gu FX, Chan JM, Wang AZ, Langer RS, Farokhzad OC. Nanoparticles in medicine: Therapeutic applications and developments. Clinical Pharmacology and Therapeutics. 2007;**83**:761-769. DOI: 10.1038/sj.clpt.6100400

[47] Gupta AK, Gupta M. Synthesis and surface engineering of iron oxide nanoparticles for biomedical applications. Biomaterials. 2005;**26**:3402-3995. DOI: 10.1016/j.biomaterials.2004.10.012

[48] Sanpo N, Tharajak J, Li Y, Berndt CC, Wen C, Wang J. Biocompatibility of transition metal-substituted cobalt ferrite nanoparticles. Journal of Nanoparticle Research. 2014;**16**:2510. DOI: 10.1007/s11051-014-2510-2513

[49] Zhang L, Pornpattananangkul D, Hu CMJ, Huang CM. Development of nanoparticles for antimicrobial drug delivery. Current Medicinal Chemistry. 2010;**17**:585-594. DOI: 10.2174/092986710790416290

[50] Sampath KA, Himanshu T, Kevin B, Singh SP. Structural, magnetic and in vitro bioactivity of co-cu ferrite and bioglass composite for hyperthermia in bone tissue engineering. Bioceramics Development and Applications. 2016;**6**:091. DOI: 0.4172/2090-5025.100091

[51] Nasrin S, Chowdhury FUZ, Hoque SM. Study of hyperthermia temperature of manganese-substituted cobalt nano ferrites prepared by chemical co-precipitation method for biomedical application. Journal of Magnetism and Magnetic Materials. 2019;**479**:129-134. DOI: 10.1016/j.jmmm.2019.02.010

[52] Giustini AJ, Petryk AA, Cassim SM, Tate JA, Baker I, Hoopes PJ. Magnetic nanoparticle hyperthermia in cancer treatment. Nano LIFE. 2010;**1**:17-32. DOI: 10.1007/s11356-019-07231-2

[53] Padmapriya G, Manikandan A, Krishnasamy V, Jaganathan SK, Arul AS. Spinel $Ni_xZn_{1-x}Fe_2O_4$ ($0.0 \leq x \leq 1.0$) nano-photocatalysts: Synthesis, characterization and photocatalytic degradation of methylene blue dye. Journal of Molecular Structure. 2016;**1119**:39-47. DOI: 10.1016/j.molstruc.2016.04.049

[54] Kazemi M, Ghobadi M, Mirzaie A. Cobalt ferrite nanoparticles ($CoFe_2O_4$ MNPs) as catalyst and support: Magnetically recoverable nanocatalysts in organic synthesis. Nanotechnology Reviews. 2018;**7**:43-68. DOI: 10.1515/ntrev-2017-0138

[55] Vozniuk O, Tabanelli T, Tanchoux N, Millet JMM, Albonetti S, Di Renzo F, et al. Mixed-oxide catalysts with spinel structure for the valorization of biomass: The chemical-loop reforming of bioethanol. Catalysts. 2018;**8**:332. DOI: 10.3390/catal8080332

[56] Arifin MN, Karim KMR, Abdullah H, Khan MR. Synthesis of titania doped copper ferrite photocatalyst and its photoactivity towards methylene blue degradation under visible light irradiation. Bulletin of Chemical Reaction Engineering & Catalysis. 2019;**14**:219-227. DOI: 10.9767/bcrec.14.1.3616.219-227

[57] Fan G, Tong J, Li F. Visible-light-induced photocatalyst based on cobalt-doped zinc ferrite nanocrystals. Industrial and Engineering Chemistry Research. 2012;**51**:13639-13647. DOI: 10.1021/ie201933g

[58] Dhiman M, Bhukal S, Chudasama B, Singhal S. Impact of metal ions (Cr^{3+}, Co^{2+}, Ni^{2+}, Cu^{2+} and Zn^{2+}) substitution on the structural, magnetic and catalytic properties of substituted Co–Mn ferrites synthesized by sol–gel route. Journal of Sol-Gel Science and Technology. 2017;**81**:831-843. DOI: 10.1007/s10971-016-4232-8

[59] Singh C, Jauhar S, Kumar V, Singh J, Singhal S. Synthesis of zinc substituted cobalt ferrites via reverse micelle technique involving in situ template formation: A study on their structural, magnetic, optical and catalytic properties. Materials Chemistry and Physics. 2015;**156**:188-197. DOI: 10.1016/j.matchemphys.2015.02.046

[60] Manikandan A, John Kennedy L, Bououdina M, Judith VJ. Synthesis, optical and magnetic properties of pure and Co-doped $ZnFe_2O_4$ nanoparticles by microwave combustion method. Journal of Magnetism and Magnetic Materials. 2004;**349**:249-258. DOI: 10.1016/j.jmmm.2013.09.013

[61] Bhukal S, Shivali, Singhal S. Magnetically separable copper substituted cobalt–zinc nano-ferrite photocatalyst with enhanced photocatalytic activity. Materials Science in Semiconductor Processing. 2014;**26**:467-476. DOI: 10.1016/j.mssp.2014.05.023

[62] Singh S, Kaur P, Bansal S, Singhal S. Enhanced photocatalytic performance of Ru-doped spinel nanoferrites for treating recalcitrant organic pollutants in wastewater. Journal of Sol-Gel Science and Technology. 2019;**92**:760-774. DOI: 10.1007/s10971-019-05142-9

[63] Sundararajan M, John Kennedy L, Nithya P, Judith Vijaya J, Bououdina M. Visible light driven photocatalytic degradation of rhodamine B using Mg doped cobalt ferrite spinel nanoparticles synthesized by microwave combustion method. Journal of Physics and Chemistry of Solids. 2017;**108**:61-75. DOI: 10.1016/j.jpcs.2017.04.002

[64] Kirankumar VS, Sumathi S. Copper and cerium co-doped cobalt ferrite nanoparticles: Structural, morphological, optical, magnetic, and photocatalytic properties. Environemental Science and Pollution Research. 2019;**26**:19189-19206. DOI: 10.1007/s11356-019-05286-9

[65] Lassoued A, Lassoued MS, Dkhil B, Ammar S, Gadri A. Improved photocatalytic activities of $Cu_xCo_{0.5-x}Ni_{0.5}Fe_2O_4$ nanoparticles through co-precipitation method in degrading methylene blue. Physica E. 2018;**101**: 29-37. DOI: 10.1016/j.physe.2018.03.015

[66] Ghadari R, Namazi H, Aghazadeh M. Nickel-substituted cobalt ferrite nanoparticles supported on arginine-modified graphene oxide nanosheets: Synthesis and catalytic activity. Applied Organometallic Chemistry. 2017;**31**:3859. DOI: 10.1002/aoc.3859

[67] Singh C, Goyal A, Singhal S. Nickel-doped cobalt ferrite nanoparticles: Efficient catalysts for the reduction of nitroaromatic compounds and photo-oxidative degradation of toxic dyes. Nanoscale. 2014;**6**:7959-7970. DOI: 10.1039/c4nr01730g

[68] Tan JB, Sahoo P, Wang JW, Hu YW, Zhang ZM, Lu TB. Highly efficient oxygen evolution electrocatalysts prepared by using reduction-engraved ferrites on graphene oxide. Inorganic Chemistry Frontiers. 2018;**5**:310-318. DOI: 10.1039/c7qi00681k

[69] Maruthapandian V, Mathankumar M, Saraswathy V, Subramanian B, Muralidharan S. A study of oxygen evolution reaction catalytic behavior of $Co_xNi_{1-x}Fe_2O_4$ in alkaline medium. ACS Applied Materials & Interfaces. 2017;**9**:13132-13141. DOI: 10.1021/acsami.6b16685

[70] Bi K, Huang K, Zeng LY, Zhou MH, Wang QM, Wang YG, et al. Tunable dielectric properties of ferrite-dielectric based metamaterial. PLOS One. 2015;**10**:e0127331. DOI: 10.1371/journal.pone.0127331

[71] Velhal NB, Patil ND, Shelke AR, Deshpande NG, Puri VR. Structural, dielectric and magnetic properties of nickel substituted cobalt ferrite nanoparticles: Effect of nickel concentration. AIP Advances. 2015;**5**:097166. DOI: 10.1063/1.4931908

[72] Bhujun B, Tan MTT, Shanmugam AA. Study of mixed ternary transition metal ferrites as potential electrodes for supercapacitor applications. Respiration Physiology. 2017;7:345-353. DOI: 10.1016/j.rinp.2016.04.010

[73] Elkholy AE, El-Taib Heakal F, Allam NK. Nanostructured spinel manganese cobalt ferrite for high-performance supercapacitors. RSC Advances. 2017;7:51888. DOI: 10.1039/c7ra11020k

Chapter 6

Twist Tetrahedral-Tilting Structure Built from Photoluminescent Cadmium Chalcogenide Clusters

Wen-Chia Wu, Chung-Sung Yang and Yan Xu

Abstract

The newly synthesized cadmium chalcogenide ternary cluster is composed by six $[S_3Se]^{2-}$ tetrahedron units, coordinated with six Cd^{2+} cations. The potential cavity, calculated by the PLATON program, occupied 38.1% of crystal cell volume. The charge of unit cell is neutral. Therefore, the unit cell formula is determinate as $[Cd_6S_{18}Se_6]$. Two strong solid-state luminescence peaks, centered at 450 nm and 498 nm, were observed from the ternary $[Cd_6S_{18}Se_6]$ clusters by λ = 370 nm radiation. The 450 nm peak is due to the porosity property of cadmium chalcogenide clusters. However, the 498 nm peak has not been reported for the cadmium chalcogenide clusters before. In this study, we demonstrate that the 498 nm peak is attributed to the embedded Se atoms confined in the $[S3Se]^{2-}$ unit of $[Cd_6S_{18}Se_6]$ cluster. The luminescent output from the ternary $[Cd_8S_{18}Se_6]$ cluster is stable in room temperature for more than 6 months.

Keywords: cadmium chalcogenides, luminescence, porosity, quantum confinement

1. Introduction

For the past decade, an intensive interest has been shown in the area of developing porous optical supertetrahedral clusters, such as for selenides [1–5]. The direct assembly of chalcogenides into the cluster framework will be a benefit approach in the development of crystalline porous chalcogenides. Recently, the Tn supertetrahedral clusters [2–4] dominate the advantages in the field for using porous materials in luminescence application. The reason is that the larger size of the Tn supertetrahedral clusters usually leads to a highly open frame work with porosity higher than 50% [1, 5, 6]. Such a property is suitable for studying optoelectronic properties induced by the quantum size cluster.

Another famous porous structure, the perovskite clusters with R3C space group, is known for its ternary ABX_3 crystal structure as a light absorbent to be used in the solar cell area [7–9]. The porous spheres of perovskite $SrTiO_3$ spheres present superior performance in photocatalytic oxygen evolution [10]. The porous structure existed in the R3C type structure could be envisioned as an alternating choice to study the optoelectronic properties due to quantum confinement. To date, the use of existed porosity in R3C structure for quantum size luminescence application is very rare [11]. Here, we employ the $Cd(SC_6H_4Me\text{-}4)_2$ as the starting material to prepare a novel ternary cadmium chalcogenide cluster $[Cd_6S_{18}Se_6]$ for studying the effect of crystalline porous frameworks in quantum size luminescence. The $[Cd_6S_{18}Se_6]$ cluster is potentially with porous R3C structure. The simpler atomic composition of R3C structure apparently allows one to manipulate the replaced site with more flexibility. Such an opportunity cannot be easily obtained from the larger Tn clusters [1, 4]. Thus, these ternary clusters will be more applicable in the study of the optoelectronic luminescent phenomena created by the embedded hetero-atoms (i.e., Se) in the quantum size cluster. Therefore, the quantum confinement luminescence induced by the Se atom confined in $[S_3Se]^{2-}$ tetrahedron is able to be observed in the ternary $[Cd_6S_{18}Se_6]$ cluster.

2. Experimental

The starting material, $Cd(SC_6H_4Me\text{-}4)_2$, was prepared according to reference [12]. The product is white precipitate with yield up to 75% in weight. The typical target cadmium chalcogenides compound, denoted as NCYU-6, was prepared as follows. $Cd(SC_6H_4Me\text{-}4)_2$ (100 mg), thiourea powder (50 mg, 99%, Alfa Aesar), tettrahydrofuran (4.5 mL, 99.9%, Pharmco), selenium powder (10 mg, 99.5%, Acros), and de-ionized H_2O (2.1 mL) were dissolved in a 33 mL stainless steel autoclave with lined Teflon. The reactants were stirred for 60 min before the sealed vessel was heated at 85°C for 28 days. When the reaction is finished, the autoclave was cooled down to room temperature for several hours in atmosphere. The product was then washed by a mixture of ethanol and methanol several times to remove any residue on crystals. The crystal is pale yellow and is stable in air, water, or polar solvents for several months. The yield is generally below 30% in weight. It is worthy to note that the weight ratio between Se and S should be carried out very precisely. If the derivation of Se/S weight ratio is higher than ±5%, no crystal could be found in the final product. For the experimental purpose, a yellow crystal, NCYU-7, was prepared via the same route as that of NCYU-6, except the addition of selenium powder in the reactants. The yield of NCYU-7 is ~50% in weight.

3. Result and discussion

The scanning electron microscopy (SEM) image of NCYU-6 is shown in **Figure 1(a)**. The geometry of NCYU-6 is tetragonal. The qualitative energy dispersive X-ray (EDX) spectrum, **Figure 1(b)**, confirms the existence of Cd ($L\alpha$ = 3.13 eV, $L\beta$ = 3.528 eV), S ($K\alpha$ = 2.31 eV), and Se ($L\ \alpha$ = 1.388 eV) in NCYU-6. The geometry of NCYU-7 is similar to NCYU-7. The EDX spectrum shows that the Se peak is lack of in NCYU-7. Crystallographic structures were solved via single-crystal X-ray diffraction. Data were collected at room temperature on a SMART CCD diffractometer with Mo-$K\alpha$ radiation. All framework atoms can be determined and a brief crystal data of NCYU-6 is given in **Table 1**. The most unique structure in NCYU-6 is the $[S_3Se]$ tetragon that is built up by three corner sulfur sites and one center Se site. Each corner sulfur site is bonded to two Cd atoms, two S atoms of tetragon, and the center Se atom. By using the bond valence sum theory [13], the bond order between corner S site and center Se site is 4/3. The bond order between S and S is 1/3. The local structure illustration of $[S_3Se]$ tetragon moiety is given in **Figure 2(a)**. On the other hand, the Cd atom is coordinated with six S atoms of $[S_3Se]^{2-}$ tetrahedron units for charge balance. Each $[S_3Se]$ tetragon is twisted 180° with adjacent $[S_3Se]^{2-}$ tetragon. The bond order between S site and

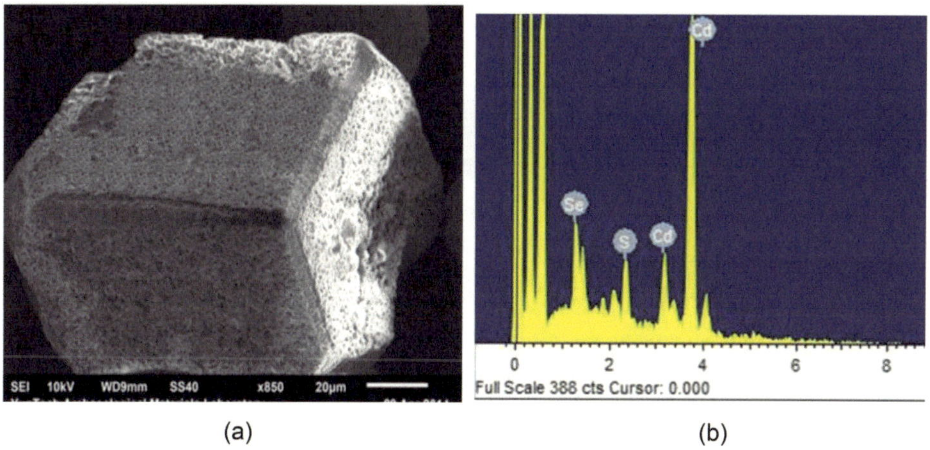

Figure 1.
(a) The SEM image of NCYU-6. The geometry of NCYU-6 is tetragonal. (b) the X-ray EDX spectrum confirms the existence of Cd (Lα = 3.13 eV, Lβ = 3.528 eV), S (Kα = 2.31 eV), and Se (L α = 1.388 eV) in NCYU-6.

Name	NCYU-6
Empirical formula	Cd_6S_{18} Se_6
Formula weight	1490.74
Temperature	295(2) K
Crystal system	Trigonal
Space group	R-3c
Unit cell dimensions	a = 4.9727(5) Å, α = 90°
	b = 4.9727(5) Å, β = 90°
	c = 16.981(3) Å, γ = 120°
Volume	363.64(9) $Å^3$
Z	1
Density (calculated)	6.807 Mg/m^3
Crystal size	0.05 × 0.03 × 0.03 mm^3
Reflections collected	949
Independent reflections	104 [R(int) = 0.0964]
Goodness-of-fit on F^2	1.331
Final R indices [I > 2sigma(I)]	R1 = 0.0735, wR2 = 0.2331
R indices (all data)	R1 = 0.0739, wR2 = 0.2344

Table 1.
Summary of crystal data and refinement for NCYU-6.

Cd site is assigned as 1/3. The local structure illustration of Cd atom coordinated with six S atoms of $[S_3Se]^{2-}$ units is given in **Figure 2(b)**. In the determination of Se location, the refined bond length data are used to verify the Se locations obtained from the R(F) value method. The bond lengths for six Cd-S locations are varied from 2.3476 Å to 2.3498 Å. The bond length of S_{1a} to center atom site (1.281 Å) is larger than the covalent radius of Se (1.16 Å) or (1.02 Å) [14]. Under the circumstance, replacement of the center site from S to Se is allowed by the architecture of

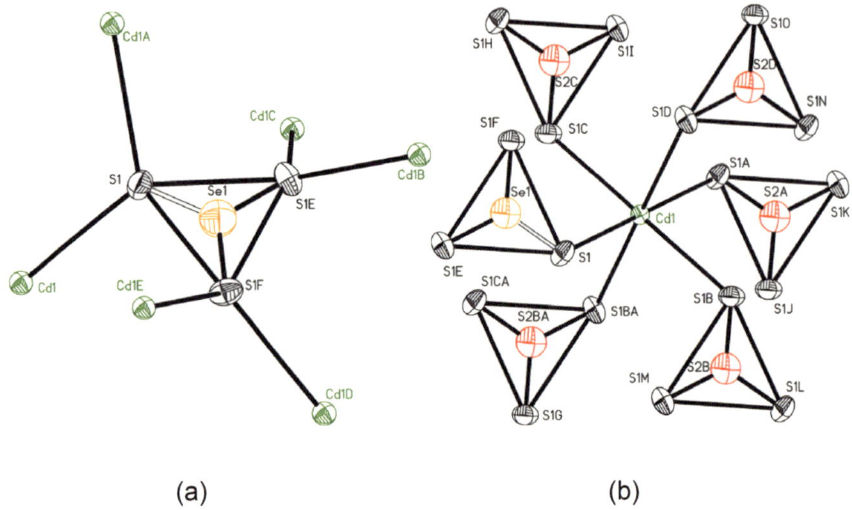

Figure 2.
(a) The local structure illustration of $[S_3Se]$ tetragon moiety. The center site is marked with tangerine ball for Se atom. (b) the local structure illustration of Cd atom coordinated with six $[S_3Se]^{2-}$ units. One $[S_3Se]$ tetragon is shown to compare the size of thermal ellipsoids in ORTEP between S and Se. If the other five center S sites are replaced by Se, the $R(F)_{all\ data}$ slightly increases from 5.27 to 7.39%.

tetragon, as shown in **Figure 2(a)**. During the crystallography determination for the location of Se, we found that the refined R(F) value is only slightly increased by an arbitrarily replacement of the center site of tetragon from S to Se atoms. For example, $R(F)_{all\ data}$ = 7.39% for $[Cd_6S_{18}Se_6]$ and 5.27% for $[Cd_6S_{24}]$. Nevertheless, the arbitrarily replacement of the corner S site of tetragon by Se atom will make the R(F) value increase more than 60%, $R(F)_{all\ data}$ ~ 11.73%. Moreover, the size of thermal ellipsoids in ORTEP for the bridging location with Cd will increase to unreasonably large size. It is clear that the bridging location occupied by S atom is reasonable [1]. On the contrary, the size of thermal ellipsoids in ORTEP for the center site in tetragon is smaller, in comparison with the other corner S sites [1]. The center site occupied by Se atom is probable. The calculated occupancy possibility of the center site for Se to S is about ~80%. The occupancy possibility of the bridging site in tetragon cluster for Se to S is about ~20%. The temperature is an important factor in the refinement of S and Se locations. The results suggest that the center site of tetragon may have an affinity for Se atom and the hetero-atom substitution leads the new tetragonal-title architecture stable.

The atomic ratio for the NCYU-6 and NCYU-7 clusters was determined by the EDX spectrum qualitatively and the ICP-MS quantitatively. The quantitative elemental microanalysis of Cd/S for NCYU-7 is about 6:24 that is consistent with the composition and structure assignment by the crystallography. As for NCYU-6, the ratio of Se/S is analyzed as a function of the atom ratio (Se/S) of Se and S by ICP-MS. Four samples are prepared for the element concentration analysis for S and Se. The 5% HNO3 solution is employed to dissolve nine single crystals for the first and second ICP-MS measurements. As for the third, and forth measurements, 12 single crystals are picked up, respectively. The quantitative microanalysis data, provided in **Table 2**, show that the atomic ratio of Se (108 ppb) to S (335 ppb) in NCYU-6 is ~32.3%, i.e., Se/S ~ 1:3. By the combination of the data obtained from the above experiments, the unit cell formula is determined as $[Cd_6S_{18}Se_6]$ for NCYU-6 (**Figure 3**). The unit cell formula for NCYU-7 is $[Cd_6S_{24}]$.

The porosity data of the NCYU-6 were calculated by the PLATON crystallographic program, a versatile SHELX97 compatible multipurpose crystallographic

Wavelength	S = 180.771 nm		Se = 196.126 nm	
Sample_Name	Cc (ppb)	RSD (%)	Cc (ppb)	RSD (%)
1	351	2.2	117	2.8
2	347	1.1	118	2.9
3	326	0.9	103	1.5
4	326	0.9	102	1.5
Avg	337		108	

Table 2.
Summary of ICP-MS data of S and Se concentration in NCYU-6.

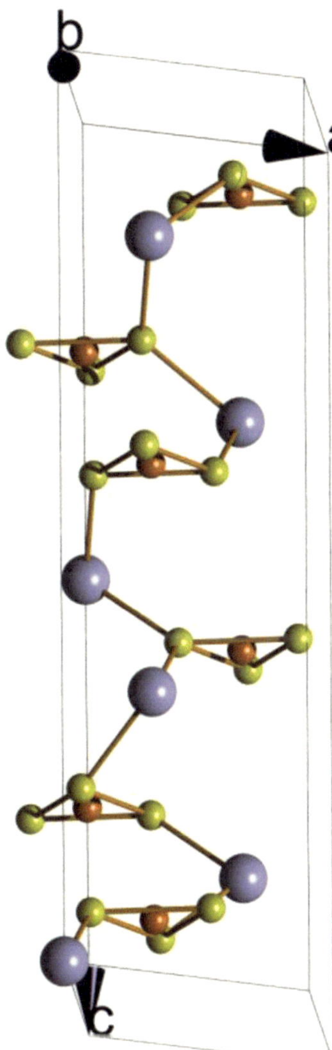

Figure 3.
The unit cell formula is determined as [$Cd_6S_{18}Se_6$] for NCYU-6. The blue ball stands for Cd atom. Dark yellow ball stands for S atom. Tangerine ball stands for Se atom.

toolkit software program [15]. In per unit cell of NCYU-6 (363.64 Å^3), the volume occupied by solvent is 138.5Å^3. The potential cavity occupied 38.1% of crystal cell volume. The percentage of cavity derived from NCYU-6 suggests that considerable

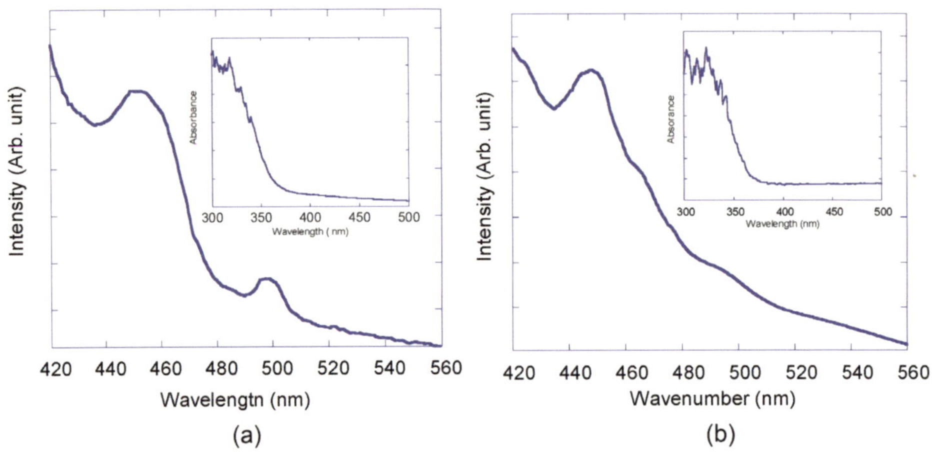

Figure 4.
Solid-state PL spectra were excited at 370 nm at RT. (a) Two emission peaks, centered at 450 nm and 498 nm, are revealed from NCYU-6. (b) Only one peak (450 nm) is observed from NCYU-7. The inset figures are DRS spectra for NCYU-6 and NCYU-7, respectively.

amount of guest molecules, i.e., tetrahydrofuran, presents in the structure of open framework. The guest molecules possibly occupied two domains to make this material with porosity. The possible domains are the in-plane pore window and the space between layers. Since the framework of the layer is wavy, the distance between layers is flexible and is adjustable by the template.

The UV–vis diffuse reflectance spectrum (DRS) of NCYU-6 and NCYU-7 are given in inset of **Figure 4(a)** and **(b)**, respectively. The bandgap of the NCYU-6 is 390 nm (3.14 eV) is 380 nm (3.26 eV) for NCYU-7. The suitable bandgaps make these two compounds possible to be used in photocatalytic reaction of hydrogen generation under UV and visible blue light irradiation [16–18].

Solid-state photoluminescence spectra (PL) were carried on Hitachi F-2500 fluorescence spectrophotometer. The emission spectra were excited at 370 nm at room temperature (RT). In the RT spectra, two emission peaks, centered at about 450 nm and 498 nm, are revealed from NCYU-6, and only one peak (450 nm) is observed for NCYU-7 (**Figure 4(a)** and **(b)**). The peak at 450 nm is generally attributed to the porosity property of open framework [1, 5]. The porosity created by the tilt-tetragonal structure allows the insertion of organic template molecule, tetrtahydrofuran [1, 5]. Thus, the conversion of UV radiation (370 nm) to visible violet light, ~450 nm, is possible. The reason is that the emission peak is induced by the interaction between the inorganic and organic phosphors [19, 20]. So far, the 498 nm emission peak has not been reported for the title-tetragonal R-3c structure before. The replacement of S atom(s) by Se atom(s) in the NCYU-6 is expected to output an absorption peak due to the quantum confinement of Se atom(s) [1, 5]. More experimental data are needed to explore the shift mechanism of the PL peak. However, it is believed that the red-shifted band-edge emission (P-type dopant) is relevant with the center site of Se atoms in tilt-tetragonal R−3c structure.

4. Conclusion

The novel ternary [$Cd_6S_{18}Se_6$] cluster is prepared via a hydrothermal method. The space group of [$Cd_6S_{18}Se_6$] is R3C. The six [S_3Se] tetragon units of [$Cd_6S_{18}Se_6$] are twisted 180 degree with each other. The cavity occupied 38.1% of [$Cd_6S_{18}Se_6$] crystal cell volume. In the solid-state PL spectra, the peak centered at 450 nm is induced by the porosity property of cadmium chalcogenides clusters, i.e., the interaction between the inorganic and organic phosphors inside the cavity of unit cell. The peak located at 498 nm is due to the quantum confinement of Se atom(s) in [S_3Se] tetragon unit.

Acknowledgements

The research is partially supported by the Ministry of Science and Technology, Taiwan, ROC (grant no. NSC 103-2622-M-415-001-CC3). Authors would like to thank Chi-Chia Chuang's help in PLATON calculation. Yan Xu would like to thank the student exchange program between National Chia Yi University and Northwest A&F University.

Author details

Wen-Chia Wu[1], Chung-Sung Yang[1*] and Yan Xu[2]

1 Department of Applied Chemistry, National Chia Yi University, Chiayi, Taiwan, ROC

2 Department of Applied Chemistry, Northwest A&F University, Xianyang, Shanxi, People's Republic of China

*Address all correspondence to: csyang@mail.ncyu.edu.tw

References

[1] Chen C-J, Yang C-S, Lin X-H. Synthesis, characterization, and photoluminescence of quaternary [Cd4In16S33 − xSex]10− supertetrahedral clusters: (0.33 < x < 0.45). Inorganic Chemistry Communications. 2005;**8**:836-840

[2] Li H, Laine A, O'Keeffe M, Yaghi OM. Supertetrahedral sulfide crystals with giant cavities and channels. Science. 1999;**1999**(283):1145-1147

[3] Bu X, Zheng N, Wang X, Wang B, Feng P. Three-dimensional frameworks of gallium selenide supertetrahedral clusters. Angewandte Chemie, International Edition. 2004;**43**:1502-1505

[4] Wang C, Lin Y, Bu X, Zheng N, Zivkovic O, Yang C-S, et al. Three-dimensional superlattices built from (M4In16S33)10- (M = Mn, Co, Zn, Cd) supertetrahedral clusters. Journal of the American Chemical Society. 2001;**123**:11506-11507

[5] Chen C-Y, Ou C-C, Huang H-F, Cheng J-H, Yang C-S. Mixed pentasupertetrahedral P1 and supertetrahedral T2 clusters as building units to create two-dimensional indium chalcogenides open framework. Inorganic Chemistry Communications. 2011;**14**:1004-1009

[6] Bu X, Zheng N, Li Y, Feng P. Pushing up the size limit of chalcogenide supertetrahedral clusters: Two- and three-dimensional photoluminescent open frameworks from $(Cu_5In_{30}S_{54})^{13-}$ clusters. Journal of the American Chemical Society. 2002;**124**:12646-12467

[7] Baikie T, Fang Y, Kadro JM, Schreyer MK, Wei F, Mhaisalkar SG, et al. Synthesis and crystal chemistry of the hybrid perovskite $(CH_3NH_3)PbI_3$ for solid-state sensitised solar cell applications. Journal of Materials Chemistry A. 2013;**1**:5628-5641

[8] Kim HS, Lee JW, Yantara N, Boix PP, Kulkarni SA, Mhaisalkar S, et al. High efficiency solid-state sensitized solar cell-based on submicrometer rutile TiO_2 nanorod and $CH_3NH_3PbI_3$ perovskite sensitizer. Nano Letters. 2013;**13**:2412-2417

[9] Burschka J, Pellet N, Moon S-J, Humphry-Baker R, Gao P, Nazeeruddin MK, et al. Sequential deposition as a route to high-performance perovskite-sensitized solar cells. Nature. 2013;**499**:316-319

[10] Pan JH, Shen C, Ivanova I, Zhou N, Wang X, Tan WC, et al. Self-template synthesis of porous perovskite titanate solid and hollow submicrospheres for photocatalytic oxygen evolution and mesoscopic solar cells. ACS Applied Materials & Interfaces. 2015;7:14859-14869

[11] Sichert JA, Tong Y, Mutz N, Vollmer M, Fischer S, Milowska KZ, et al. Quantum size effect in organometal halide perovskite nanoplatelets. Nano Letters. 2015;**15**:6521-6527

[12] Dance I, Garbutt RG, Craig DC, Scudder ML. The different nonmolecular polyadamantanoid crystal structures of cadmium benzenethiolate and 4-methylbenzenethiolate. Analogies with microporous aluminosilicate frameworks. Inorganic Chemistry. 1987;**26**:4057-4064

[13] Altermatt D, Brown ID. Bond-valence parameters obtained from a systematic analysis of the inorganic crystal structure database. Acta Crystallographica. Section B. 1985;**41**:244-247

[14] Data adopted from Cambridge Structural Database (CSD)

[15] Speck AL. Single-crystal structure validation with the program PLATON. Journal of Applied Crystallography. 2003;**36**:7-13

[16] Ou C-C, Yang C-S, Lin S-H. Selective photo-degradation of Rhodamine B over zirconia incorporated titania nanoparticles: A quantitative approach. Catalysis Science & Technology. 2011;**1**:295-307

[17] Lin S-H, Ou C-C, Su MD, Yang C-S. Photo-catalytic behavior of vanadia incorporated titania nanoparticles. Catalysis Science & Technology. 2013;**3**:2081-2091

[18] Chang Y-H, Ou C-C, Yeh H-W, Yang C-S. Photo-catalytic selectivity of anthranilic acid over iron oxide incorporated titania nanoparticles: Influence of the Fe_2+/Fe_3+ ratio of iron oxide. Journal of Molecular Catalysis A: Chemical. 2016;**412**:67-77

[19] Feng P. Photoluminescence of open-framework phosphates and germanates. Chemical Communications. 2001:1668-1669

[20] Xu X, Wang W, Liu D, Hu D, Wu T, Bu X, et al. Pushing up the size limit of metal chalcogenide supertetrahedral nanocluster. Journal of the American Chemical Society. 2018;**140**:888-891

Chapter 7

Transition Metal Chalcogenides for the Electrocatalysis of Water

Chi-Ang Tseng and Chuan-Pei Lee

Abstract

Sustainable energy technology has received enormous attention in recent years. Specifically, electrochemical water splitting is considered to be the cleanest technique for the production of promising fuels, for example, hydrogen and oxygen, where transition metal (di)chalcogenides (TMCs) as electrocatalysts for hydrogen evolution reaction (HER) and oxygen evolution reaction (OER) have been a growing interest. In this chapter, the typical preparation methods of TMCs such as chemical vapor phase deposition (CVD) and solvothermal synthesis are introduced. Then, several TMC materials for catalyzing HER and OER are reviewed. Most importantly, this chapter also introduced some *in situ* approaches to realize the mechanism of electrocatalytic behavior toward HER and OER. Finally, the conclusion and futuristic prospects of TMCs in HER and OER are discussed.

Keywords: energy conversion, hydrogen evolution reaction (HER), oxygen evolution reaction (OER), transition metal chalcogenides (TMCs)

1. Introduction

Energy crisis and environmental pollution arising from the burning of carbon-based fossil fuel in the past decades facilitate people to reconsider the way we utilized the resource on earth. Hydrogen as an ideal energy source came up to the stage due to its high energy density and environmental benignity [1, 2]. Electrochemical water splitting is not only regarded as the cleanest technique for hydrogen generation but also suitable to perform on a large scale. The appropriate electrocatalysts are developed to boost the cathodic hydrogen evolution reaction (HER) and anodic oxygen evolution reaction (OER) because of the sluggish kinetics of these two core reactions [overpotential (η), **Figure 1a**]. Although platinum-based and iridium/ruthenium-based catalysts, respectively, have shown very promising performance in HER and OER, the industrial application is restricted due to their high cost and limited availability [3, 4]. Thus, developing alternative electrocatalysts based on non-noble metals or earth-abundant elements is still highly demanded.

The realization of HER and OER mechanisms plays a crucial role to design the efficient electrocatalysts, so the brief discussions on their mechanisms were introduced.

Two kinds of HER mechanisms have been completely studied and widely accepted [5, 6]. The first step, a proton from the solution adsorbs onto the catalytic sites of the electrode with a reduction process, is called as Volmer step:

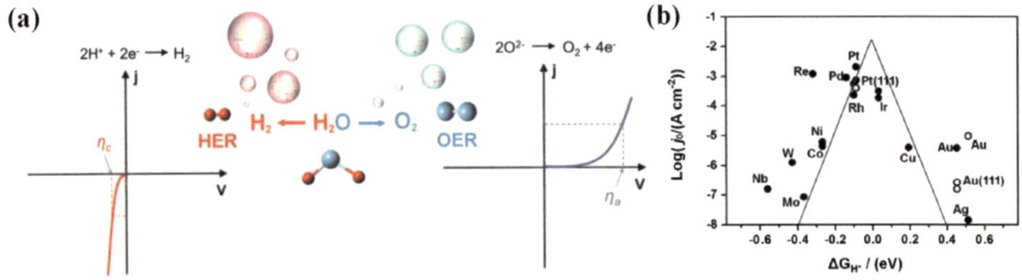

Figure 1.
(a) Polarization curves for HER and OER [9] and (b) volcano plot of exchange current density (j_o) as a function of DFT-calculated Gibbs free energy (ΔG_H) of adsorbed atomic hydrogen on pure metals [7].

$$H^+_{(aq)} + e^- \rightarrow H^{\bullet}_{(ads)} \tag{1}$$

Then, either the recombination of two adsorbed hydrogen atoms on the electrode surface is called as Tafel step:

$$H^{\bullet}_{(ads)} + H^{\bullet}_{(ads)} \rightarrow H_{2(g)} \tag{2}$$

or the reaction of an adsorbed hydrogen atom with the hydrated proton, which proceeds with an electron transfer from the electrode surface, is called as Heyrovsky step:

$$H^{\bullet}_{(ads)} + H^+_{(aq)} + e^- \rightarrow H_{2(g)} \tag{3}$$

Nørskov et al. proposed a Volcano curve showing that experimental exchange currents of materials are as a function of the Gibbs free energy of the adsorbed hydrogen (ΔG_H) (**Figure 1b**) [7, 8]. Ideally, the interaction of hydrogen with the electrode surface should be thermoneutral ($\Delta G_H \cong 0$), otherwise either Heyrovsky or Tafel step (strong bonding) or Volmer step would become the rate-determining step. In the case of the OER, many researchers have proposed possible mechanisms at the anode in acidic electrolyte (Eqs. (4)–(8)). However, there are some differences around the reaction of forming oxygen. One route of forming oxygen is through the direct recombination of two MO (M represents Mn, Fe, Co, and Ni metals) (Eq. (6)), while the other route of forming oxygen is through the decomposition of the MOOH intermediate (Eq. (8)).

$$M + H_2O_{(l)} \rightarrow MOH + H^+ + e^- \tag{4}$$

$$MOH + OH^- \rightarrow MO + H_2O_{(l)} + e^- \tag{5}$$

$$2MO \rightarrow 2M5 + O_{2(g)} \tag{6}$$

$$MO + H_2O_{(l)} \rightarrow MOOH + H^+ + e^- \tag{7}$$

$$MOOH + H_2O_{(l)} \rightarrow M + O_{2(g)} + H^+ + e^- \tag{8}$$

2. Preparations of TMCs

2.1 CVD synthesis of TMCs

The preparation methods of TMCs layers can be categorized into two main approaches: top-down and bottom-up methods (**Figure 2**). The TMC layers

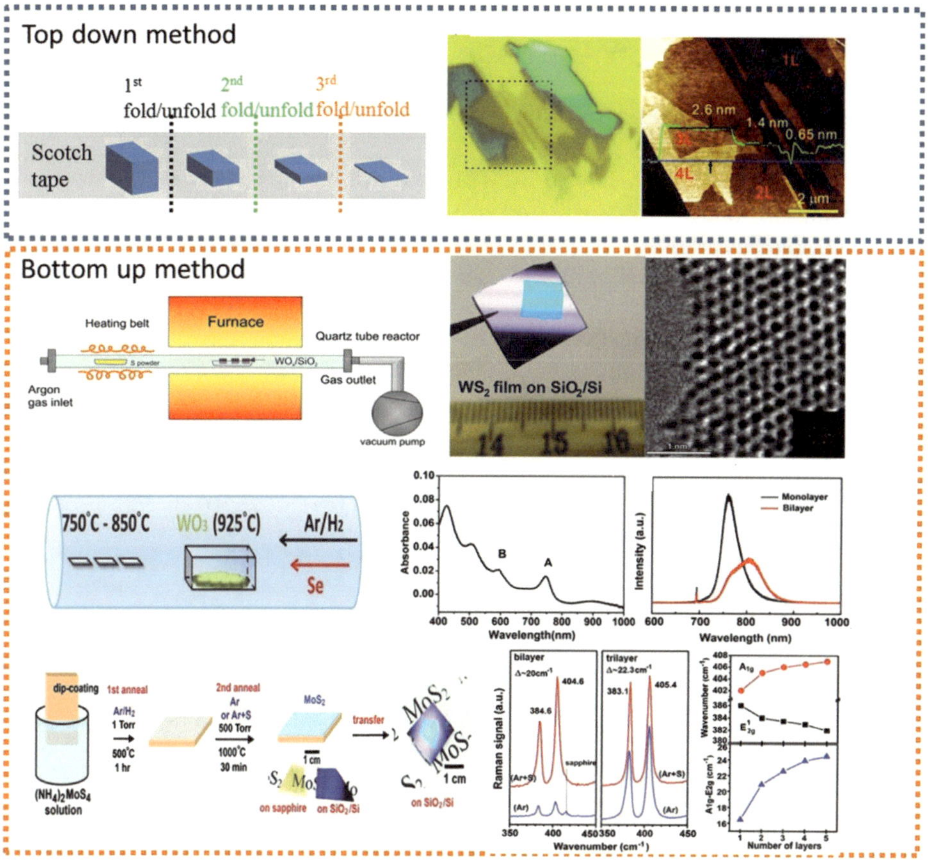

Figure 2.
Two routes of the TMC layer preparation: (top) a schematic representation of the top-down method – mechanical exfoliation and (bottom) a schematic illustration of the bottom-up method – CVD [22].

prepared from their bulk material by mechanical exfoliation are called the top-down method, while the TMC layers were produced from the elemental precursors on the target substrate in the bottom-up method, such as chemical vapor deposition (CVD) approach. The CVD method for TMC synthesis has been widely used due to its unique advantages such as scalable size, high crystallinity, and controllable thickness of TMCs [10–13]. One of the most classic TMCs, molybdenum sulfide (MoS_2), has been developed via CVD synthesis to replace the zero bandgap graphene. [14] There are two routes to synthesize MoS_2 layer by CVD (**Figure 3a**). One is a two-step growth route, where the Mo-based precursors are first deposited and then conducted the sulfurization or decomposition process to form MoS_2 (route 1, **Figure 3a**). The other growth way of MoS_2 layer is an one-step growth, where the gaseous Mo- and S-based precursors are simultaneously introduced and react to form MoS_2 on a substrate (route 2, **Figure 3a**). Recent reports have also demonstrated to grow MoS_2 on other kinds of insulating substrates such as quartz, mica, and sapphire (**Figure 3b**) [15, 16]. **Figure 3c** shows a typical experimental setup for the low-pressure CVD synthesis of MoS_2. The formation of entire MoS_2 would be prevented according to the ternary Mo—O—S phase diagram (**Figure 3d**) if the reducing atmosphere is too weak [17]. **Figure 3e** displays two possible mechanisms for the growth of MoS_2. The well-established CVD synthesis of MoS_2 is regarded as a prototype for the synthesis of other TMCs such as $MoSe_2$ and WS_2 [18, 19]. Except from the VI B group metals such as molybdenum and tungsten, the research works of TMCs based on VIII B group metals such as cobalt [20] and nickel [21] are also widely studied in recent years.

2.2 Solvothermal synthesis of TMCs

Wet chemical synthesis, a bottom-up process, is broadly used to synthesize TMC nanosheets with tunable thickness and size [24–26]. The desired nanosheets form directly in solution or onto substrate in high yield. The synthesis of TMCs via wet chemical method usually relies on the chemical reaction of metal salts and sulfur/selenide-based materials as precursors. The solvothermal method is one of the most typical routes of wet chemical method, in which the reaction is conducted in a sealed autoclave at suitable temperature (**Figure 4**) [27]. As a representative example, Xie et al. [28] synthesized oxygen incorporated MoS_2 nanosheets from the

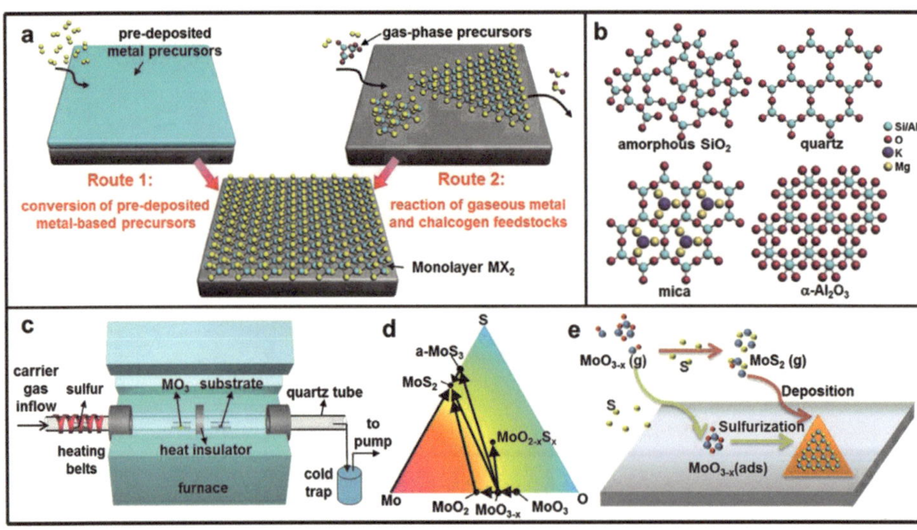

Figure 3.
The general introduction to the CVD synthesis, CVD setup, and growth mechanism of MoS_2 layer [23].

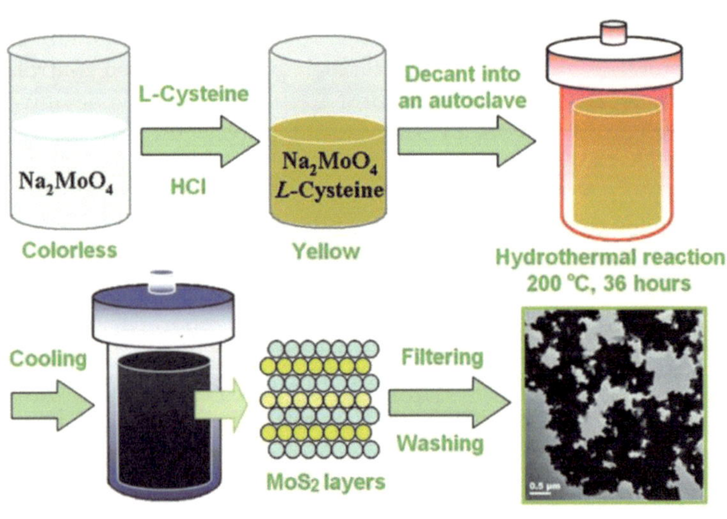

Figure 4.
The synthesis process of MoS_2 nanosheets via a simple hydrothermal method [27].

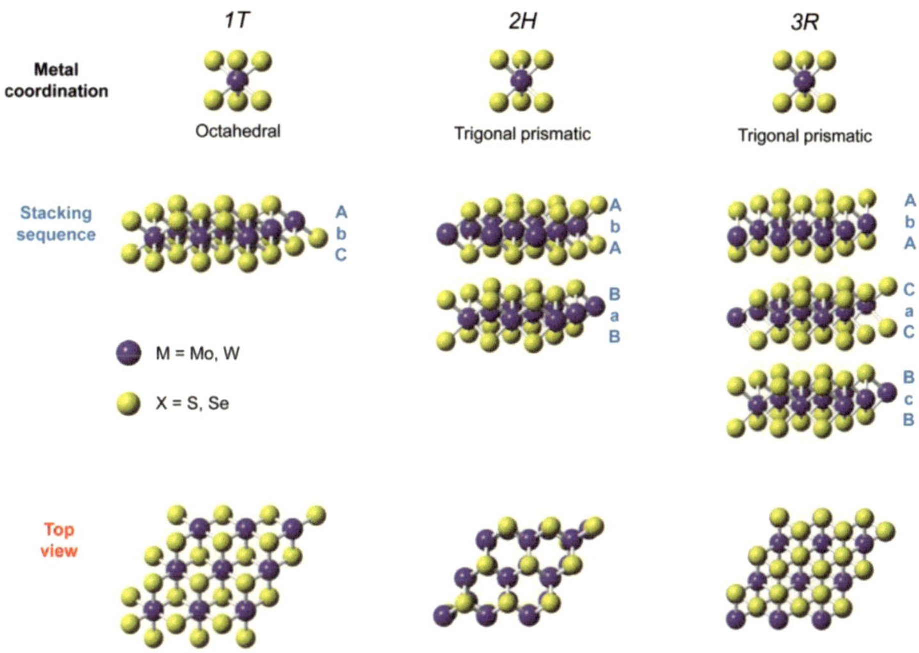

Figure 5.
Different metal coordination and stacking sequence in TMC unit cells [10].

precursors of $(NH_4)_6Mo_7O_{24}·4H_2O$ and thiourea through a solvothermal method. By tuning the synthetic temperatures, the disordered structure and intrinsic conductivity of MoS_2 could be controlled to a moderate degree. Thus, the optimal catalyst for electrocatalytic hydrogen evolution reaction exhibits onset overpotential as low as 120 mV. It is well known that the materials with different crystal orientations have the anisotropic properties. This phenomenon appears in TMCs depending on stacking sequence of the chalcogen and transition-metal coordination. The commonly found polymorphs in TMC materials are the so-defined 1 T, 2H, and 3R, where the number and letter, respectively, indicate the layer number in the unit cell and the type of symmetry with T, H, and R representing for tetragonal, hexagonal, and rhombohedral, respectively (**Figure 5**). For example, Zhang et al. synthesized metallic CoS_2 nanopyramid array on carbon fiber paper by an one-step solvothermal synthesis [29]. In brief, the carbon fiber papers were transferred to the precursor solution containing $CoCl_2·6H_2O$ and $CS(NH_2)_2$, which was maintained at 180°C for 12 hours. The unique 3D nanostructure and intrinsic metallic properties of the CoS_2 by this method contribute to the ultrahigh activity toward hydrogen evolution reaction.

3. Hydrogen evolution reaction (HER) based on TMC electrodes

In order to improve HER performance, three main factors including the number of active sites, intrinsic catalytic activity, and the conductivity of TMCs play crucial roles. In 2005, Hinnemann et al. suggested the active sites for HER only exist in the edges of TMCs by density functional theory (DFT) calculation [30]. Thus, many research works are dedicated to enrich the defect sites and/or active sites by nanostructural engineering [31–33]. Zhang et al. synthesized the edge-rich 3D MoS_2 coupling with conductive polymer polyaniline (PANI) as catalyst (**Figure 6**) [34].

The MoS_2 grown on 3D PANI substrate tends to grow vertically and expose abundant edge sites for HER. Consequently, excellent HER performance can be achieved with a low onset potential of 100 mV and a small Tafel slop of 45 mV dec^{-1}.

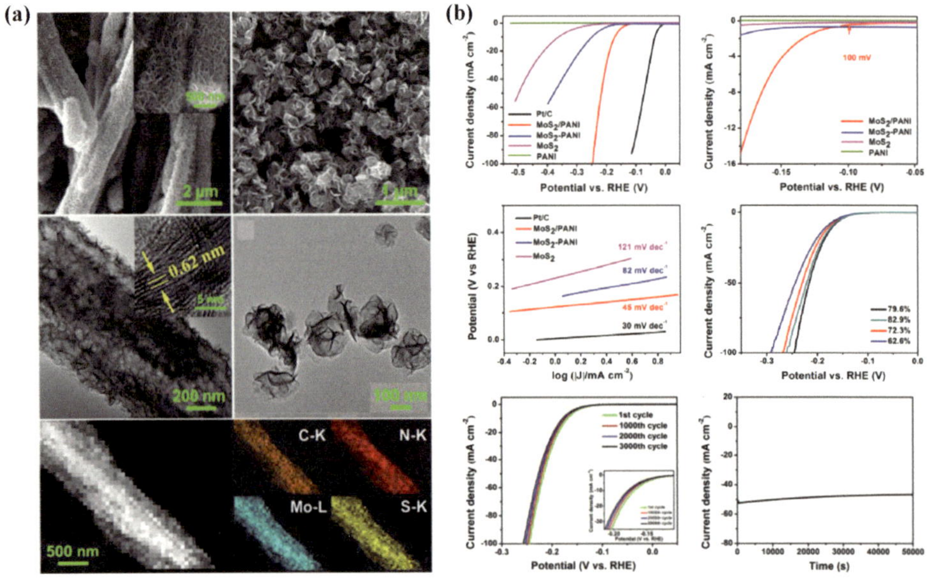

Figure 6.
The defect-rich 3D MoS_2/PANI catalyst for HER (a) the morphologies of synthesized 3D MoS_2/PANI and (b) the electrochemical performance of 3D MoS_2/PANI catalyst [34].

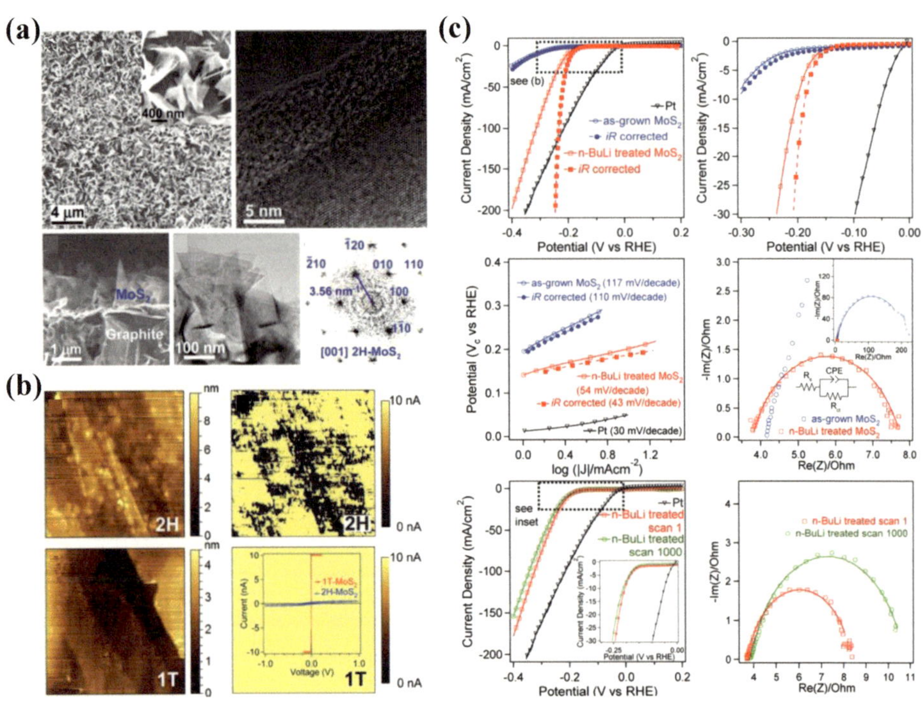

Figure 7.
(a) Electron microscopy characterization of as-grown 2H-MoS_2 nanostructures. (b) Comparison of as-grown and exfoliated MoS_2 nanosheets. (c) Electrocatalytic performance of chemically exfoliated and as-grown MoS_2 nanosheets [36].

Additionally, MoS_2/PANI achieved superior stability for HER electrocatalysis. Although the promising catalytic activity of MoS_2 for HER was achieved by creating abundant edge sites and/or active sites, the performance was still limited by its intrinsic properties such as poor electrical transport and inefficient electrical contact to the catalyst [35]. Lukowski et al. reported that the metallic nanosheets of 1T-MoS_2, which were chemically exfoliated by lithium intercalation of semiconducting 2H-MoS_2 nanostructures, dramatically enhanced HER performance (**Figure** 7) [36]. The current density of 10 mA cm^{-1} can be reached at a low overpotential of −187 mV with a Tafel slope of 43 mV dec^{-1}. The excellent performance can be attributed to the favorable kinetics, metallic conductivity, and increasing number of active sites in the metallic 1T-MoS_2 nanosheets, which was proven by the dramatically decrease of charge-transfer resistance from 232 Ω of 2H-MoS_2 nanostructures to 4 Ω of metallic 1T-MoS_2 nanosheets. This finding proves that the metallic 1T polymorph of TMCs is competitive to earth-abundant catalysts in heterogeneous catalysis.

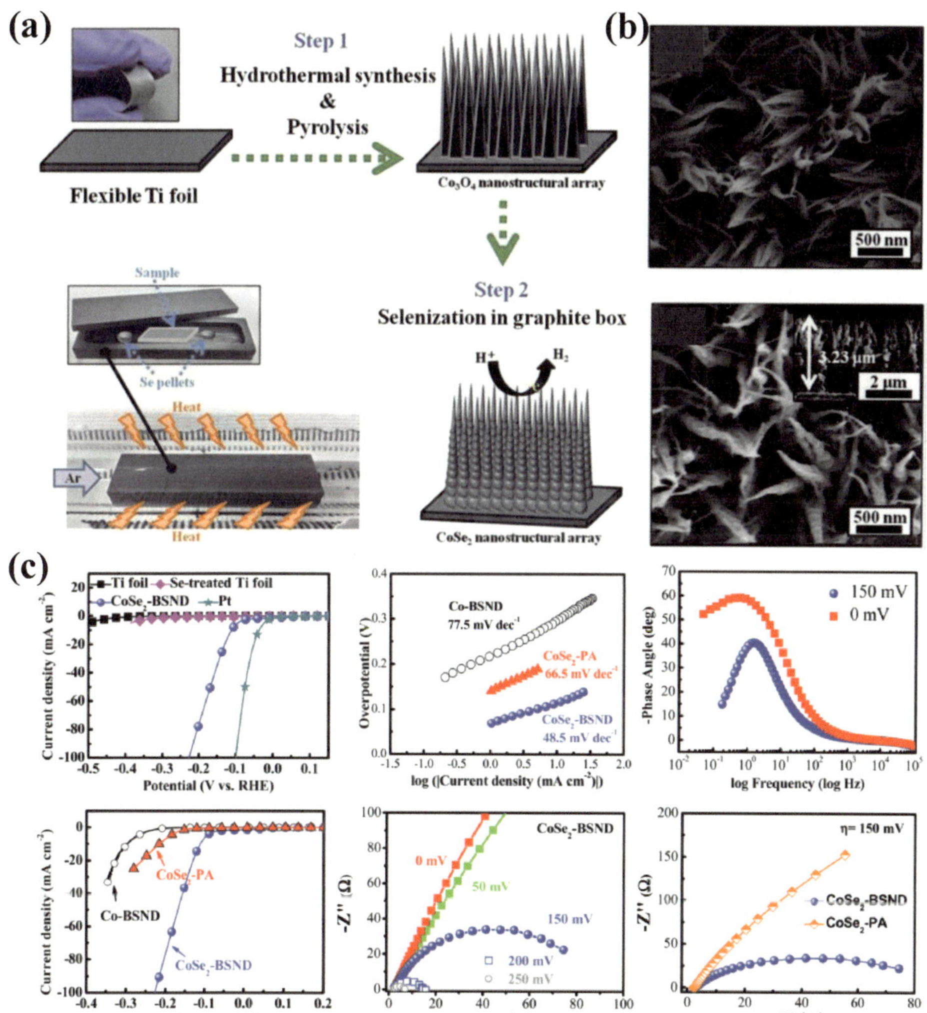

Figure 8.
(a) The schematic illustration for the synthesis of beaded stream-like $CoSe_2$ nanoneedles, (b) the SEM images of beaded stream-like ($CoSe_2$) nanoneedle array, and (c) the electrochemical measurements of Co-BSND, $CoSe_2$-PA, and $CoSe_2$-BSND electrodes [37].

Lee et al. developed earth-abundant nanostructuring beaded stream-like cobalt diselenide ($CoSe_2$) nanoneedles ($CoSe_2$-BSND) as electrocatalyst for HER [37]. The $CoSe_2$ nanoneedles derived from the cobalt oxide (Co_3O_4) nanoneedle array directly formed on flexible titanium foils after selenization treatment (**Figure 8**). The $CoSe_2$-BSND can drive the HER at a current density of 20 mA cm^{-2} with a small overpotential of 125 mV. Also, it possesses a small Tafel slope of 48.5 mV dec^{-1} suggesting that the HER follows the Volmer-Heyrovsky mechanism where a fast discharge of protons is followed by rate-determining electrochemical desorption. Moreover, the $CoSe_2$-BSND electrode achieved great stability in an acidic electrolyte for 3000 cycles. The enhanced electrochemical activity is attributed to the highly accessible surface active sites, the improved charge transfer kinetics, and the super hydrophilic surface of $CoSe_2$-BSND electrode.

4. Oxygen evolution reaction (OER) based on TMC electrodes

The TMC materials have shown promising performance toward HER as we mentioned above. Researchers are searching for the possibility of bifunctional electrocatalysts for both HER and OER to perform the overall water-splitting reaction. Moreover, TMC electrocatalysts have attracted tremendous attentions since Alonso-vante and coworker discovered that $Mo_4Ru_2Se_8$ had a ORR activity comparable to platinum [38]. As compared to ruthenium (Ru) and Rhenium (Rh), the low cost and earth abundant transition metals such as iron-, nickel- and cobalt-based TMCs have much attention for OER [39–41]. Liu and coworkers discovered that electrodeposited CoS nanosheet films on Ti mesh show high activity toward OER (**Figure 9**) [42]. The CoS nanosheets tend to drive a current density of 10 mA cm^{-2} with an overpotential of 361 mV. In addition, this electrode maintains highly catalytic activity for at least 20 hours. The superior catalytic activity along with excellent stability of CoS nanosheets offers the good opportunity to become a cost-effective and industry-feasible electrode toward OER. On the other hand, Swesi et al. first reported that the catalytic activity of OER was observed by the nickel selenide (Ni_3Se_2) in alkaline condition (**Figure 10**) [43]. The low overpotential

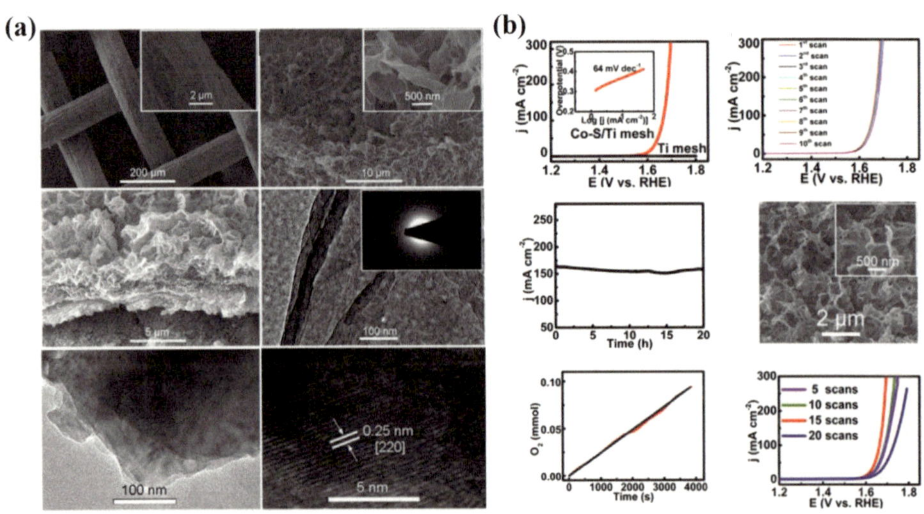

Figure 9.
(a) The surface morphology and crystal structure of electrodeposited CoS nanosheets and (b) the electrochemical performance and stability toward OER on CoS nanosheets [42].

required to reach 10 mA cm^{-2} was 290 mV, suggesting that this catalyst exhibits its competitivity among the oxide-based electrocatalysts. The catalytic ability of Ni_3Se_2 can be further improved through the modification of Se-deficient phase in Ni_3Se_2. Moreover, electrodeposited Ni_3Se_2 catalysts exhibited exceptional stability under OER for 42 hours. The effect of the underlying substrates such as glassy carbon, ITO-coated glass, and Ni foam on OER was also investigated. The results revealed that the glassy carbon substrate exhibited the lowest onset potential and the highest

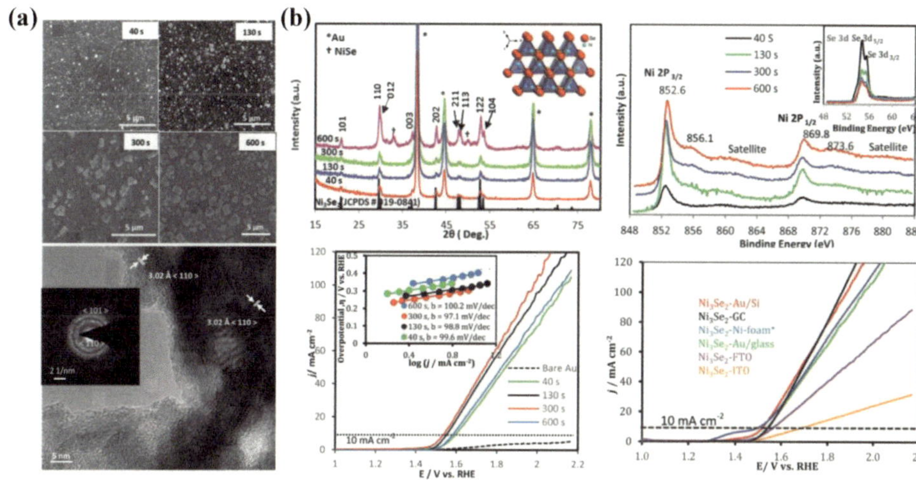

Figure 10.
(a) The morphological characterization of Ni_3Se_2 grown by electrochemical deposition and (b) the crystal structure identification and OER performance of Ni_3Se_2 electrocatalyst [43].

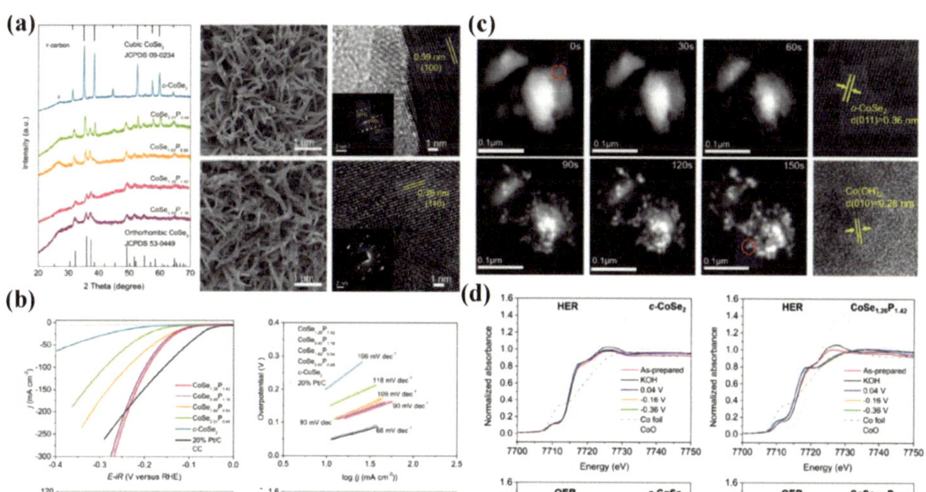

Figure 11.
(a) The structural and morphological characterizations of different P-doped $CoSe_2$, (b) electrochemical OER activities of the different P-doped $CoSe_2$ and standard RuO_2 electrodes, (c) in situ *STEM images of the P-doped $CoSe_2$ catalyst taken at different times after immersing in the KOH solution, and (d)* in situ *Co K-edge XANES spectra of different P-doped $CoSe_2$ electrodes for HER and OER processes [3].*

current density, suggesting that the interaction between the underlying substrate and the Ni_3Se_2 may play a role in O_2 evolution reaction.

Although the electrocatalytic performance of TMCs toward OER was significantly enhanced, the reaction mechanism and actual active sites responsible for the reaction were in dispute. Recently, operando or *in situ* experiments like *in situ* Fourier transform infrared spectroscopy, *in situ* Raman spectroscopy, and *in situ* X-ray absorption/diffraction are commonly carried out to provide atomic-level information [3, 44, 45]. Zhu and coworkers [3] conducted *in situ* X-ray absorption spectroscopy, *in situ* liquid-phase TEM, and *in situ* Raman spectroscopy, revealing that P-doped $CoSe_2$ in an alkaline solution was acting as the "pre-catalyst" rather than the real reactive species, which has been debated for a while (**Figure 11**). They found that the introduction of phosphorus would generate more vacancies, which facilitated the structural transformation into the real active electrocatalyst, such as metallic cobalt for HER and cobalt oxyhydroxide (CoOOH) for OER. $CoSe_{1.26}P_{1.42}$ shows the best OER performance among all catalysts, which requires an overpotential of 255 mV to reach the current density of 10 mA cm^{-2}. Furthermore, the $CoSe_{1.26}P_{1.42}$ catalyst with a Tafel slop of 87 mV dec^{-1} exhibits a slightly lower than those of the other P-doped $CoSe_2$ catalysts. Such performance is comparable to many leading earth-abundant HER and OER catalysts in alkaline electrolyte.

5. Conclusion

Electrocatalysts for HER and OER play an important role for sustainable energy, which require converting renewable energy to storable chemical fuels or employing clean energy. In this article, we have partially reviewed the promising candidates, transition metal (di)chalcogenides (TMCs), from preparation methods to electrochemical measurements toward HER and OER. It can be concluded that a good electrocatalyst should possess good conductivity and moderate adsorption energy to reactive species and/or intermediates. In addition, the large number of active/defect sites would be favorable for catalytic ability. The future challenges for TMCs toward HER and OER are summarized as follows. The realization of the mechanisms toward the HER and OER plays a key role to design a perfect electrocatalyst. The development of *in situ* or operando techniques shines the light on this difficulty. Most importantly, the understanding of HER and OER mechanisms offers the significant information to many other fields due to the similar operating concepts in electrocatalytic and photocatalytic applications.

Acknowledgements

This work was supported by the Ministry of Science and Technology (MOST) of Taiwan, under grant numbers 107-2113-M-845-001-MY3.

Author details

Chi-Ang Tseng[1] and Chuan-Pei Lee[2]*

1 Department of Chemistry, National Taiwan University, Taiwan

2 Department of Applied Physics and Chemistry, University of Taipei, Taiwan

*Address all correspondence to: cplee@utaipei.edu.tw

References

[1] Bockris JOM. The origin of ideas on a hydrogen economy and its solution to the decay of the environment. International Journal of Hydrogen Energy. 2002;**27**:731-740. DOI: 10.1016/S0360-3199(01)00154-9

[2] Dresselhaus MS, Thomas IL. Alternative energy technologies. Nature. 2001;**414**:332-337. DOI: 10.1038/35104599

[3] Zhu Y, Chen HC, Hsu CS, Lin TS, Chang CJ, Chang SC, et al. Operando unraveling of the structural and chemical stability of P-substituted $CoSe_2$ electrocatalysts toward hydrogen and oxygen evolution reactions in alkaline electrolyte. ACS Energy Letters. 2019;**4**:987-994. DOI: 10.1021/acsenergylett.9b00382

[4] Zhang J, Zhao Z, Xia Z, Dai L. A metal-free bifunctional electrocatalyst for oxygen reduction and oxygen evolution reactions. Nature Nanotechnology. 2015;**10**:444-452. DOI: 10.1038/nnano.2015.48

[5] Li Y, Wang H, Xie L, Liang Y, Hong G, Dai H. MoS_2 nanoparticles grown on graphene: An advanced catalyst for the hydrogen evolution reaction. Journal of the American Chemical Society. 2011;**133**:7296-7299. DOI: 10.1021/ja201269b

[6] Conway BE, Tilak BV. Interfacial processes involving electrocatalytic evolution and oxidation of H_2, and the role of chemisorbed H. Electrochimica Acta. 2002;**47**:3571-3594. DOI: 10.1016/S0013-4686(02)00329-8

[7] Zeng M, Li Y. Recent advances in heterogeneous electrocatalysts for the hydrogen evolution reaction. Journal of Materials Chemistry A. 2015;**3**:14942-14962. DOI: 10.1039/c5ta02974k

[8] Nørskov JK, Bligaard T, Logadottir A, Kitchin JR, Chen JG, Pandelov S, et al. Trends in the exchange current for hydrogen evolution. Journal of the Electrochemical Society. 2005;**152**:J23. DOI: 10.1149/1.1856988

[9] Suen NT, Hung SF, Quan Q, Zhang N, Xu YJ, Chen HM. Electrocatalysis for the oxygen evolution reaction: Recent development and future perspectives. Chemical Society Reviews. 2017;**46**:337-365. DOI: 10.1039/c6cs00328a

[10] Chia X, Eng AYS, Ambrosi A, Tan SM, Pumera M. Electrochemistry of nanostructured layered transition-metal dichalcogenides. Chemical Reviews. 2015;**115**:11941-11966. DOI: 10.1021/acs.chemrev.5b00287

[11] Huang JK, Pu J, Hsu CL, Chiu MH, Juang ZY, Chang YH, et al. Large-area synthesis of highly crystalline WSe_2 monolayers and device applications. ACS Nano. 2014;**8**:923-930. DOI: 10.1021/nn405719x

[12] Liu G, Li Z, Hasan T, Chen X, Zheng W, Feng W, et al. Vertically aligned two-dimensional SnS_2 nanosheets with a strong photon capturing capability for efficient photoelectrochemical water splitting. Journal of Materials Chemistry A. 2017;**5**:1989-1995. DOI: 10.1039/c6ta08327g

[13] Chang YH, Zhang W, Zhu Y, Han Y, Pu J, Chang JK, et al. Monolayer $MoSe_2$ grown by chemical vapor deposition for fast photodetection. ACS Nano. 2014;**8**:8582-8590. DOI: 10.1021/nn503287m

[14] Lee YH, Zhang XQ, Zhang W, Chang MT, TeLin C, DiChang K, et al. Synthesis of large-area MoS_2 atomic layers with chemical vapor deposition. Advanced Materials. 2012;**24**:2320-2325. DOI: 10.1002/adma.201104798

[15] Ji Q, Zhang Y, Gao T, Zhang Y, Ma D, Liu M, et al. Epitaxial monolayer

MoS_2 on mica with novel photoluminescence. Nano Letters. 2013; **13**:3870-3877. DOI: 10.1021/nl401938t

[16] Yu Y, Li C, Liu Y, Su L, Zhang Y, Cao L. Controlled scalable synthesis of uniform, high-quality monolayer and few-layer MoS_2 films. Scientific Reports. 2013;**3**:1-6. DOI: 10.1038/srep01866

[17] Feldman Y, Tenne R, Wasserman E, Srolovitz DJ. High-rate, gas-phase growth of MoS_2 nested inorganic fullerenes and nanotubes. Science. 1995; **267**:222

[18] Zhao Y, Lee H, Choi W, Fei W, Lee CJ. Large-area synthesis of monolayer $MoSe_2$ films on SiO_2/Si substrates by atmospheric pressure chemical vapor deposition. RSC Advances. 2017;**7**:27969-27973. DOI: 10.1039/c7ra03642f

[19] Okada M, Sawazaki T, Watanabe K, Taniguch T, Hibino H, Shinohara H, et al. Direct chemical vapor deposition growth of WS_2 atomic layers on hexagonal boron nitride. ACS Nano. 2014;**8**:8273-8277. DOI: 10.1021/nn503093k

[20] Wang B, Wang Z, Wang X, Zheng B, Zhang W, Chen Y. Scalable synthesis of porous hollow $CoSe_2$-$MoSe_2$/carbon microspheres for highly efficient hydrogen evolution reaction in acidic and alkaline media. Journal of Materials Chemistry A. 2018;**6**: 12701-12707. DOI: 10.1039/c8ta03523g

[21] An T, Wang Y, Tang J, Wei W, Cui X, Alenizi AM, et al. Interlaced NiS_2-MoS_2 nanoflake-nanowires as efficient hydrogen evolution electrocatalysts in basic solutions. Journal of Materials Chemistry A. 2016; **4**:13439-13443. DOI: 10.1039/c6ta05022k

[22] Dong R, Kuljanishvili I. Review article: Progress in fabrication of transition metal dichalcogenides heterostructure systems. Journal of Vacuum Science and Technology B. 2017;**35**:030803. DOI: 10.1116/1.4982736

[23] Ji Q, Zheng Y, Zhang Y, Liu Z. Chemical vapour deposition of group-VIB metal dichalcogenide monolayers: Engineered substrates from amorphous to single crystalline. Chemical Society Reviews. 2015;**44**:2587-2602. DOI: 10.1039/c4cs00258j

[24] Wang S, Chen Y, Li X, Gao W, Zhang L, Liu J, et al. Injectable 2D MoS_2-integrated drug delivering implant for highly efficient NIR-triggered synergistic tumor hyperthermia. Advanced Materials. 2015;**27**:7117-7122. DOI: 10.1002/adma.201503869

[25] Cheng L, Yuan C, Shen S, Yi X, Gong H, Yang K, et al. Bottom-up synthesis of metal-ion-doped WS_2 nanoflakes for cancer theranostics. ACS Nano. 2015;**9**:11090-11101. DOI: 10.1021/acsnano.5b04606

[26] Shen S, Chao Y, Dong Z, Wang G, Yi X, Song G, et al. Bottom-up preparation of uniform ultrathin rhenium disulfide nanosheets for image-guided photothermal radiotherapy. Advanced Functional Materials. 2017;**27**: 1-9. DOI: 10.1002/adfm.201700250

[27] Zhu S, Gong L, Xie J, Gu Z, Zhao Y. Design, synthesis, and surface modification of materials based on transition-metal dichalcogenides for biomedical applications. Small Methods. 2017;**1**:1700220. DOI: 10.1002/smtd.201700220

[28] Xie J, Zhang J, Li S, Grote F, Zhang X, Zhang H, et al. Controllable disorder engineering in oxygen-incorporated MoS_2 ultrathin nanosheets for efficient hydrogen evolution. Journal of the American Chemical Society. 2013; **135**:17881-17888. DOI: 10.1021/ja408329q

[29] Zhang H, Li Y, Zhang G, Xu T, Wan P, Sun X. A metallic CoS_2 nanopyramid array grown on 3D carbon

fiber paper as an excellent electrocatalyst for hydrogen evolution. Journal of Materials Chemistry A. 2015; **3**:6306-6310. DOI: 10.1039/c5ta00707k

[30] Hinnemann B, Moses PG, Bonde J, Jørgensen KP, Nielsen JH, Horch S, et al. Biomimetic hydrogen evolution: MoS_2 nanoparticles as catalyst for hydrogen evolution. Journal of the American Chemical Society. 2005;**127**:5308-5309. DOI: 10.1021/ja0504690

[31] Ren X, Pang L, Zhang Y, Ren X, Fan H, Liu S. One-step hydrothermal synthesis of monolayer MoS_2 quantum dots for highly efficient electrocatalytic hydrogen evolution. Journal of Materials Chemistry A. 2015;**3**:10693-10697. DOI: 10.1039/c5ta02198g

[32] Du H, Gu S, Liu R, Li CM. Tungsten diphosphide nanorods as an efficient catalyst for electrochemical hydrogen evolution. Journal of Power Sources. 2015;**278**:540-545. DOI: 10.1016/j.jpowsour.2014.12.095

[33] Zhang X, Li L, Guo Y, Liu D, You T. Amorphous flower-like molybdenum-sulfide-at-nitrogen-doped-carbon-nanofiber film for use in the hydrogen-evolution reaction. Journal of Colloid and Interface Science. 2016;**472**:69-75. DOI: 10.1016/j.jcis.2016.03.041

[34] Zhang N, Ma W, Wu T, Wang H, Han D, Niu L. Edge-rich MoS_2 naonosheets rooting into polyaniline nanofibers as effective catalyst for electrochemical hydrogen evolution. Electrochimica Acta. 2015;**180**:155-163. DOI: 10.1016/j.electacta.2015.08.108

[35] Laursen AB, Kegnæs S, Dahl S, Chorkendorff I. Molybdenum sulfides-efficient and viable materials for electro- and photoelectrocatalytic hydrogen evolution. Energy & Environmental Science. 2012;**5**: 5577-5591. DOI: 10.1039/c2ee02618j

[36] Lukowski MA, Daniel AS, Meng F, Forticaux A, Li L, Jin S. Enhanced hydrogen evolution catalysis from chemically exfoliated metallic MoS_2 nanosheets. Journal of the American Chemical Society. 2013;**135**: 10274-10277. DOI: 10.1021/ja404523s

[37] Lee CP, Chen WF, Billo T, Lin YG, Fu FY, Samireddi S, et al. Beaded stream-like $CoSe_2$ nanoneedle array for efficient hydrogen evolution electrocatalysis. Journal of Materials Chemistry A. 2016;**4**:4553-4561. DOI: 10.1039/c6ta00464d

[38] Vante NA, Tributsch H. Energy conversion catalysis using semiconducting transition metal cluster compounds. Nature. 1986;**323**:431-432

[39] Feng Y, He T, Alonso-vante N. In situ free-surfactant synthesis and ORR-electrochemistry of carbon-supported Co_3S_4 and $CoSe_2$ nanoparticles. Chemistry of Materials. 2008;**20**:26-28

[40] Gao MR, Jiang J, Yu SH. Solution-based synthesis and design of late transition metal chalcogenide materials for oxygen reduction reaction (ORR). Small. 2012;**8**:13-27. DOI: 10.1002/smll.201101573

[41] Gao MR, Xu YF, Jiang J, Yu SH. Nanostructured metal chalcogenides: Synthesis, modification, and applications in energy conversion and storage devices. Chemical Society Reviews. 2013;**42**:2986-3017. DOI: 10.1039/c2cs35310e

[42] Liu T, Liang Y, Liu Q, Sun X, He Y, Asiri AM. Electrodeposition of cobalt-sulfide nanosheets film as an efficient electrocatalyst for oxygen evolution reaction. Electrochemistry Communications. 2015;**60**:92-96. DOI: 10.1016/j.elecom.2015.08.011

[43] Swesi AT, Masud J, Nath M. Nickel selenide as a high-efficiency catalyst for oxygen evolution reaction. Energy & Environmental Science. 2016;**9**: 1771-1782. DOI: 10.1039/C5EE02463C

[44] Deng Y, Handoko AD, Du Y, Xi S, Yeo BS. In situ Raman spectroscopy of copper and copper oxide surfaces during electrochemical oxygen evolution reaction: Identification of CuIII oxides as catalytically active species. ACS Catalysis. 2016;**6**: 2473-2481. DOI: 10.1021/acscatal.6b00205

[45] Zhang M, DeRespinis M, Frei H. Time-resolved observations of water oxidation intermediates on a cobalt oxide nanoparticle catalyst. Nature Chemistry. 2014;**6**:362-367. DOI: 10.1038/nchem.1874

Chapter 8

A Novel MDD-Based BEM Model for Transient 3T Nonlinear Thermal Stresses in FGA Smart Structures

Mohamed Abdelsabour Fahmy

Abstract

The main objective of this chapter is to introduce a novel memory-dependent derivative (MDD) model based on the boundary element method (BEM) for solving transient three-temperature (3T) nonlinear thermal stress problems in functionally graded anisotropic (FGA) smart structures. The governing equations of the considered study are nonlinear and very difficult if not impossible to solve analytically. Therefore, we develop a new boundary element scheme for solving such equations. The numerical results are presented highlighting the effects of the MDD on the temperatures and nonlinear thermal stress distributions and also the effect of anisotropy on the nonlinear thermal stress distributions in FGA smart structures. The numerical results also verify the validity and accuracy of the proposed methodology. The computing performance of the proposed model has been performed using communication-avoiding Arnoldi procedure. We can conclude that the results of this chapter contribute to increase our understanding on the FGA smart structures. Consequently, the results also contribute to the further development of technological and industrial applications of FGA smart structures of various characteristics.

Keywords: boundary element method, memory-dependent derivative, three-temperature, nonlinear thermal stresses, FGA smart structures

1. Introduction

Smart materials, which are also called intelligent materials, are engineered materials that have the ability to respond to the changes that occur around them in a controlled fashion by external stimuli, such as stress, heat, light, ultraviolet, moisture, chemical compounds, mechanical strength, and electric and magnetic fields. We can simply define smart materials as materials which adapt themselves as per required condition. The history of the discovery of these materials dates back to the 1880s when Jacques and Pierre Curie noticed a phenomenon that pressure generates electrification around a number of minerals such as quartz and tourmaline, and this phenomenon is called piezoelectric effect, so the piezoelectric materials are the oldest type of smart materials, which are utilized extensively in the fabrication of various devices such as transducers, sensors, actuators, surface acoustic wave devices, frequency control, etc. There are a lot of smart material types like piezoelectric materials, thermochromic pigments, shape memory alloys, magnetostrictive, shape memory polymers, hydrogels, electroactive polymers and bi-component fibers, etc.

Anisotropic smart structures (ASSs) are getting great attention of researchers due to their applications in textile, aerospace, mass transit, marine, automotive, computers and other electronic industries, consumer goods applications, mechanical and civil engineering, infertility treatment, micropumps, medical equipment applications, ultrasonic micromotors, microvalves and photovoltaics, rotating machinery applications, and much more [1–12].

The classical thermoelasticity (CTE) theory of Duhamel [13] and Neumann [14] has two shortcomings based on parabolic heat conduction equation of this theory: the first does not involve any elastic terms, while the second has infinite propagation speeds of thermoelastic waves. In order to overcome the first shortcoming, Biot [15] proposed the classical coupled thermoelasticity (CCTE). But CTE and CCTE have the second shortcoming. So, several generalized thermoelasticity theories have been developed to overcome the second shortcoming of CTE. Among of these theories are Lord and Shulman (LS) [16], Green and Lindsay (GL) [17], and Green and Naghdi [18, 19] theories of thermoelasticity with and without energy dissipation, dual-phase-lag thermoelasticity (DPLTE) [20, 21] and three-phase-lag thermoelasticity (TPLTE) [22]. Although thermoelastic phenomena in the majority of practical applications are adequately modeled with the classical Fourier heat conduction equation, there are an important number of problems that require consideration of nonlinear heat conduction equation. It is appropriate in these cases to apply the nonlinear generalized theory of thermoelasticity; great attention has been paid to investigate the nonlinear generalized thermoelastic problems by using numerical methods [23–34]. Fahmy [35–39] introduced the mathematical foundations of three-temperature (3T) field to thermoelasticity.

The fractional calculus is the mathematical branch that used to study the theory and applications of derivatives and integrals of arbitrary non-integer order. This branch has emerged in recent years as an effective tool for modeling and simulation of various engineering and industrial applications [40, 41]. Due to the nonlocal nature of fractional order operators, they are useful for describing the memory and hereditary properties of various materials and processes. Also, the fractional calculus has drawn wide attention from the researchers of various countries in recent years due to its applications in solid mechanics, fluid dynamics, viscoelasticity, heat conduction modeling and identification, biology, food engineering, econophysics, biophysics, biochemistry, electrochemistry, electrical engineering, finance and control theory, robotics and control theory, signal and image processing, electronics, electric circuits, wave propagation, nanotechnology, etc. [42–44].

Several mathematics researchers have contributed to the history of fractional calculus, where Euler mentioned interpolating between integral orders of a derivative in 1730. Then, Laplace defined a fractional derivative by means of an integral in 1812.

Lacroix presented the first formula for the fractional order derivative appeared in 1819, where he introduced the nth derivative of the function $y = x^m$ as follows:

$$\frac{d^n}{dx^n} = \frac{\Gamma(m+1)}{\Gamma(m\text{-}n+1)} x^{m\text{-}n} \tag{1}$$

Liouville supposed that $\frac{d^v}{dx^v}(e^{ax}) = a^v e^{ax}$ for $v > 0$ to get the following fractional order derivative:

$$\frac{d^v x^{-a}}{dx^v} = (-1)^v \frac{\Gamma(a+v)}{\Gamma(a)} x^{-a-v} \tag{2}$$

By using Cauchy's integral formula for complex valued analytical functions, Laurent defined the integration of arbitrary order $v > 0$ as

$${}_cD_x^v f(x) = {}_cD_x^{m-\rho} f(x) = \frac{d^m}{dx^m}\left[\frac{1}{\Gamma(\rho)}\int_c^x (x\text{-}t)^{\rho\text{-}1} f(t)dt\right], 0 < \rho \le 1 \tag{3}$$

where ${}_cD_x^v$ denotes differentiation of order v of the function f along the x-axis. Cauchy presented the following fractional order derivative:

$$f_+^{(\alpha)} = \int f(\tau)\frac{(t-\tau)^{-\alpha-1}}{\Gamma(-a)} d\tau \tag{4}$$

In 1967, the Italian mathematician Caputo presented his fractional derivative of order $\alpha > 0$ as

$$D_*^\alpha f(t) = \frac{1}{\Gamma(m-\alpha)}\int_0^t \frac{f^{(m)}(\tau)}{(t-\tau)^{\alpha+1-m}} d\tau, m-1 < \alpha < m, \quad \alpha < 0 \tag{5}$$

Diethelm [45] has suggested the Caputo derivative to be in the following form:

$$D_a^\zeta f(\tau) = \int_a^\tau K_\zeta(\tau\text{-}\xi) f^{(m)}(\xi)\, d\xi \tag{6}$$

where $f^{(m)}$ is the mth order derivative and m is an integer such that $m\text{-}1 < \zeta \le m$

$$K_\zeta(\tau\text{-}\xi) = \frac{(\tau\text{-}\xi)^{m\text{-}\zeta\text{-}1}}{\Gamma(m\text{-}\zeta)} \tag{7}$$

Wang and Li [46] have introduced a memory-dependent derivative (MDD)

$$D_\omega^\zeta f(\tau) = \frac{1}{\omega}\int_{\tau\text{-}\omega}^\tau K_\zeta(\tau\text{-}\xi) f^{(m)}(\xi)\, d\xi \tag{8}$$

where the first-order ($\zeta = 1$) of MDD for a differentiable function $f(\tau)$ can be expressed as

$$D_\omega f(\tau) = \frac{1}{\omega}\int_{\tau\text{-}\omega}^\tau K(\tau\text{-}\xi) f'(\xi)\, d\xi \tag{9}$$

Based on several practical applications, the memory effect needs weight $0 \le K(\tau\text{-}\xi) < 1$ for $\xi \in [\tau\text{-}\omega, \tau]$, so the MDD magnitude $D_\omega f'(\tau)$ is usually smaller than $f'(\tau)$, where the time delay ($\omega > 0$) and the kernel function ($0 \le K(\tau\text{-}\xi) \le 1$ for $\xi \in [\tau\text{-}\xi, \tau]$) can be chosen arbitrarily on the delayed interval $[\tau\text{-}\omega, \tau]$, the practical kernel functions are 1, $[1-(\tau\text{-}\xi)]$ and $\left[1-\frac{\tau\text{-}\xi}{\omega}\right]^p$, $p = \frac{1}{4}, 1, 2$, etc. These functions are monotonically increasing with $K = 0$ for the past time $\tau\text{-}\xi$ and $K = 1$ for the present time τ. The main feature of MDD is that the real-time functional value depends also on the past time $[\tau\text{-}\xi\text{-}\tau[$. So, D_ω depends on the past time (nonlocal operator), while the integration does not depend on the past time (local operator).

As a special case $K(\tau\text{-}\xi) \equiv 1$, we have

$$D_\omega f(\tau) = \frac{1}{\omega}\int_{\tau-\omega}^{\tau} f'(\xi)d\xi = \frac{f(\tau) - f(\tau - \omega)}{\omega} \to f'(\tau) \tag{10}$$

The above equation shows that the common derivative $\frac{d}{d\tau}$ is the limit of D_ω as $\omega \to 0$. That is,

$$D_\omega f(\tau) \le \left|\frac{\partial f}{\partial \tau}\right| = \lim_{\omega \to 0} \frac{f(\tau + \omega) - f(\tau)}{\omega} \tag{11}$$

Now, the boundary element method (BEM) [47–80] is widely adopted for solving several engineering problems due to its easy implementation. In the BEM, only the boundary of the domain needs to be discretized, so it has a major advantage over other methods requiring full domain discretization [81–87] such as finite difference method (FDM), finite element method (FEM), and finite volume method (FVM) in engineering applications. This advantage of BEM over domain methods has significant importance for modeling of nonlinear generalized thermoelastic problems which can be implemented using BEM with little cost and less input data. Previously scientists have proven that FEM covers more engineering applications than BEM which is more efficient for infinite domain problems. But currently BEM scientists have changed their thinking and vision on BEM, where the BEM researchers developed the BEM technique for solving inhomogeneous and nonlinear problems involving infinite and semi-infinite domains by using a lot of software like FastBEM and ExaFMM.

The main objective of this chapter is to introduce a novel memory-dependent derivative model for solving transient three-temperature nonlinear thermal stress problems in functionally graded anisotropic (FGA) smart structures. The governing equations of the considered model are nonlinear and very difficult if not impossible to solve analytically. Therefore, we develop a new efficient boundary element technique for solving such equations. Numerical results show the effects of MDD on the three-temperature distributions and the influence of MDD and anisotropy on the nonlinear thermal stresses of FGA smart structures. Also, numerical results demonstrate the validity and accuracy of the proposed model.

A brief summary of the chapter is as follows: Section 1 introduces the background and provides the readers with the necessary information to books and articles for a better understanding of smart material problems, memory-dependent derivative history, and their applications. Section 2 describes the physical modeling of memory-dependent derivative problems of three-temperature nonlinear thermal stresses in FGA structures. Section 3 outlines the BEM implementation for obtaining the temperature field of the considered problem. Section 4 outlines the BEM implementation for obtaining the dispacement field of the considered problem. Section 5 introduces computing performance of the proposed model. Section 6 presents the new numerical results that describe the effects of memory-dependent derivative and anisotropy on the problem's field variations. Lastly, Section 7 outlines the significant findings of this chapter.

2. Formulation of the problem

With reference to a Cartesian system (x_1, x_2, x_3) with a configuration R bounded by a closed surface S as shown in **Figure 1**.

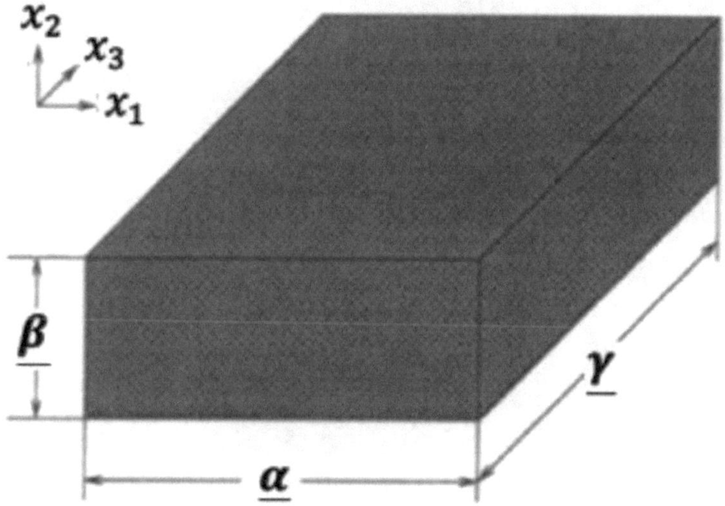

Figure 1.
Computational domain of the considered smart structure.

The governing equations for the transient three-temperature nonlinear thermal stresses problems of FGA smart structures with memory-dependent derivatives can be written as [35].

$$\sigma_{ij,j} + \rho F_i = \ddot{\pi} \mathrm{u}_i \tag{12}$$

$$D_{i,i} = 0 \tag{13}$$

where

$$\sigma_{ij} = (x+1)^m \left[C_{ijkl} e \delta_{ij} \text{-} \beta_{ab} \left(T_\alpha \text{-} T_{\alpha 0} + \tau_1 \dot{T}_\alpha\right)\right] \tag{14}$$

$$D_i = (x+1)^m \left[e_{ijk}\varepsilon_{jk} + \mathrm{f}_{ik} E_k\right] \varepsilon_{ij} = \frac{1}{2}\left(u_{i,j} + u_{\,j,i}\right), \tag{15}$$

where σ_{ij}, F_i, ε_{ij}, ε_{ijk}, u_i, and ρ are the force stress tensor, mass force vector, strain tensor, alternate tensor, displacement vector, and density, respectively, $C_{ijkl}\left(C_{ijkl} = C_{klij} = C_{jikl}\right)$ is the constant elastic moduli, e is the dilatation, $\beta_{ij}\left(\beta_{ij} = \beta_{ji}\right)$ are the stress-temperature coefficients, D_i is the electric displacement, e_{ijk} is the piezoelectric tensor, f_{ik} is the permittivity tensor, and E_k is the electric field vector.

The two-dimensional three-temperature (2D-3T) radiative heat conduction equations can be expressed as

$$c_e \frac{\partial T_e(r,\tau)}{\partial \tau} - \frac{1}{\rho}\nabla[K_e \nabla T_e(r,\tau)] = -W_{ei}(T_e - T_i) - W_{ep}\left(T_e - T_p\right) \tag{16}$$

$$c_i \frac{\partial T_i(r,\tau)}{\partial \tau} - \frac{1}{\rho}\nabla[K_i \nabla T_i(r,\tau)] = W_{ei}(T_e - T_i) \tag{17}$$

$$c_p T_p^3 \frac{\partial T_p(r,\tau)}{\partial \tau} - \frac{1}{\rho}\nabla\left[K_p \nabla T_p(r,\tau)\right] = W_{ep}\left(T_e - T_p\right) \tag{18}$$

where $e, i, \wedge p$ denote electron, ion, and phonon, respectively; $\left(c_e, c_i, c_p\right)$, $\left(K_e, K_i, K_p\right)$, and $\left(T_e, T_i, T_p\right)$ are specific heat capacities, conductive coefficients, and temperature functions, respectively; W_{ei} is the electron-ion coefficient; and W_{ep} is the electron-phonon coefficient.

3. BEM solution of temperature field

This section concerns using a boundary element method to solve the temperature model.

The above 2D-3T radiative heat conduction Eqs. (16)-(18) can be expressed in the context of nonlinear thermal stresses of FGA smart structures as in [36].

$$\nabla\left[\left(\delta_{1j}K_\alpha + \delta_{2j}K_\alpha^*\right)\nabla T_\alpha(r,\tau)\right] - \acute{W}(r,\tau) = c_\alpha\rho\delta_1\delta_{1j}D_{\omega_\alpha}T_\alpha(r,\tau) \tag{19}$$

which can be written in the following form:

$$L_{ab}T_\alpha(r,\tau) = f_{ab} \tag{20}$$

where

$$L_{ab} = \nabla\left[\left(\delta_{1j}K_\alpha + \delta_{2j}K_\alpha^*\right)\nabla\right] \tag{21}$$

$$f_{ab} = \acute{W}(r,\tau) + \acute{W}(r,\tau) \tag{22}$$

where

$$\acute{W}(r,\tau) = \begin{cases} \rho W_{ei}(T_e - T_i) + \rho W_{er}(T_e - T_p) + \acute{W}, \alpha = e, \delta_1 = 1 \\ -\rho W_{ei}(T_e - T_i) + \acute{W}, \qquad \alpha = i, \delta_1 = 1 \\ -\rho W_{er}(T_e - T_p) + \acute{W}, \qquad \alpha = p, \delta_1 = T_p^3 \end{cases} \tag{23}$$

$$\begin{aligned} \acute{W}(r,\tau) = F(r,\tau) &- \frac{\delta_{2j}K_\alpha}{\omega_\alpha}\int_{\tau-\omega_\alpha}^{\tau} K(\tau-\xi)\frac{\partial}{\partial\xi}\left(\nabla^2 T_\alpha(r,\tau)\right)d\xi \\ &+ \frac{\rho C_\alpha\delta_1\delta_{1j}}{\omega_\alpha}\int_{\tau-\omega_\alpha}^{\tau} K(\tau-\xi)\frac{\partial}{\partial\xi}\left(T_\alpha(r,\tau)\right)d\xi \\ &+ \frac{\rho C_\alpha\left(\tau_0 + \delta_{1j}\tau_2 + \delta_{2j}\right)}{\omega_\alpha}\int_{\tau-\omega_\alpha}^{\tau} K(\tau-\xi)\frac{\partial^2}{\partial\xi^2}\left(T_\alpha(\mathrm{r},\tau)\right)d\xi \end{aligned} \tag{24}$$

$$F(r,\tau) = \beta_{ab}T_{\alpha 0}\left[\text{Å}\ \delta_{1j}\acute{u}_{a,b} + \left(\tau_0 + \delta_{2j}\right)\acute{u}_{a,b}\right] \tag{25}$$

and

$$W_{ei} = \rho A_{ei}T_e^{-2/3}, W_{er} = \rho A_{er}T_e^{-1/2}, K_\alpha = A_\alpha T_\alpha^{5/2}, \alpha = e, i, K_p = A_p T_p^{3+B} \tag{26}$$

where δ_{ij}, $(i, j = 1, 2)$, $\omega_\alpha(0)(\alpha = e, i \wedge p)$, and $K(\tau - \xi)$ are the Kronecker delta, delay times, and kernel function, respectively.

The total energy can be expressed as

$$P = P_e + P_i + P_{p'}P_e = c_eT_e, P_i = c_iT_i, P_p = \frac{1}{4}c_pT_p^4 \tag{27}$$

Initial and boundary conditions can be expressed as

$$T_\alpha(x,y,0) = T_\alpha^0(x,y) = g_1(x,\tau) \tag{28}$$

$$\mathbb{K}_\alpha \frac{\partial T_\alpha}{\partial n}\bigg|_{\Gamma_1} = 0, \alpha = e, i, T_p\big|_{\Gamma_1} = g_2(x, \tau) \tag{29}$$

$$\mathbb{K}_\alpha \frac{\partial T_\alpha}{\partial n}\bigg|_{\Gamma_2} = 0, \alpha = e, i, p \tag{30}$$

By using the fundamental solutions T_α^* that satisfies the following differential equation:

$$L_{ab} T_\alpha^* = f_{ab} \tag{31}$$

Now, by implementing the technique of Fahmy [35], we can write (19) as

$$CT_\alpha = \frac{D}{\mathbb{K}_\alpha}\int_O^\tau \int_S [T_\alpha q^* \text{-} T_\alpha^* q]\, dS d\tau + \frac{D}{\mathbb{K}_\alpha}\int_O^\tau \int_R b T_\alpha^*\, dR d\tau + \int_R T_\alpha^i T_\alpha^* \big|_{\tau=0} dR \tag{32}$$

which can be written in the absence of heat sources as follows:

$$CT_\alpha = \int_S [T_\alpha q^* - T_\alpha^* q]\, dS \text{-} \int_R \frac{\mathbb{K}_\alpha}{D}\frac{\partial T_\alpha^*}{\partial \tau} T_\alpha dR \tag{33}$$

In order to transform the domain integral in (33) to the boundary, we approximate the temperature time derivative as

$$\frac{\partial T_\alpha}{\partial \tau} \cong \sum_{i=1}^{N} f^j(r)^j a^j(\tau) \tag{34}$$

where $f^j(r)$ are known functions and $a^j(\tau)$ are unknown coefficients.

We assume that $\widehat{T}_\alpha^j$ is a solution of

$$\nabla^2 \widehat{T}_\alpha^j = f^j \tag{35}$$

Then, Eq. (33) leads to the following boundary integral equation

$$CT_\alpha = \int_S [T_\alpha q^* \text{-} T_\alpha^* q]\, dS + \sum_{i=1}^{N} a^j(\tau) D^{-1}\left(C\widehat{T}_\alpha^j \text{-} \int_S [T_\alpha^j q^* \text{-} \hat{q}^j T_\alpha^*]\, dS \right) \tag{36}$$

where

$$\hat{q}^j = \text{-}\mathbb{K}_\alpha \frac{\partial \widehat{T}_\alpha^j}{\partial n} \tag{37}$$

and

$$a^j(\tau) = \sum_{i=1}^{N} f_{ji}^{-1} \frac{\partial T_\alpha(r_i, \tau)}{\partial \tau} \tag{38}$$

where f_{ji}^{-1} are the coefficients of F^{-1} which are defined as [58].

$$\{F\}_{ji} = f^j(r_i) \tag{39}$$

By discretizing Eq. (36) and using Eq. (38), we get [35].

$$C\dot{T}_{\alpha}+HT_{\alpha}=GQ \tag{40}$$

where Q is the heat flux vector and H and G are matrices.
The diffusion matrix can be defined as

$$C=-\left[H\widehat{T}_{\alpha}-G\widehat{Q}\right]F^{-1}D^{-1} \tag{41}$$

where

$$\left\{\widehat{T}\right\}_{ij}=\widehat{T}^{j}(x_i) \tag{42}$$

$$\left\{\widehat{Q}\right\}_{ij}=\hat{q}^{j}(x_i) \tag{43}$$

To solve numerically Eq. (41), the functions T_{α} and q were interpolated as

$$T_{\alpha}=(1-\theta)\,T_{\alpha}^{m}+\theta T_{\alpha}^{m+1} \tag{44}$$

$$q=(1-\theta)q^{m}+\theta q^{m+1} \tag{45}$$

where $0\leq 0=\frac{\tau-\tau^{m}}{\tau^{m+1}\text{-}\tau^{m}}\leq 1$ determines the practical time τ of the current time step. By time differentiation of Eq. (44), we obtain

$$\dot{T}_{\alpha}=\frac{dT_{\alpha}}{d\theta}\frac{d\theta}{d\tau}=\frac{T_{\alpha}^{m+1}\text{-}T_{\alpha}^{m}}{\tau^{m+1}\text{-}\tau^{m}}=\frac{T_{\alpha}^{m+1}\text{-}T_{\alpha}^{m}}{\Delta\tau^{m}} \tag{46}$$

By substitution from (44)–(46) into (40), we get

$$\left(\frac{C}{\Delta\tau^{m}}+\theta H\right)T_{\alpha}^{m+1}-\theta GQ^{m+1}=\left(\frac{C}{\Delta\tau^{m}}\text{-}(1-\theta)H\right)T_{\alpha}^{m}+(1-\theta)\,GQ^{m} \tag{47}$$

By considering the initial and boundary conditions, we can write the following system of equations

$$\text{a}X=\text{b} \tag{48}$$

We apply an explicit staggered algorithm to solve the system (48) and obtain the temperature in terms of the displacement field.

4. BEM solution of displacement field

By using the weighted residual method, we can write (12) and (13) in the following form:

$$\int_{R}\left(\sigma_{ij,j}+U_i\right)u_i^{*}\,dR=0 \tag{49}$$

$$\int_{R}\left(D_{,i}\right)\Phi_i^{*}\,dR=0 \tag{50}$$

where

$$U_i = \rho F_i \text{-} \rho \ddot{u}_i, \tag{51}$$

where u_i^* and Φ_i^* are weighting functions and u_i and Φ_i are approximate solutions.

Now, we assume the following boundary conditions:

$$u_i = \overline{u}_i \quad on\ \mathrm{S}_1 \tag{52}$$

$$\lambda_i = \sigma_{ij} n_j = \overline{\lambda}_i \quad on\ \mathrm{S}_2 \tag{53}$$

$$\Phi = \overline{\Phi} \quad on\ \mathrm{S}_5 \tag{54}$$

$$Q = \frac{\partial \Phi}{\partial n} = \overline{Q} \quad on\ \mathrm{S}_6 \tag{55}$$

By integration by parts for the first term of Eqs. (49) and (50), we have

$$-\int_R \sigma_{ij} u_{i,j}^* dR + \int_R U_i u_i^* dR = -\int_{S_2} \lambda_i u_i^* dS \tag{56}$$

$$-\int_R D \Phi_{i,i}^* dR = -\int_{S_6} Q_i \Phi_i^* dS \tag{57}$$

Based on Huang and Liang [88], the boundary integral equation can be expressed as

$$\begin{aligned} -\int_R \sigma_{ij,j} u_i^* \, dR + \int_R U_i u_i^* \, dR - \int_R D \Phi_{i,i}^* dR = & \int_{S_2} (\lambda_i - \overline{\lambda}_i) u_i^* \, dS + \int_{S_1} (\overline{u}_i - u_i) \lambda_i^* \, dS \\ & + \int_{S_6} (Q_i - \overline{Q}_i) \Phi_i^* \, dS \\ & + \int_{S_5} (\overline{\Phi}_i - \Phi_i) Q_i^* \, dS \end{aligned} \tag{58}$$

By integrating by parts for the left-hand side of (58), we get

$$\begin{aligned} -\int_R \sigma_{ij} \varepsilon_{ij}^* \, dR + \int_R U_i u_i^* \, dR - \int_R D \Phi_{i,i}^* dR = & -\int_{S_2} \overline{\lambda}_i u_i^* \, dS - \int_{S_1} \lambda_i u_i^* \, dS] \\ & + \int_{S_1} (\overline{u}_i - u_i) \lambda_i^* \, dS - \int_{S_6} \overline{Q}_i \Phi_i^* \, dS \\ & - \int_{S_5} Q_i \Phi_i^* \, dS + \int_{S_5} (\overline{\Phi}_i - \Phi_i) Q_i^* \, dS \end{aligned} \tag{59}$$

Based on Eringen [89], the elastic stress can be expressed as

$$\sigma_{ij} = \mathbb{A}_{ijkl} \, \varepsilon_{kl}, \tag{60}$$

where

$$\mathbb{A}_{ijkl} = \mathbb{A}_{klij} \tag{61}$$

Hence, Eq. (59) can be rewritten as

$$-\int_R \sigma^{ij*}\,\varepsilon_{ij}\,dR + \int_R U_i\,u_i^*\,dR - \int_R D\,\Phi_{i,i}^*\,dR = -\int_{S_2} \overline{\lambda}_i\,u_i^*\,dS - \int_{S_1} \lambda_i\,u_i^*\,dS + \int_{S_1} (\overline{u}_i - u_i)\,\lambda_i^*\,dS - \int_{S_6} \overline{Q}_i\,\Phi_i^*\,dS - \int_{S_5} Q_i\,\Phi_i^*\,dS + \int_{S_5} (\overline{\Phi}_i - \Phi_i)\,Q_i^*\,dS \quad (62)$$

By integration by parts again, we obtain

$$\int_R \sigma_{ij,i}^* u_i dR = -\int_S u_i^*\,\lambda_i\,dS - \int_S \Phi_i^*\,Q_i\,dS + \int_S \lambda_i^*\,u_i\,dS + \int_S Q_i^*\,\Phi_i\,dS \quad (63)$$

The weighting functions of $U_i = \Delta^n$ and $V_i = 0$ along e_1 can be obtained as follows:

$$\sigma_{1j,j}^* + \Delta^n e_1 = 0 \quad (64)$$

According to Dragos [90], the fundamental solution can be written as

$$u_i^* = u_{1i}^*\,e_1, \Phi_i^* = \Phi_{1i}^*\,e_1, \lambda_i^* = \lambda_{1i}^*\,e_1, Q_i^* = Q_{1i}^* e_1 \quad (65)$$

The weighting functions of $U_i = 0$ and $V_i = \Delta^n$ along e_1 can be written as follows:

$$\sigma_{ij,j}^* = 0 \quad (66)$$

Based on Dragos [90], the fundamental solution can be obtained analytically as

$$u_i^* = u_{1i}^{**}\,e_1, \Phi_i^* = \Phi_{1i}^{**}\,e_1, \lambda_i^* = \lambda_{1i}^{**}\,e_1, Q_i^* = Q_{1i}^{**}\,e_1 \quad (67)$$

By using the weighting functions of (65) and (67) into (63), we have

$$C_{1i}^n u_i^n = -\int_S \lambda_{1i}^*\,u_i\,dS - \int_S Q_{1i}^*\,\Phi_i\,dS + \int_S u_{1i}^*\,\lambda_i\,dS + \int_S \Phi_{1i}^*\,Q_i\,dS \quad (68)$$

$$C_{1i}^n \omega_i^n = -\int_S \lambda_{1i}^{**}\,u_i\,dS - \int_S Q_{1i}^{**}\,\Phi_i\,dS + \int_S u_{1i}^{**}\,\lambda_i\,dS + \int_S \Phi_{1i}^{**}\,Q_i\,dS \quad (69)$$

Thus, we can write

$$C^n \mathrm{q}^n = -\int_S \mathrm{p}^*\,\mathrm{q}\,dS + \int_S \mathrm{q}^*\,\mathrm{p}\,dS + \int_S \mathrm{d}^*\,\Phi\,ds + \int_S \mathrm{f}^*\,\frac{\partial \Phi}{\partial n}\,dS \quad (70)$$

where

$$C^n = \begin{bmatrix} C_{11} & C_{12} \\ C_{21} & C_{22} \end{bmatrix}, \mathrm{q}^* = \begin{bmatrix} u_{11}^* & u_{12}^* & 0 \\ u_{21}^* & u_{22}^* & 0 \\ u_{31}^{**} & u_{32}^{**} & 0 \end{bmatrix}, \mathrm{p}^* = \begin{bmatrix} \lambda_{11}^* & \lambda_{12}^* & 0 \\ \lambda_{21}^* & \lambda_{22}^* & 0 \\ \lambda_{31}^{**} & \lambda_{32}^{**} & 0 \end{bmatrix}, \mathrm{q} = \begin{bmatrix} u_1 \\ u_2 \\ \omega_3 \end{bmatrix},$$

$$\mathrm{p} = \begin{bmatrix} \lambda_1 \\ \lambda \\ \mu_3 \end{bmatrix}, \mathrm{d}^* = \begin{bmatrix} \mathrm{d}_1^* \\ \mathrm{d}_2^* \\ 0 \end{bmatrix}, \mathrm{f}^* = \begin{bmatrix} \mathrm{f}_1^* \\ \mathrm{f}_2^* \\ 0 \end{bmatrix} \quad (71)$$

In order to solve (70) numerically, we suppose the following definitions:

$$\mathrm{q} = \psi\, \mathrm{q}^j, \mathrm{p} = \psi\, \mathrm{p}^j, \Phi = \psi_0\, \Phi^j, \frac{\partial \Phi}{\partial n} = \psi_0 \left(\frac{\partial \Phi}{\partial n}\right)^j \tag{72}$$

Substituting from (72) into (70) and discretizing the boundary, we obtain

$$\begin{aligned} C^n\, \mathrm{q}^n = & \sum_{j=1}^{N_e} \left[-\int_{\Gamma_j} \mathrm{p}^*\, \psi\, d\Gamma \right] \mathrm{q}^j + \sum_{j=1}^{N_e} \left[\int_{\Gamma_j} \mathrm{q}^*\, \psi\, d\Gamma \right] \mathrm{p}^j + \sum_{j=1}^{N_e} \left[\int_{\Gamma_j} \mathrm{d}^*\, \psi_0\, d\Gamma \right] \Phi^j \\ & + \sum_{j=1}^{N_e} \left[\int_{\Gamma_j} \mathrm{f}^*\, \psi_0\, d\Gamma \right] \left(\frac{\partial \Phi}{\partial n}\right)^j \end{aligned} \tag{73}$$

Equation after integration can be written as

$$C^i \mathrm{q}^i = -\sum_{j=1}^{N_e} \hat{\mathbb{H}}^{ij}\, \mathrm{q}^j + \sum_{j=1}^{N_e} \hat{\mathbb{G}}^{ij}\, \mathrm{p}^j + \sum_{j=1}^{N_e} \hat{\mathbb{D}}^{ij}\, \Phi^i + \sum_{j=1}^{N_e} \hat{\mathbb{F}}^{ij} \left(\frac{\partial \Phi}{\partial n}\right)^j \tag{74}$$

By using the following representation:

$$\mathbb{H}^{ij} = \begin{cases} \hat{\mathbb{H}}^{ij} & \text{if } i \neq j \\ \hat{\mathbb{H}}^{ij} + C^i & \text{if } i = j \end{cases} \tag{75}$$

Thus, we can write (74) as follows:

$$\sum_{j=1}^{N_e} \mathbb{H}^{ij}\, \mathrm{q}^j = \sum_{j=1}^{N_e} \hat{\mathbb{G}}^{ij} \mathrm{p}^j + \sum_{j=1}^{N_e} \hat{\mathbb{D}}^{ij} \Phi^j + \sum_{j=1}^{N_e} \hat{\mathbb{F}}^{ij} \left(\frac{\partial \Phi}{\partial n}\right)^j \tag{76}$$

The global matrix equation for all **i** nodes can be expressed as

$$\mathbb{H}\mathbb{Q} = \mathbb{G}\mathbb{P} + \mathbb{D}\Theta + \mathbb{F}\mathbb{S} \tag{77}$$

where $\mathbb{Q}$ is the displacement vector, $\mathbb{P}$ is the traction vector, Θ is the electric potential vector, and $\mathbb{S}$ is the electric potential gradient vector.

Substituting the boundary conditions into (77), we obtain the following system of equations:

$$\mathbb{A}\mathbb{X} = \mathbb{B} \tag{78}$$

We apply an explicit staggered algorithm to solve the system (78) and obtain the temperature and displacement fields as follows:

1. From Eq. (48) we obtain the temperature field in terms of the displacement field.

2. We predict the displacement field and solve the resulted equation for the temperature field.

3. We correct the displacement field using the computed temperature field for Eq. (78).

An explicit staggered algorithm based on communication-avoiding Arnoldi as described in Hoemmen [91] is very suitable for efficient implementation in Matlab (R2019a) with the aim of specifically improving its performance for the solution of the resulting linear algebraic systems.

5. Computational performance of the problem

According to Fahmy [35], the computer performance with simulation can be computed based on account and communication process, elements underlying the hardware and functional computation. The main objective of our proposed technique during simulation process is to use the preconditioners which are efficient to improve the overall CPU utilization of the cluster, accelerate the iterative method, and reduce the input/output and the interprocessor communication costs. Also, Fahmy [35] compared the communication-avoiding Krylov methods that are based on the s-step Krylov methods such as communication-avoiding generalized minimal residual (CA-GMRES) of Saad and Schultz [92], communication-avoiding Arnoldi (CA-Arnoldi) of the Arnoldi [93] and communication-avoiding Lanczos (CA-Lanczos) of Lanczos [94], with their corresponding standard Krylov methods. CA-Arnoldi which is also called Arnoldi (s, t) algorithm is different from standard Arnoldi (s) $(s, t = 1)$, where s is the number of inner iteration steps and t is the number of outer iteration steps. According to [35], the CA-Arnoldi has numerical stability, convergence, and performance due to the implementation of algorithm shown in **Figure 2**, which is based on the QR factorization update and block classical Gram-Schmidt (block CGS) approach or block modified Gram-Schmidt (block MGS) approach where

$$V_k = [v_{sk+1}, v_{sk+2}, \dots, v_{sk+s}] \tag{79}$$

Require: $n \times n$ matrix A and $n \times 1$ starting vector v

$h_0 := \|r\|_2$, $q_1 := v/h_0$

for $k = 0$ to $t - 1$ **do**

  Fix basis conversion matrix $\underline{B_k}$

  Compute $\underline{\acute{V}_k}$ using matrix powers kernel

  if $k = 0$ **then**

    Compute QR factorization $\underline{V_0} = \underline{Q_0 R}$

    $\underline{\mathfrak{Q}_0} := \underline{Q_0}$

    $\underline{\mathfrak{H}_0} := \underline{R_0 B_0} R_0^{-1}$

  else

    $\underline{\acute{\mathfrak{R}}_{k-1,k}} := \underline{\mathfrak{Q}_{k-1}^*} \underline{\acute{V}_k}$

    $\underline{\acute{V}'_k} := \underline{\acute{V}_k} - \underline{\mathfrak{Q}_{k-1}} \underline{\acute{\mathfrak{R}}_{k-1,k}}$

    Compute QR factorization $\underline{\acute{V}'_k} = \underline{\acute{Q}_k \acute{R}_k}$

    Compute $\underline{\mathfrak{H}_{k-1,k}} := -\mathfrak{H}_{k-1} \mathfrak{R}_{k-1,k} R_k^{-1} + \underline{\mathfrak{R}_{k-1,k} B_k} R_k^{-1}$

    Compute H_k via Equation (3.21)

    Compute h_k via Equation (3.22)

    $$\underline{\mathfrak{H}_k} := \begin{pmatrix} \mathfrak{H}_{k-1} & \underline{\mathfrak{H}_{k-1,k}} \\ h_{k-1} e_1 e_{s(k-1)}^T & H_k \\ 0_{1,s(k-1)} & h_k e_s^T \end{pmatrix}$$

  end if

end for

Figure 2.
CA-Arnoldi iteration algorithm.

and

$$\mathcal{Q}_k = [\mathrm{Q}_0, \mathrm{Q}_1, \ldots, \mathrm{Q}_{k-1}] \tag{80}$$

The generalized minimal residual (GMRES) method of Saad and Schultz [92] is a Krylov subspace method for solving nonsymmetric linear systems. The CA-GMRES algorithm is based on Arnoldi (s, t) and equivalent to standard GMRES in exact arithmetic. Also, the GMRES or CA-GMRES are convergent at the same rate for problems, but Hoemmen [91] proved that CA-GMRES algorithm shown in **Figure 3** converges for the s-step basis lengths and restart lengths used for obtaining maximum performance. Lanczos method can be considered as a special case of Arnoldi method for symmetric and real case of A or Hermitian and complex case of A. Symmetric Lanczos which is also called Lanczos is different from nonsymmetric Lanczos. We implemented a communication-avoiding version of symmetric Lanczos (CA-Lanczos) for solving symmetric positive definite (SPD) eigenvalue problems. Also, we implement CA-Lanczos iteration algorithm shown in **Figure 4**, which is also called Lanczos (s, t), where s is the s-step basis length and t is the outer iterations number before restart. This algorithm is based on using rank revealing-tall skinny QR-block Gram-Schmidt (RR-TSQR-BGS) orthogonalization method

Require: $n \times n$ linear system $Ax = b$, and initial guess x_0
$r_0 := b - Ax_0$, $\beta := \|r_0\|_2$, $q_1 := r_0/\beta$
for $k = 0$ to $t - 1$ **do**
  Fix basis conversion matrix $\underline{B}_k$
  Compute $\underline{\acute{V}}_k$ using matrix powers kernel
  if $k = 0$ **then**
    Compute QR factorization $\underline{V}_0 = \underline{Q}_0\underline{R}$
    Set $\underline{\mathfrak{Q}}_0 := \underline{Q}_0$ and $\underline{\mathfrak{H}}_0 := \underline{R}_0\underline{B}_0R_0^{-1}$
    Reduce $\underline{H_0}$ from upper Hessenberg to upper triangular form using s Givens rotations $G_1, G_2, \ldots, G_s$. Apply the same rotations in the same order to βe_1, resulting in the length $s+1$ vector ζ_0.
  else
    $\acute{\mathfrak{R}}_{k-1,k} := \mathfrak{Q}_{k-1}^{*}\underline{\acute{V}}_k$
    $\underline{\acute{V}}_k' := \underline{\acute{V}}_k - \underline{\mathfrak{Q}}_{k-1}\acute{\mathfrak{R}}_{k-1,k}$
    Compute QR factorization $\underline{\acute{V}}_k' = \underline{\acute{Q}}_k\underline{\acute{R}}_k$
    Compute $\underline{\mathfrak{H}}_{k-1,k} := -\mathfrak{H}_{k-1}\mathfrak{R}_{k-1,k}R_k^{-1} + \mathfrak{R}_{k-1,k}\underline{B}_kR_k^{-1}$
    Compute H_k via Equation (3.21)
    Compute h_k via Equation (3.22)
    $\underline{\mathfrak{H}}_k := \begin{pmatrix} \mathfrak{H}_{k-1} & \underline{\mathfrak{H}}_{k-1,k} \\ h_{k-1}e_1e_{s(k-1)}^T & H_k \\ 0_{1,s(k-1)} & h_ke_s^T \end{pmatrix}$
    Apply Givens rotations $G_1, \ldots, G_{sk}$ in order to $\begin{pmatrix} \mathfrak{H}_{k-1,k} \\ \underline{H}_k \end{pmatrix}$.
    Reduce $\underline{H}_k$ to upper triangular form using s Givens rotations $G_{sk+1}, \ldots, G_{s(k+1)}$. Apply the rotations in the same order to $\begin{pmatrix} \zeta_{k-1} \\ 0_s \end{pmatrix}$, resulting in the length $s(k+1)+1$ vector ζ_k.
  end if
  Element $s(k+1)+1$ of ζ_k is the 2-norm (in exact arithmetic) of the current residual $r_{k+1} = b - Ax_{k+1}$ of the current solution x_{k+1}.
  if converged **then**
    Use the above reduction of $\underline{\mathfrak{H}}_k$ to upper triangular form, and ζ_k, to solve $y_k := \operatorname{argmin}_y \|\underline{\mathfrak{H}}_k y - \beta e_1\|_2$
    Set $x_k := x_0 + \underline{\mathfrak{Q}}_k y_k$, and exit
  end if
end for

Figure 3.
CA-GMRES iteration algorithm.

Input: $n \times n$ symmetric matrix A and $n \times 1$ starting vector r

$\beta_0 := \|r\|_2$, $q_1 := r/\beta_0$
for $k = 0$ to $t - 1$ **do**
  Fix basis conversion matrix $\underline{B_k}$
  Compute $\underline{\acute{V}_k}$ from q_{sk+1} and A using matrix powers kernel
  if $k = 0$ **then**
    $\underline{\mathfrak{B}_0} := \underline{B_0}$
    Compute (via TSQR) the QR factorization $\underline{V_0} = \underline{Q_0}\underline{R_0}$
    $\underline{\mathfrak{Q}_0} := \underline{Q_k}, \underline{\mathfrak{R}_0} := \underline{R_0}$
    $\underline{T_0} := \underline{R_0 B_0 R_0^{-1}}, \underline{\mathfrak{T}_0} := \underline{T_0}$
  else
    $\underline{\mathfrak{B}_k} := \begin{pmatrix} \mathfrak{T}_{k-1} & 0 \cdot e_1 e_s^T \\ \beta_{sk+1} e_1 e_{s(k-1)}^T & \underline{B_k} \end{pmatrix}$
    $\underline{\acute{R}_{k-1,k}} := \underline{Q_{k-1}^*}\underline{\acute{V}_k}$
    $\underline{\acute{V}_k} := \underline{\acute{V}_k} - \underline{Q_{k-1}}\underline{\acute{R}_{k-1,k}}$
    Compute (via TSQR) the QR factorization $\underline{\acute{V}_k} = \underline{\acute{Q}_k}\underline{\acute{R}_k}$
    $\underline{\mathfrak{Q}_k} := [\underline{\mathfrak{Q}_{k-1}}, \underline{\acute{Q}_k}]$
    Compute T_k (Eq. (4.10)) and $\beta_{s(k+1)+1}$ (Eq. (4.11))
    $\underline{\mathfrak{T}_k} := \begin{pmatrix} \mathfrak{T}_{k-1} & \beta_{sk+1} e_{s(k-1)} e_1^T \\ \beta_{sk+1} e_1 e_{s(k-1)}^T & T_k \\ 0_{1,s(k-1)} & \beta_{s(k+1)+1} e_s^T \end{pmatrix}$
  end if
end for

Figure 4.
CA-Lanczos iteration algorithm.

which connects between TSQR and block Gram-Schmidt, where we have been using the right-shifted basis matrix at outer iteration k as follows:

$$V'_k = [V_{sk+2}, \dots, v_{sk+s}] \tag{81}$$

and

$$V'_k = \left[V'_k, v_{sk+s+1}\right] \tag{82}$$

For more details about the considered preconditioners and algorithms, we refer the interested readers to [91].

The main objective of this section is to implement an accurate and robust preconditioning technique for solving the dense nonsymmetric algebraic system of linear equations arising from the BEM. So, a communication-avoiding Arnoldi of the Arnoldi [93] has been implemented for solving the resulting linear systems in order to reduce the iteration number and CPU time. The BEM discretization is employed in 1280 quadrilateral elements, with 3964 degrees of freedom (DOF). A comparative performance of preconditioned Krylov subspace solvers (CA-Arnoldi, CA-GMRES, and CA-Lanczos) has been shown in **Table 1**, where the number of DOF is 3964 and "–" was defined as the divergence process. From the results of **Table 1**. The CA-Arnoldi, CA-GMRES, and CA-Lanczos are more cost-effective than the other Krylov subspace methods Arnoldi, GMRES, and Lanczos, respectively. Also, CA-Arnoldi, CA-GMRES, and CA-Lanczos have been compared with each other in **Table 2**. It can be seen from this table that the performance of CA-Arnoldi is superior than the other iterative methods.

Methods	Preconditioning techniques	Iterations	Residual	Time of each iterative step (s)	Time of solution
Direct methods	NO	—	—	—	9 min 50 s
Arnoldi	NO	174	7.21E–07	3.85	11 min 25 s
	JOBI	26	5.22E–07	3.86	2 min 38 s
	BJOB	22	1.34E–06	3.86	2 min 23 s
	ILU3	47	1.66E–06	3.84	4 min 2 s
	ILU5	48	1.38E–06	3.89	4 min 6 s
	DILU	48	1.53E–06	5.45	4 min 18 s
CA–Arnoldi	NO	360	6.96E–07	1.95	11 min 53 s
	JOBI	20	4.42E–07	1.96	1 min 30 s
	BJOB	20	2.30E–08	1.96	1 min 30 s
	ILU3	40	7.87E–07	1.96	2 min 11 s
	ILU5	60	1.28E–08	1.96	2 min 48 s
	DILU	60	1.59E–07	3.07	4 min 1 s
GMRES	NO	280	2.36E–08	1.90	6 min 20 s
	JOBI	40	5.01E–13	1.91	2 min 10 s
	BJOB	40	2.05E–11	1.91	2 min 10 s
	ILU3	40	4.70E–08	1.91	2 min 10 s
	ILU5	40	3.13E–08	2.60	2 min 10 s
	DILU	40	6.19E–08	3.07	2 min 48 s
CA–GMRES	NO	120	6.89E–07	3.78	7 min 57 s
	JOBI	12	1.00E–05	3.76	1 min 41 s
	BJOB	12	2.22E–06	3.76	1 min 42 s
	ILU3	26	3.63E–06	3.75	2 min 34 s
	ILU5	22	4.05E–06	3.75	2 min 20 s
	DILU	25	5.19E–06	5.93	3 min 18 s
Lanczos	NO	135	7.24E–07	3.80	8 min 41 s
	JOBI	22	4.87E–07	3.75	2 min 33 s
	BJOB	18	9.27E–07	5.18	3 min 2 s
	ILU3	42	2.41E–07	3.81	3 min 48 s
	ILU5	36	6.41E–07	3.78	3 min 18 s
	DILU	38	2.04E–07	5.00	3 min 32 s
CA–Lanczos	NO	129	1.30E–04	3.75	9 min 22 s
	JOBI	16	8.64E–07	3.76	2 min 3s
	BJOB	14	1.69E–07	3.77	2 min 0 s
	ILU3	24	9.29E–07	3.87	2 min 31 s
	ILU5	31	1.91E–07	3.90	3 min 1 s
	DILU	27	8.11E–07	5.95	3 min 31 s

Table 1.
Performances of preconditioned Krylov subspace iterative methods for DOF 3964.

Solvers		DOF				
		965	1505	3380	3964	6005
CA–Arnoldi	Residual	6.81E–12	5.38E–12	4.13E–11	4.17E–11	7.57E–11
	CPU time (s)	4.96	10.78	99.24	134.26	293.29
	Iterations	25	25	25	25	25
CA–GMRES	Residual	2.98E–12	1.90E–12	1.28E–11	1.36E–11	1.22E–11
	CPU time (s)	5.06	11.49	126.38	164.09	445.51
	Iterations	50	50	50	50	50
CA– Lanczos	Residual	7.20E–11	3.35E–11	2.72E–11	3.97E–11	8.33E–11
	CPU time (s)	5.05	11.47	139.07	180.49	514.72
	Iterations	22	26	28	30	32

Table 2.
The CPU time and the number of iterations for some communication–avoiding Krylov subspace solvers.

6. Numerical results and discussion

In order to illustrate the numerical results of this study, we consider a monoclinic graphite-epoxy as an anisotropic smart material which has the following constants [35].

The elasticity tensor is expressed as

$$C_{pjkl} = \begin{bmatrix} 430.1 & 130.4 & 18.2 & 0 & 0 & 201.3 \\ 130.4 & 116.7 & 21.0 & 0 & 0 & 70.1 \\ 18.2 & 21.0 & 73.6 & 0 & 0 & 2.4 \\ 0 & 0 & 0 & 19.8 & -8.0 & 0 \\ 0 & 0 & 0 & -8.0 & 29.1 & 0 \\ 201.3 & 70.1 & 2.4 & 0 & 0 & 147.3 \end{bmatrix} \text{GPa} \tag{83}$$

The mechanical temperature coefficient is

$$\beta_{pj} = \begin{bmatrix} 1.01 & 2.00 & 0 \\ 2.00 & 1.48 & 0 \\ 0 & 0 & 7.52 \end{bmatrix} \cdot 10^6 \text{ N/km}^2 \tag{84}$$

The thermal conductivity tensor is

$$k_{pj} = \begin{bmatrix} 5.2 & 0 & 0 \\ 0 & 7.6 & 0 \\ 0 & 0 & 38.3 \end{bmatrix} \text{W/km} \tag{85}$$

Mass density $\rho = 7820 \text{ kg/m}^2$ and heat capacity $c = 461 \text{ J/kg k}$.

The technique that has been proposed in the current chapter can be applicable to a wide range of three-temperature nonlinear thermal stress problems of FGA structures. The main aim of this chapter is to assess the impact of MDD and anisotropy on the three-temperature nonlinear thermal stress distributions.

The proposed technique that has been implemented in the current study can be applicable to a wide variety of FGA smart structure problems involving three temperatures. All the physical parameters satisfy the initial and boundary conditions. The efficiency of our BEM modeling technique has been improved using an explicit staggered algorithm based on communication-avoiding Arnoldi procedure to decrease the computation time.

Figure 5 shows the variations of the three temperatures T_e, T_i and T_p with the time τ in the presence of MDD. **Figure 6** shows the variations of the three temperatures T_e, T_i and T_p with the time τ in the presence of MDD. It can be seen from **Figures 5** and **6** that the MDD has a significant effect on the temperature distributions.

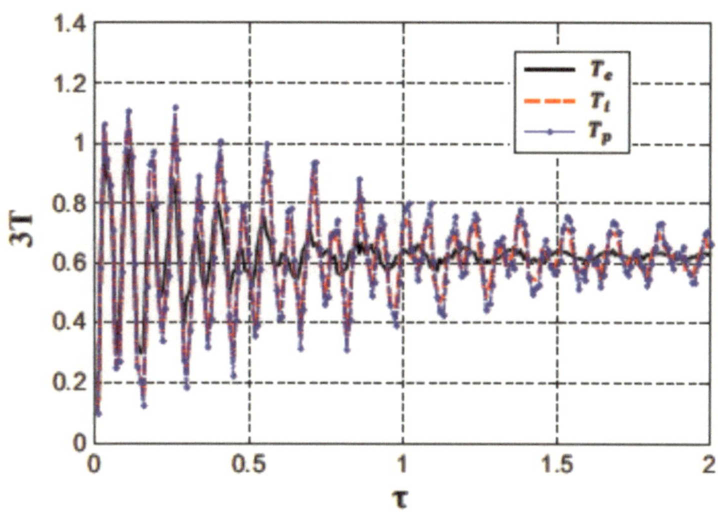

Figure 5.
Variation of the three-temperature (with memory) with time τ.

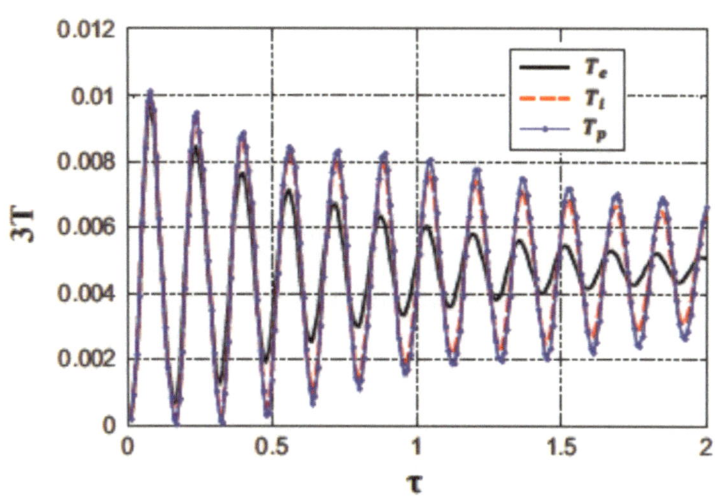

Figure 6.
Variation of the three-temperature (without memory) with time τ.

In order to study the anisotropy and MDD effects on the nonlinear thermal stresses, we assume the following four cases: A, B, C, and D, where case A denotes the nonlinear thermal stress distribution in the isotropic material without MDD effect, case B denotes the nonlinear thermal stress distribution in isotropic material with MDD effect, case C denotes the nonlinear thermal stress distribution in anisotropic material without MDD effect, and case D denotes nonlinear thermal stress distribution in anisotropic material with MDD effect.

Figures 7–9 show the variation of the nonlinear thermal stresses σ_{11}, σ_{12} and σ_{22} with the time τ. It is clear from these figures that both anisotropy and MDD have a significant influence on the nonlinear thermal stress distributions.

Since there are no available results for the considered problem in the literature. Therefore, we only considered the one-dimensional special case for the variations of the nonlinear thermal stress σ_{11} with the time τ as shown in **Figure 10**. The validity and accuracy of our proposed technique was confirmed by comparing graphically

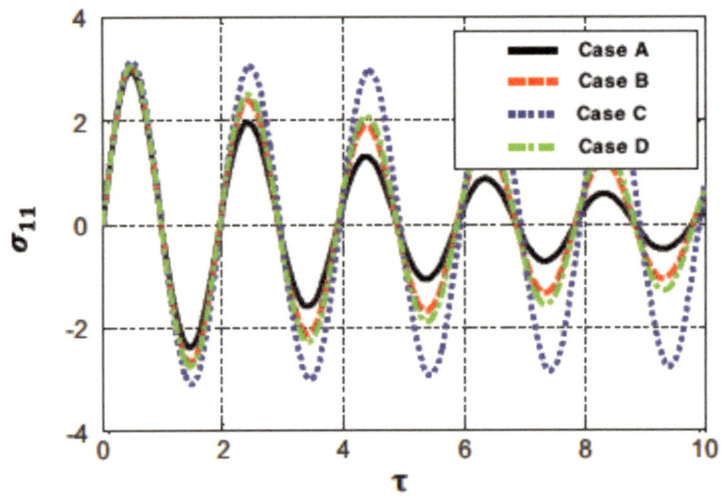

Figure 7.
Variation of the nonlinear thermal stress σ_{11} with time τ.

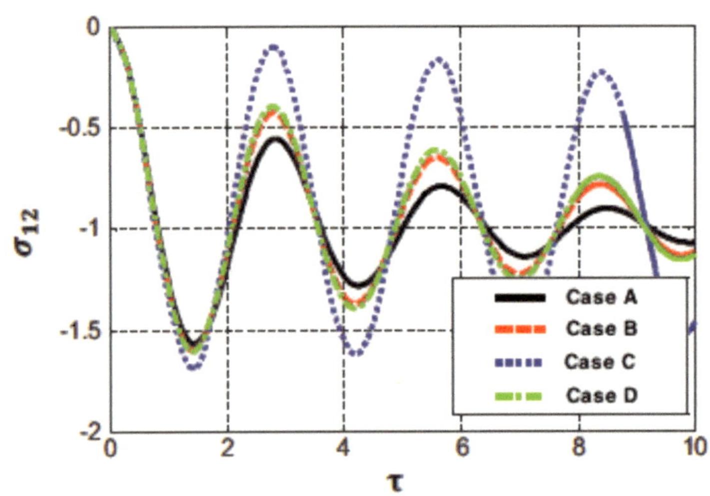

Figure 8.
Variation of the nonlinear thermal stress σ_{12} with time τ.

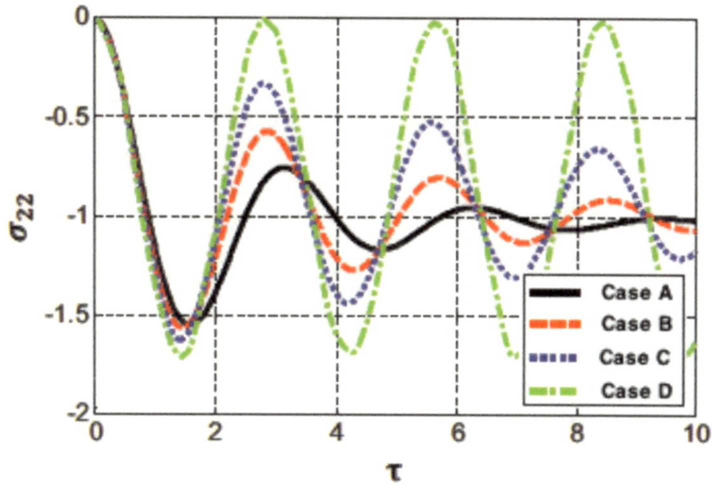

Figure 9.
Variation of the nonlinear thermal stress σ_{22} with time τ.

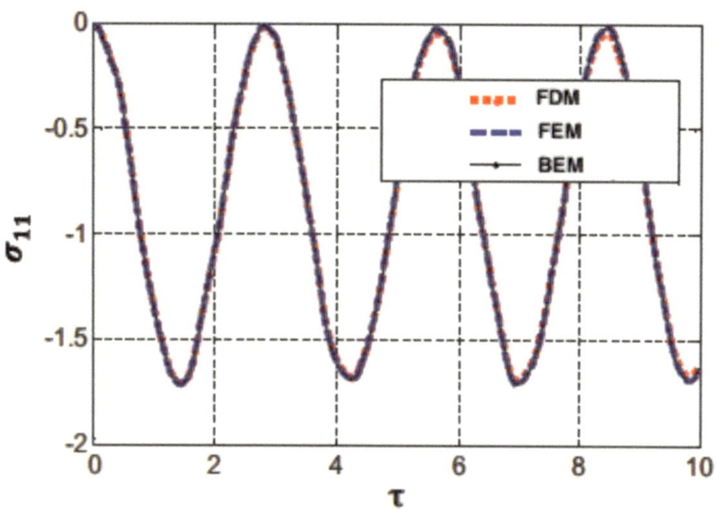

Figure 10.
Variation of the nonlinear thermal stress σ_{11} with time τ.

our BEM results with those obtained using the FDM of Pazera and Jędrysiak [95] and FEM of Xiong and Tian [96] results based on replacing one-temperature heat conduction with the total three-temperature T $(T = T_e + T_i + T_r)$ heat conduction. It can be noticed that the BEM results are found to agree very well with the FDM and FEM results.

7. Conclusion

The main aim of this chapter is to introduce a new MDD model based on BEM for obtaining the transient three-temperature nonlinear thermal stresses in FGA smart structures. The governing equations of this model are very hard to solve analytically because of nonlinearity and anisotropy. To overcome this, we propose a new boundary element formulation for solving such equations. Since the CA kernels of the s-step Krylov methods are faster than the kernels of standard Krylov methods. Therefore, we used an explicit staggered algorithm based on CA-Arnoldi procedure to solve the resulted linear equations. The computational performance of the proposed technique has been performed using communication-avoiding Arnoldi procedure. The numerical results are presented highlighting the effects of MDD on the three-temperature distributions and the influence of MDD and anisotropy on the nonlinear thermal stresses of FGA smart structures. The numerical results also demonstrate the validity and accuracy of the proposed technique. It can be concluded from numerical results of our current general problem that all generalized and nonlinear generalized thermoelasticity theories can be combined with the three-temperature radiative heat conduction to describe the deformation of FGA smart structures in the context of memory-dependent derivatives. From the research that has been performed, it is possible to conclude that the proposed BEM technique is effective and stable for transient three-temperature thermal stress problems in FGA smart structures.

The numerical results for our complex and general problem can provide data references for computer scientists and engineers, geotechnical and geothermal engineers, designers of new materials, and researchers in material science as well as for those working on the development of anisotropic smart structures. In the application of three-temperature theories in advanced manufacturing technologies, with the development of soft machines and robotics in biomedical engineering and advanced manufacturing, transient thermal stresses will be encountered more often where three-temperature radiative heat conduction will turn out to be the best choice for thermomechanical analysis in the design and analysis of advanced smart materials and structures.

Author details

Mohamed Abdelsabour Fahmy
Faculty of Computers and Informatics, Suez Canal University, Ismailia, Egypt

*Address all correspondence to: mohamed_fahmy@ci.suez.edu.eg

References

[1] Fahmy MA. A time-stepping DRBEM for 3D anisotropic functionally graded piezoelectric structures under the influence of gravitational waves. In: Proceedings of the 1st GeoMEast International Congress and Exhibition (GeoMEast 2017); 15–19 July 2017; Sharm El Sheikh, Egypt. Facing the Challenges in Structural Engineering, Sustainable Civil Infrastructures. 2017. pp. 350-365

[2] Fahmy MA. 3D DRBEM modeling for rotating initially stressed anisotropic functionally graded piezoelectric plates. In: Proceedings of the 7th European Congress on Computational Methods in Applied Sciences and Engineering (ECCOMAS 2016); 5–10 June 2016; Crete Island, Greece. pp. 7640-7658

[3] Fahmy MA. Boundary element solution of 2D coupled problem in anisotropic piezoelectric FGM plates. In: Proceedings of the 6th International Conference on Computational Methods for Coupled Problems in Science and Engineering (Coupled Problems 2015); 18–20 May 2015; Venice, Italy. 2015. pp. 382-391

[4] Fahmy MA. The DRBEM solution of the generalized magneto-thermo-viscoelastic problems in 3D anisotropic functionally graded solids. In: Proceedings of the 5th International Conference on Coupled Problems in Science and Engineering (Coupled Problems 2013); 17–19 June 2013; Ibiza, Spain. 2013. pp. 862-872

[5] Fahmy MA. A computerized boundary element model for simulation and optimization of fractional-order three temperatures nonlinear generalized piezothermoelastic problems based on genetic algorithm. In: AIP Conference Proceedings 2138 of Innovation and Analytics Conference and Exhibition (IACE 2019); 25–28 March 2019; Sintok, Malaysia. 2019. p. 030015

[6] Huang R, Zheng SJ, Liu ZS, Ng TY. Recent advances of the constitutive models of smart materials—Hydrogels and shape memory polymers. International Journal of Applied Mechanics. 2020;**12**:2050014

[7] Fahmy MA. Shape design sensitivity and optimization of anisotropic functionally graded smart structures using bicubic B-splines DRBEM. Engineering Analysis with Boundary Elements. 2018;**87**:27-35

[8] Sigmund O, Torquato S. Design of smart composite materials using topology optimization. Smart Materials and Structures. 1999;**8**:365-379

[9] Jin B, Zhong Z. A moving mode-III crack in functionally graded piezoelectric material: Permeable problem. Mechanics Research Communications. 2002;**29**:217-224

[10] Lin S, Narita F, Shindo Y. Electroelastic analysis of a penny-shaped crack in a piezo-electric ceramic under mode I loading. Mechanics Research Communications. 2003;**30**:371-386

[11] Liu W, Ma S, Wu H. Three-dimensional analysis of functionally graded piezoelectric plate with arbitrarily distributed material properties. Journal of Wuhan University of Technology. 2014;**29**(August):712-720

[12] Stanak P. Three-dimensional meshless modelling of functionally graded piezoelectric sensor. J Mech Phys Solids Mech. 2014;**2013**:425-432

[13] Duhamel J. Some memoire sur les phenomenes thermo-mechanique. Journal de l'École Polytechnique. 1837;**15**: 1-57

[14] Neumann F. Vorlesungen Uber die theorie der elasticitat. Brestau: Meyer; 1885

[15] Biot M. Thermoelasticity and irreversible thermo-dynamics. Journal of Applied Physics. 1956;**27**:249-253

[16] Lord HW, Shulman Y. A generalized dynamical theory of thermoelasticity. Journal of the Mechanics and Physics of Solids. 1967;**15**:299-309

[17] Green AE, Lindsay KA. Thermoelasticity. Journal of Elasticity. 1972;**2**:1-7

[18] Green AE, Naghdi PM. On undamped heat waves in an elastic solid. Journal of Thermal Stresses. 1992;**15**: 253-264

[19] Green AE, Naghdi PM. Thermoelasticity without energy dissipation. Journal of Elasticity. 1993; **31**:189-208

[20] Tzou DY. A unified field approach for heat conduction from macro to micro scales. ASME Journal of Heat Transfer. 1995;**117**:8-16

[21] Chandrasekharaiah DS. Hyperbolic thermoelasticity: A review of recent literature. Applied Mechanics Reviews. 1998;**51**:705-729

[22] Roychoudhuri SK. On a thermoelastic three-phase-lag model. Journal of Thermal Stresses. 2007;**30**: 231-238

[23] Fahmy MA. A time-stepping DRBEM for magneto-thermo-viscoelastic interactions in a rotating nonhomogeneous anisotropic solid. International Journal of Applied Mechanics. 2011;**3**:1-24

[24] Fahmy MA. A time-stepping DRBEM for the transient magneto-thermo-visco-elastic stresses in a rotating non-homogeneous anisotropic solid. Engineering Analysis with Boundary Elements. 2012;**36**:335-345

[25] Fahmy MA. Numerical modeling of transient magneto-thermo-viscoelastic waves in a rotating nonhomogeneous anisotropic solid under initial stress. International Journal of Modeling. Simulation and Scientific Computing. 2012;**3**:1250002

[26] Fahmy MA. Transient magneto-thermo-viscoelastic stresses in a rotating nonhomogeneous anisotropic solid with and without a moving heat source. Journal of Engineering Physics and Thermophysics. 2012;**85**:950-958

[27] Fahmy MA. Transient magneto-thermo-elastic stresses in an anisotropic viscoelastic solid with and without moving heat source. Numerical Heat Transfer, Part A: Applications. 2012;**61**: 547-564

[28] Fahmy MA. Transient magneto-thermoviscoelastic plane waves in a non-homogeneous anisotropic thick strip subjected to a moving heat source. Applied Mathematical Modelling. 2012; **36**:4565-4578

[29] Fahmy MA. The effect of rotation and inhomogeneity on the transient magneto-thermoviscoelastic stresses in an anisotropic solid. ASME Journal of Applied Mechanics. 2012;**79**:1015

[30] Fahmy MA. Finite difference algorithm for transient magneto-thermo-elastic stresses in a non-homogeneous solid cylinder. International Journal of Materials Engineering and Technology. 2010;**3**:87-93

[31] Fahmy MA. Boundary element algorithm for nonlinear modeling and simulation of three temperature anisotropic generalized micropolar piezothermoelasticity with memory-dependent derivative. International Journal of Applied Mechanics. 2020;**12**: 2050027

[32] Fahmy MA. Thermal stresses in a spherical shell under three thermoelastic

models using FDM. International Journal of Numerical methods and Applications. 2009;**2**:123-128

[33] Ezzat MA, El-Karamany AS, Alaa A, El-Bary AA. On dual-phase-lag thermoelasticity theory with memory-dependent derivative. Mechanics of Advanced Materials and Structures. 2017;**24**:908-916

[34] Ezzat MA, El-Karamany AS, Alaa A, El-Bary AA. Generalized thermoelasticity with memory-dependent derivatives involving two temperatures. Mechanics of Advanced Materials and Structures. 2016;**23**: 545-553

[35] Fahmy MA. A new boundary element strategy for modeling and simulation of three temperatures nonlinear generalized micropolar-magneto-thermoelastic wave propagation problems in FGA structures. Engineering Analysis with Boundary Elements. 2019;**108**: 192-200

[36] Fahmy MA. A new computerized boundary element model for three-temperature nonlinear generalized thermoelastic stresses in anisotropic circular cylindrical plate structures. In: Awrejcewicz J, Grzelczyk D, editors. Dynamical Systems Theory. London, UK: IntechOpen; 2019. pp. 1-17

[37] Fahmy MA. Boundary element model for nonlinear fractional-order heat transfer in magneto-thermoelastic FGA structures involving three temperatures. In: Ebrahimi F, editor. Mechanics of Functionally Graded Materials and Structures. IntechOpen: London, UK; 2019. pp. 1-22

[38] Fahmy MA. Boundary element mathematical modelling and boundary element numerical techniques for optimization of micropolar thermoviscoelastic problems in solid deformable bodies. In: Sivasankaran S, Nayak PK, Günay E, editors. Mechanics of Solid Deformable Bodies. IntechOpen: London, UK; 2020. pp. 1-21

[39] Fahmy MA. Boundary element modeling and optimization based on fractional-order derivative for nonlinear generalized photo-thermoelastic stress wave propagation in three-temperature anisotropic semiconductor structures. In: Sadollah A, Sinha TS, editors. Recent Trends in Computational Intelligence. IntechOpen: London, UK; 2020. pp. 1-16

[40] Cattaneo C. Sur une forme de i'equation de la chaleur elinant le paradox d'une propagation instantanc. Comptes Rendus de l'Académie des Sciences. 1958;**247**:431-433

[41] Oldham KB, Spanier J. The Fractional Calculus: Theory and Applications of Differentiation and Integration to Arbitrary Order. Mineola: Dover Publication; 2006

[42] Soukkou A, Belhour MC, Leulmi S. Review, design, optimization and stability analysis of fractional-order PID controller. International Journal of Intelligent Systems Technologies and Applications. 2016;**8**:73-96

[43] Kilbas AA, Srivastava HM, Trujillo JJ. Theory and applications of fractional differential equations. In: Vol. 204 of North-Holland Mathematics Studies. Amsterdam, The Netherlands: Elsevier Science; 2006

[44] Sabatier J, Agrawal OP, Machado JAT, editors. Advances in Fractional Calculus: Theoretical Developments and Applications in Physics and Engineering. Dordrecht, The Netherlands: Springer; 2007

[45] Diethelm K. Generalized compound quadrature formulae for finite-part integrals. IMA Journal of Numerical Analysis. 1997;**17**:479-493

[46] Wang JL, Li HF. Surpassing the fractional derivative: Concept of the memory-dependent derivative. Computers and Mathematics with Applications. 2011;**62**:1562-1567

[47] Fahmy MA. A three-dimensional generalized magneto-thermo-viscoelastic problem of a rotating functionally graded anisotropic solids with and without energy dissipation. Numerical Heat Transfer, Part A: Applications. 2013;**63**:713-733

[48] Fahmy MA. A 2-D DRBEM for generalized magneto-thermo-viscoelastic transient response of rotating functionally graded anisotropic thick strip. International Journal of Engineering and Technology Innovation. 2013;**3**:70-85

[49] Fahmy MA, Salem AM, Metwally MS, Rashid MM. Computer implementation of the DRBEM for studying the generalized thermoelastic responses of functionally graded anisotropic rotating plates with one relaxation time. International Journal of Applied Science and Technology. 2013; **3**:130-140

[50] Fahmy MA, Salem AM, Metwally MS, Rashid MM. Computer implementation of the DRBEM for studying the classical uncoupled theory of thermoelasticity of functionally graded anisotropic rotating plates. International Journal of Engineering Research and Applications. 2013;**3**: 1146-1154

[51] Fahmy MA. A computerized DRBEM model for generalized magneto-thermo-visco-elastic stress waves in functionally graded anisotropic thin film/substrate structures. Latin American Journal of Solids and Structures. 2014;**11**:386-409

[52] Fahmy MA, Salem AM, Metwally MS, Rashid MM. Computer implementation of the DRBEM for studying the classical coupled thermoelastic responses of functionally graded anisotropic plates. Physical Science International Journal. 2014;**4**: 674-685

[53] Fahmy MA, Salem AM, Metwally MS, Rashid MM. Computer implementation of the DRBEM for studying the generalized thermo elastic responses of functionally graded anisotropic rotating plates with two relaxation times. British Journal of Mathematics & Computer Science. 2014;**4**:1010-1026

[54] Fahmy MA. Computerized Boundary Element Solutions for Thermoelastic Problems: Applications to Functionally Graded Anisotropic Structures. Saarbrücken: LAP Lambert Academic Publishing; 2017

[55] Fahmy MA. Boundary Element Computation of Shape Sensitivity and Optimization: Applications to Functionally Graded Anisotropic Structures. Saarbrücken: LAP Lambert Academic Publishing; 2017

[56] Fahmy MA. A new computerized boundary element algorithm for cancer modeling of cardiac anisotropy on the ECG simulation. Asian Journal of Research in Computer Science. 2018;**2**: 1-10

[57] Brebbia CA, Telles JCF, Wrobel L. Boundary Element Techniques in Engineering. New York: Springer-Verlag; 1984

[58] Wrobel LC, Brebbia CA. The dual reciprocity boundary element formulation for nonlinear diffusion problems. Computer Methods in Applied Mechanics and Engineering. 1987;**65**:147-164

[59] Partridge PW, Brebbia CA. Computer implementation of the BEM dual reciprocity method for the solution of general field equations.

Communications in Applied Numerical Methods. 1990;**6**:83-92

[60] Partridge PW, Brebbia CA, Wrobel LC. The Dual Reciprocity Boundary Element Method. Southampton: Computational Mechanics Publications; 1992

[61] Gaul L, Kögl M, Wagner M. Boundary Element Methods for Engineers and Scientists. Berlin: Springer-Verlag; 2003

[62] Zirakashvili N. Solution of contact problems for half-space by boundary element methods based on singular solutions of flamant and boussinesq's problems. International Journal of Applied Mechanics. 2020;**12**:2050015

[63] Fahmy MA. Implicit-explicit time integration DRBEM for generalized magneto-thermoelasticity problems of rotating anisotropic viscoelastic functionally graded solids. Engineering Analysis with Boundary Elements. 2013; **37**:107-115

[64] Fahmy MA. Generalized magneto-thermo-viscoelastic problems of rotating functionally graded anisotropic plates by the dual reciprocity boundary element method. Journal of Thermal Stresses. 2013;**36**:1-20

[65] Fahmy MA. A 2D time domain DRBEM computer model for magneto-thermoelastic coupled wave propagation problems. International Journal of Engineering and Technology Innovation. 2014;**4**:138-151

[66] Fahmy MA, Al-Harbi SM, Al-Harbi BH. Implicit time-stepping DRBEM for design sensitivity analysis of magneto thermo-elastic FGA structure under initial stress. American Journal of Mathematical and Computational Sciences. 2017;**2**:55-62

[67] Fahmy MA. The effect of anisotropy on the structure optimization using golden-section search algorithm based on BEM. Journal of Advances in Mathematics and Computer Science. 2017;**25**:1-18

[68] Fahmy MA. DRBEM sensitivity analysis and shape optimization of rotating magneto-thermo-viscoelastic FGA Structures using golden-section search algorithm based on uniform bicubic B-splines. Journal of Advances in Mathematics and Computer Science. 2017;**25**:1-20

[69] Fahmy MA. A predictor-corrector time-stepping DRBEM for shape design sensitivity and optimization of multilayer FGA structures. Transylvanian Review. 2017;**XXV**: 5369-5382

[70] Fahmy MA. Shape design sensitivity and optimization for two-temperature generalized magneto-thermoelastic problems using time-domain DRBEM. Journal of Thermal Stresses. 2018;**41**: 119-138

[71] Fahmy MA. Boundary element algorithm for modeling and simulation of dual-phase lag bioheat transfer and biomechanics of anisotropic soft tissues. International Journal of Applied Mechanics. 2018;**10**:1850108

[72] Fahmy MA. Modeling and optimization of anisotropic viscoelastic porous structures using CQBEM and moving asymptotes algorithm. Arabian Journal for Science and Engineering. 2019;**44**:1671-1684

[73] Fahmy MA. Boundary element modeling and simulation of biothermomechanical behavior in anisotropic laser-induced tissue hyperthermia. Engineering Analysis with Boundary Elements. 2019;**101**: 156-164

[74] Fahmy MA, Al-Harbi SM, Al-Harbi BH, Sibih AM. A computerized boundary element algorithm for

modeling and optimization of complex magneto-thermoelastic problems in MFGA structures. Journal of Engineering Research and Reports. 2019;**3**:1-13

[75] Fahmy MA. A new LRBFCM-GBEM modeling algorithm for general solution of time fractional order dual phase lag bioheat transfer problems in functionally graded tissues. Numerical Heat Transfer, Part A: Applications. 2019;**75**:616-626

[76] Fahmy MA. Design optimization for a simulation of rotating anisotropic viscoelastic porous structures using time-domain OQBEM. Mathematics and Computers in Simulation. 2019;**66**: 193-205

[77] Fahmy MA. A new convolution variational boundary element technique for design sensitivity analysis and topology optimization of anisotropic thermo-poroelastic structures. Arab Journal of Basic and Applied Sciences. 2020;**27**:1-12

[78] Fahmy MA. Thermoelastic stresses in a rotating non-homogeneous anisotropic body. Numerical Heat Transfer, Part A: Applications. 2008;**53**: 1001-1011

[79] Abd-Alla AM, Fahmy MA, El-Shahat TM. Magneto-thermo-elastic problem of a rotating non-homogeneous anisotropic solid cylinder. Archives of Applied Mechanics. 2008;**78**:135-148

[80] Fahmy MA, El-Shahat TM. The effect of initial stress and inhomogeneity on the thermoelastic stresses in a rotating anisotropic solid. Archives of Applied Mechanics. 2008; **78**:431-442

[81] Soliman AH, Fahmy MA. Range of applying the boundary condition at fluid/porous interface and evaluation of Beavers and Joseph's slip coefficient using finite element method. Computation. 2020;**8**:14

[82] Eskandari AH, Baghani M, Sohrabpour S. A time-dependent finite element formulation for thick shape memory polymer beams considering shear effects. International Journal of Applied Mechanics. 2019;**10**:1850043

[83] Othman MIA, Khan A, Jahangir R, Jahangir A. Analysis on plane waves through magneto-thermoelastic microstretch rotating medium with temperature dependent elastic properties. Applied Mathematical Modelling. 2019;**65**:535-548

[84] El-Naggar AM, Abd-Alla AM, Fahmy MA, Ahmed SM. Thermal stresses in a rotating non-homogeneous orthotropic hollow cylinder. Heat and Mass Transfer. 2002;**39**:41-46

[85] El-Naggar AM, Abd-Alla AM, Fahmy MA. The propagation of thermal stresses in an infinite elastic slab. Applied Mathematics and Computation. 2003;**12**:220-226

[86] Abd-Alla AM, El-Naggar AM, Fahmy MA. Magneto-thermoelastic problem in non-homogeneous isotropic cylinder. Heat and Mass Transfer. 2003; **39**:625-629

[87] Hu Q, Zhao L. Domain decomposition preconditioners for the system generated by discontinuous Galerkin discretization of 2D-3T heat conduction equations. Communications in Computational Physics. 2017;**22**: 1069-1100

[88] Huang FY, Liang KZ. Boundary element method for micropolar thermoelasticity. Engineering Analysis with Boundary Elements. 1996;**17**:19-26

[89] Eringen AC. Theory of micropolar elasticity. In: Liebowitz H, editor. Fracture. New York: Academic Press; 1968

[90] Dragos L. Fundamental solutions in micropolar elasticity. International Journal of Engineering Science. 1984;**22**: 265-275

[91] Hoemmen M. Communication-Avoiding Krylov Subspace Methods. Berkeley: University of California; 2010

[92] Saad Y, Schultz MH. GMRES: A generalized minimal residual algorithm for solving nonsymmetric linear systems. SIAM Journal on Scientific and Statistical Computing. 1986;**7**:856-869

[93] Arnoldi WE. The principle of minimized iterations in the solution of the matrix eigenvalue problem. Quarterly of Applied Mathematics. 1951; **9**:17-29

[94] Lanczos C. An iteration method for the solution of the eigenvalue problem of linear differential and integral operators. Journal of Research of the National Bureau of Standards. 1950;**45**: 255-282

[95] Pazera E, Jędrysiak J. Effect of microstructure in thermoelasticity problems of functionally graded laminates. Composite Structures. 2018; **202**:296-303

[96] Xiong QL, Tian XG. Generalized magneto-thermo-microstretch response during thermal shock. Latin American Journal of Solids and Structures. 2015; **12**:2562-2580

Section 2

Advanced Functional Materials for Health

Chapter 9

Self-Healing in Titanium Alloys: A Materials Science Perspective

Paul Sunday Nnamchi and Camillus Sunday Obayi

Abstract

Self-healing materials (SHM's) is an emerging class of smart materials, which are capable of autonomous or spontaneous repair of their damage under external stimuli, such as heat, light, and solvent, to the original or near original functionalities much like the biological organisms. The emergence of self-healing in metallic materials presents an exciting paradigm for an ideal combination of metallic and biological properties. The driving force behind this effort is to decrease the consequences of accidents, reduction of cost and extending the service life of metallic components. While previous reviews have focused on self-healing in polymers, composite, concrete and cementous materials, and ceramic, discussions about self-healing in metallic materials remains scarce and the survey of literatures suggests Ti-based self-healing materials known to be biocompatible in human body is rare. The present chapter examines the art of self-healing in titanium-based alloys with the scope to provide an overview of recent advancements and to highlight current problems and perspectives with respect to potential application.

Keywords: self-healing metals, self-healing coating, Ti-based alloys, shape memory effect (SME), autonomous repair, damage, design strategy

1. Introduction and background

Historically, solving material reliability issues has been an old and long term quest of material scientist and engineers, due to their implications for material safety. Considering the fact that structural materials degrade irreversibly over time owing to proliferation of damage like microscopic cracks: the growth of which eventually results in failure. And most times, these internal defects or damage are deep inside materials and difficult to perceive and repair. Recently, there has been a huge interest in materials that can self-heal, as this property can potentially extend materials lifetime, minimize replacement costs, and improve product safety and reliability [1]. Thus, having materials with intrinsic self-repair capabilities—a sort of biomimetic healing functionality, may then allow failures to be averted and the useful lives of components and structures to be extended [2, 3].

Although self-healing is an exclusive specialty of living organisms of biological origin and not easy to put in place in non-biological materials, continuous efforts are now being made to mimic natural materials and to integrate self-healing capability into polymers and polymer composites. Self-engineered healing properties, which are applied in closing and healing crack initiated in a material during its utilization, have been described in cementous [4] and polymer materials [5]. Self-healing approaches mostly gained by surface modification [6–9] or by the creation

IntechOpen

of a composite material with some other smart material like NiTi [10, 11] are been utilized in metals and other inanimate materials. For example, damage to oxide films, which normally protect the surfaces of metals such as aluminum (Al) and titanium (Ti) from corrosion, can be repaired by reoxidation in air, which can be seen as a form of self-repair. Also identified are the self-healing properties obtained by encapsulating a solder material into a metallic matrix [11–13]. Self-healing behavior was also observed in a commercial Al alloy after suitable heat treatment [14] and some other precipitation-forming systems [15, 16]. Healing can be initiated by means of an external source of energy as was shown in the case of a bullet penetration [17] where the ballistic impact caused local heating of the material by allowing self-healing of ionomers.

There are several different strategies to impart self-healing functionality that have been developed and the number of publications dealing with various aspects of self-healing materials has increased markedly in recent years. On the whole, the vast majority of the articles deal with polymer composites and cementous materials. Research in the field of metallic systems is still in its infancy. However, the emergence of self-healing in metallic materials, such as titanium adjured to be biocompatible and explored here presents an exciting paradigm for an ideal combination of metallic and biological properties in application traditionally dominated by metallic materials. Depending on the method of healing, self-healing in metallic system can be classified into two categories: (i) intrinsic ones that are able to heal cracks or repair damage by the metals themselves and (ii) extrinsic in which healing agent has to be pre-embedded.

This chapter begins with an overview on the importance of titanium as an engineering of self-healing materials. Since all processes of self-repair, including healing in living bodies depends on rapid transportation of repair substance to the injured part and reconstruction of the tissues, Therefore, the knowledge of basic principle of solid state diffusion is essential for understanding the self-repair processes, such as phase transformation, precipitation and shape memory effects taking place titanium and other alloys, were briefly discussed. The chapter concludes by considering future research.

2. Titanium: A special engineering material

Titanium has been an important development in the history of non-ferrous industry. Titanium is an attractive material with excellent corrosion resistance and high strength-to-weight ratio. It combines the strength of iron and steel with the light weight of aluminum, which accounts for its widespread use. Industrial applications of titanium materials have recently expanded widely in many areas such as the aerospace, chemical plants, automobiles, and aviation industries, and even in high performance sports equipment, and in the medical field for bone. Their biological compatibility is particularly of interest to the medical industry implants and replacement devices [17]. Currently, the chemical industry is the largest user of titanium due to its excellent corrosion resistance, particularly in the presence of oxidizing acids. The ballistic properties of titanium are also excellent on a density-normalized basis. Some physical properties as compared with other engineering materials by Hanson are presented in **Table 1** [18]. Detailed discussions on other applications of titanium in other areas can be found elsewhere [18, 19].

Besides the areas mentioned above, building applications such as exterior walls and roofing material have emerged as a new market for titanium. Using CP titanium as building material has become especially popular in Japan [20]. One example is the Fukuoka Dome, built in 1993, which is covered with titanium roofing, retractable for multi-role and all-weather purposes (**Figure 1**) [20]. Each of these building projects uses large quantities of CP titanium leading to the increased usage in the civil engineering area in Japan. Another "new area" in which titanium use is growing is the area of consumer products, such as spectacle frames, cameras, watches, jewelry, and various kinds of sporting goods. The largest application in the area of sporting goods is golf club heads. Other examples are tennis rackets, bicycle

Physical Property	Titanium	Aluminium	Copper	Iron	Magnesium
Melting point (°C)	1665	660	1083	1535	650
Density (Mg/m^3)	4.51	2.70	8.94	7.86	1.74
Thermal conductivity (W/mK)	16	239	384	71	147
Electrical resistivity (nΩm)	482	26.8	17.2	97.1	44
Specific heat (J/KgK)	528	883	389	456	1026
Magnetic susceptibility (X10^{-6})	+ 3.4	+ 0.65	-0.086	Ferro magnetic	+ 0.55
Elastic modulus (GPa)	110	70	120	200	45

Table 1.
Physical properties of titanium compared with other metals [18].

Figure 1.
(a) Arial approach view of the Fukuoka dome, built in 1993, which is covered with (b) titanium roofing, retractable for multi-role and all-weather purposes.

Alloy	Nominal Composition	Density	Coefficient of Expansion (20-100°C)	Thermal Conductivity	Resistivity (20°C)	Temperature Coefficient of Resistivity (20-100°C)	Specific Heat (50°C)	Magnetic Susceptibility	Elastic Modulus
		(Mg/m^3)	(x10^{-6}/K)	(W/[mK])	(nΩm)		(J/[KgK])	(x10^{-6})	(GPa)
IMI 115-160	Commercially pure	4.51	7.6	16	482	0.0036	528	+3.4	110
IMI 260/262	Ti-0.15Pd	4.52	7.6	16	482	0.0036	528		
IMI 230	Ti-2.5Cu	4.56	9.0	13	700	0.0026			110
IMI 318	Ti-6.A1-4V	4.42	8.0	5.8	1680	0.0004	610	+3.3	110-125
IMI 550	Ti-4A1-4Mo-2Sn-0.25Si	4.60	8.6	7.9	1590	0.0004			117
IMI 551	Ti-4A1-4Mo-4Sn-0.25Si	4.62	8.4	5.7	1700	0.0003	400	+3.1	112
IMI 685	Ti-6Al-5Zr-0.5Mo-0.25Si	4.45	9.8	4.8	1670	0.0004			124

Table 2.
Physical properties of titanium and some of its alloys [18].

frames, spikes in sprinters. Their low coefficient of thermal expansion is also an important factor. The ballistic properties of titanium are also excellent on a density-normalized basis. Future applications are likely to be in the areas of steam turbine blading, flue gas desulphurization plant consumer products and many marine applications. Some of the basic characteristics of titanium and its alloys are listed in **Table 2** in [18] and compared to those of other structural metallic materials based on Fe, Ni, and Al. Detailed discussions on other applications of titanium in other areas can be found elsewhere [4].

3. Crystallographic structures in titanium metal and alloys

Generally, all property of materials depends directly or indirectly on the type of crystallographic phase and its constructions. The stable structure of pure Ti is the hexagonal close-packed (hcp) structure (α phase) at room temperature, which transforms to the body-centered cubic (bcc) structure (β phase) at high temperature. Apart from these stable phases, other metastable phases can emerge in a quenched alloy such as (α′) martensite with hexagonal structure, martensite with orthorhombic (α″) structure or the β phase [20] or an omega (ω) phase. There are two types of omega ω phases with hcp structure, one is athermal, which forms during quenching from β-phase at high temperature and this type is cooling rate dependent. The other is isothermal ω phases, which precipitate during aging at certain temperatures. However, the exact relationship between the two omega ω phases and the α″ martensite phase for reversible transformation is still a subject of many research studies.

For this, interests in titanium-base alloys as structural materials has inspired several studies of their phase relationships. These have provided the background essential to the development of commercial alloys, but have also revealed some unexpected, and still to some extent unexplained, aspects that are of considerable scientific interest [2]. Although the exact transus point is dependent on the composition and processing treatment for the alloy, for alloyed Ti based materials, the transformation of crystallographic phase could also be driven by alloying elements.

Titanium alloys are stabilized by solute elements that have strong effect on the transformation temperature. Alloying elements of titanium are typically grouped based on their effect on the beta-transus temperature. They are often termed as neutral, alpha stabilizers, or beta-stabilizers. According to this classification scheme the alpha stabilizing alloying elements extend the room temperature hexagonal alpha phase field to elevated temperatures, while beta-stabilizing elements shift the high temperature beta phase field to lower temperatures. Neutral elements have only minor influence on the beta-transus temperature. Among the alpha stabilizing elements aluminum is by far the most important alloying element of titanium. The interstitial elements oxygen O_2, nitrogen N, and carbon C also belong to this category, which are referred to as α-stabilizers [20] (see **Figure 2**).

Some of the high temperature β phase stabilizing elements are Nb Mo; V; W; Fe; Cr; Mn; Co; Cu; Si, H and Ta at room temperature. For β-stabilizers, a minimum concentration β element is required to fully stabilize the β-phase following a quench from the high temperature. The β stabilizing elements are categorized into two groups, namely:

i. β eutectoid stabilizers are elements (which lead to a partially stabilized β phase) such as Fe; Cr; Mn; Co; Cu; Si and H.

ii. β isomorphous forming elements are heavy refractory BCC elements such as Mo; V; W; Nb and Ta.

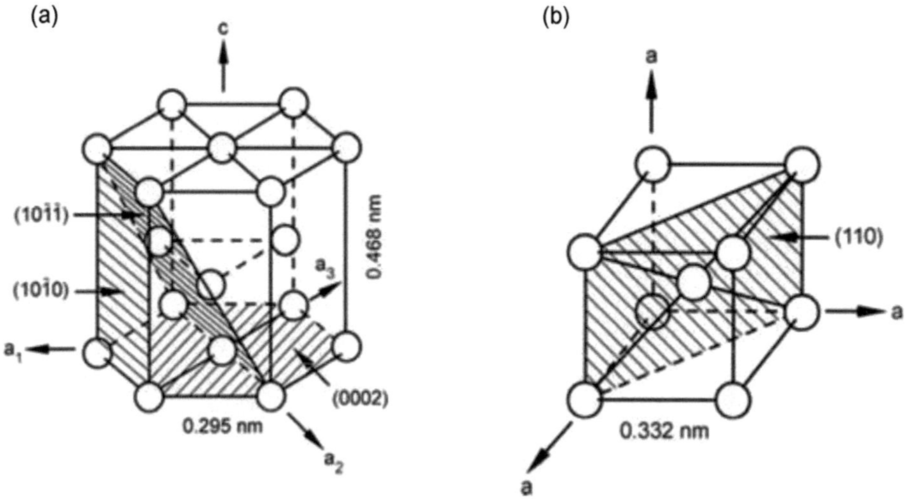

Figure 2.
(a) Unit cell of α type titanium phase (b) unit cell of β type titanium phase [21].

The β phase can be easily transformed to hexagonal (a′ and ω) or orthorhombic (α″) phases depending on the contents of the β stabilizing elements at room temperature. The (α″) phase has more slip systems than that in the hexagonal phases but fewer than in the β phase.

Although it might be that no metallic element is purely neutral, some elements are classified as neutral because they have a minor influence on the transus temperature. They can lower the β transus slightly, but again increase it at a higher concentration. These elements include Sn, Zr and Hf, which may slightly lower the α/β transformation temperatures after certain threshold concentrations. Zr and Sn are the commonly used neutral stabilizing elements. Zr and Hf are isomorphous with titanium and therefore exhibit the same allotropic phase transformation from β to α and are completely soluble in both the α and β phases. Zr also substitutes titanium in a multicomponent alloy and thereby indirectly has a α stabilizing effect [20].

4. Some self-healing assisting phenomena in titanium metal and alloys

4.1 Phase transformation in titanium metal and alloys

Phase transformation occurs whenever a materials system is not at equilibrium, or changes its microstate, as a result of external constraints such as pressure or temperature. In effect, these materials adopt different crystal structures favorable for the minimization of their free energy. In general, the microstructural features and the order in the system changes, leading to variations in most of the important properties. By so doing, phase transformation provides an effective way to modify the microstructure of solids. If it can be activated by a mechanical or other physical force, it becomes part of the deformation process and directly affects the properties of materials as well [22].

In CP titanium and titanium alloys, the most common equilibrium phases are those of α and β, phases. The transformation of high temperature phase can occur by martensitic or by a diffusion controlled nucleation and growth process depending on cooling rate and alloy composition. Their relationship was confirmed for Zirconium by Burger [23, 24] and later for titanium by [22]. This Burgers relationship is closely obeyed for both the martensite transformation and the diffusional transformations (see **Figure 3**).

Thus, during phase transition from β to α, several slip systems operate within the α titanium alloys system. The most common slip direction is $\langle 11\bar{2}0\rangle$. The $\vec{a}$ direction slip occurs in one (0002) basal plane, three $\{10\bar{2}0\}$ prism planes, and six $\{10\bar{1}1\}$ pyramidal planes [21, 26–29]. They are therefore responsible for the four independent slip systems. The fifth slip system required for homogeneous deformation of polycrystals (Von-Mises criteria) is provided by the two basal slips that occur in the $\langle 11\bar{2}3\rangle$ direction and in the $\{11\bar{2}2\}$ plane. If the slip systems are unable to operate, twinning occurs in α titanium. The main twinning modes are $\{10\bar{1}2\}$ and $\{11\bar{2}1\}$ in tension and {1122} in tension and {1122} in compression loading [21, 30–32].

4.2 Martensitic transformation in titanium metal and alloys

Phase transformation provides an effective way to modify microstructure and property of solids. It becomes part of a deformation process, when it can be activated by mechanical or other physical forces. Titanium and its alloy undergo a series of stable and metastable allotropic transformations, depending on alloy elements and process. The exact transus point is dependent on the composition and processing treatment for the alloy. Apart from the stable phase(s), other metastable phases can emerge in a quenched alloy such as martensite with hexagonal structure, martensite with orthorhombic structure or the metastable β phase (see **Figure 4**).

Some striking characteristics distinguish martensite transformations from any other [34]. First, the martensite phase is either a substitutional or interstitial solid solution. Secondly, the transformation takes place in a very short time (i.e.) very rapid. This can only be measured by high speed cameras. The complexity inherent in its measurements is an added problem to its study. The third is that it is accompanied by a shape change (surface relief) of a definite value. This has been confirmed metallographic ally by scratch line test [34]. The fourth is that martensite crystal has a specific habit plane; interfacing between the parent phase and martensite phase

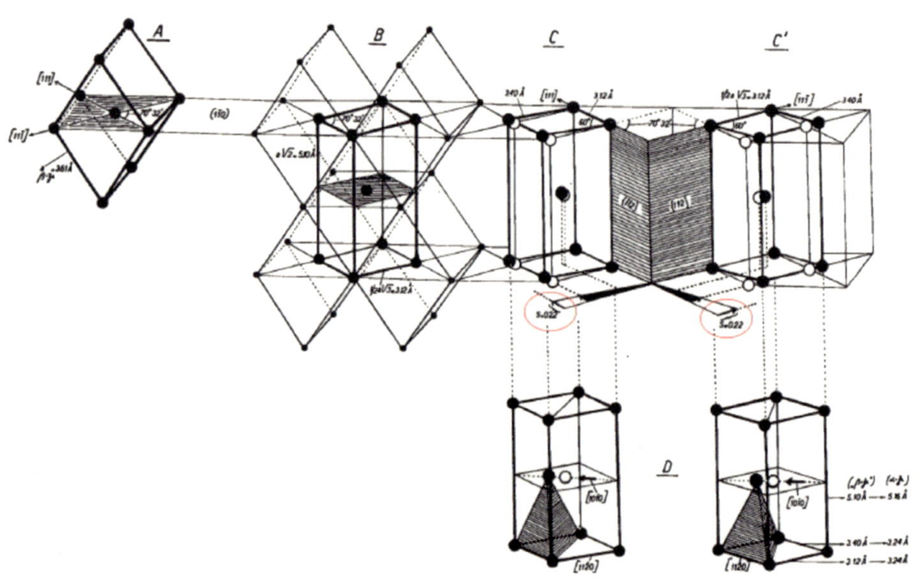

Figure 3.
Schematic illustration of the Burgers' lattice correspondence model of bcc-hcp transformation. The two simultaneous shears are marked by the red ellipses [25].

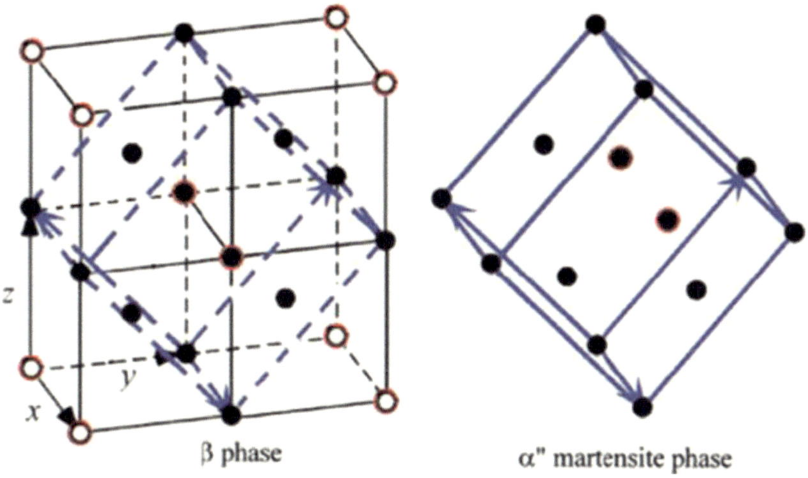

Figure 4.
A schematic illustration showing the lattice correspondence between the β and α" phases, after Kym et al. [33].

which lie along the shear plane during the transformation. This implies that an orientation relationship exist between the two phase lattices. The presence of lattice defects will necessarily exist in martensite crystal due to shearing. The list of alloy components where martensitic transformation can be exploited are not exhaustible and still growing with research in the field of Titanium [35–37].

The recent innovative drive for Ti-alloys has drifted to many other areas such as the formation of SIM for electrical appliances, GUM metal for industrial applications, bio-implants resulting from its excellent compatibility with body tissues, better the mechanical and physical properties. There is an agreement among researchers that the formation is due to deformation by twinning (A twinning process is shown in **Figure 5** as culled adapted from [38]) of β-phase material to a type of martensite α″ with

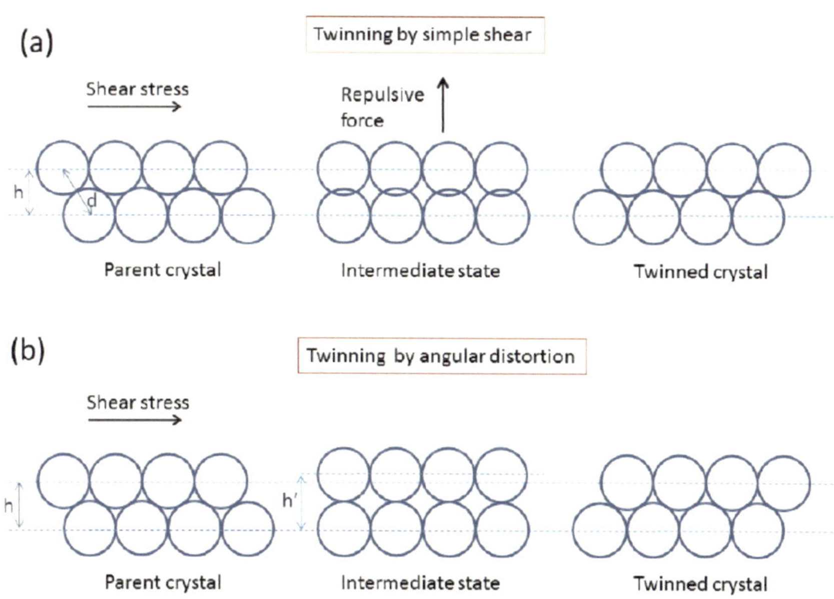

Figure 5.
Schematic 2D representation of a collective displacement of the atoms during deformation twinning according to (a) simple shear, (b) angular distortion.

orthorhombic structure. It is also thought that over a narrow compositional average the β-phase material is said to transform to martensite under an applied stress.

4.3 Shape memory effect and super-elasticity in titanium alloys

Shape Memory Alloys (SMAs) are special materials with great potential in various engineering applications since they possess a number of unique characteristics, including superior energy dissipation capacity compared to normal metallic materials [39]. Other beneficial properties, apart from SMEs, including superelasticity, favorable damping ability and other important characteristics of shape memory alloys, allow it to be applied in a wide range of fields, including electronic, chemical, medical devices, electricity, aerospace, etc. [40].

The value and demand of SMAs was not positively understood for most engineering and technological applications until William Buehler and Frederick Wang discovered the shape memory effect (SME) in a nickel-titanium (NiTi) alloy in 1962 [41]. Thereafter, the use of SMAs, has expanded and the research interests and patents have become quite large. Examples of the possible beneficiaries of these materials abound in a variety of fields, such as automobile and mechanical engineering applications [42, 43], automotive [42], aerospace [44], mini actuators and micro-electromechanical systems (MEMS) [45], robotics [35], biomedical [36] and even in clothing/fashion industries [37]. Titanium (Ti) alloys are one of the most important SMAs and until now, development of new Ti-based SMAs is still one of the most important directions of metal intelligent materials. The Ti-Nb based [35–37, 45], Ti-Ta based [46], Ti-Mo based [46–48] and Ti-Zr based [49] SMAs are developed in recent years.

The interest in using shape memory alloys (SMAs) stems from the fact that they can "remember" their original shape. When subjected to an external force above a threshold, they exhibit stress-induced martensitic transformation from austenite into martensite through twinning, and can recover the apparent permanent strains, returning to the original form. An Illustration of the superelastic response in shape memory alloys (deformation at a temperature N austenite finish temperature Af). This important attribute exhibited by many titanium based alloy can be exploited to accelerate the self-healing process in metallic materials. In addition by adjusting the hysteresis width can allow materials scientists to precisely adjust temperature change ΔT during the self-healing process. One of the probable set back is the functional degradation in properties that manifested as a reduction in the superelastic strains (εSE) and accumulation of residual strains (εresidual).

4.4 Diffusion in titanium metals and alloys

It is well known that atoms in almost all metals and alloys crystallize or has a tendency to pack in dense structural arrangement at room temperature, due to the strong bond that bind atom together in a metallic substance. This architecture determines how fast or how slow a healing mechanism would autonomously respond when a metallic component fails. Therefore, it is rational to deduce that triggering autonomous self-healing should be easier if the rate of diffusion of the part is high enough to be transported to the point where it fails. The molecular diffusion or atomic transport of matter by diffusion is represented ideally by the net flux, J, of atoms per second per unit area of reference plane in opposite directions (±x) in the presence of a concentration gradient, dc/dx, as given by Fick's first law:

$$J = -D(dc/dx). \tag{1}$$

where D is the diffusion coefficient, given by:

$$D = Do\ \exp.(-Q/RT). \tag{2}$$

Do (the frequency factor in cm^2/s), Q (the activation energy for diffusion in kJ/mol) and R (the gas constant, 8.314510 J/Kmol) are all constants, so the only variable is the temperature T, in Kelvin. In other words, In Dvs. 1/T forms a single straight line. In the case of Ti, higher temperature induces thermal diffusion needed to increase the kinetic energy needed to overcome the binding energy of the metallic substance. In addition to Vacant lattice sites or other in homogeneities within a metal, molecular transport can also be influenced by pressure and electrical and magnetic activations, chemical process and mechanical agitation of atoms.

There have been several improvement in the study of diffusivity in the case of in titanium alloys, beginning with the work documented in German by Zwicker in 1974 [50]. On transformation from the α (hcp) to the β(bcc) phase, the diffusivity shows some changes. With respect to the widely used Ti-6Al-4V alloy. Liu and Welsch in 1987 studied the diffusivities of oxygen, aluminum, and vanadium in α and β titanium [33]. Zwicker observed in the plot that the self-diffusion of titanium in the β phase is about three orders of magnitude faster than the self-diffusion in the α phase [50] (see **Figure 6**). The diffusion rates of substitutional elements in the β phase can be either slower or faster than the self-diffusion of titanium [33].

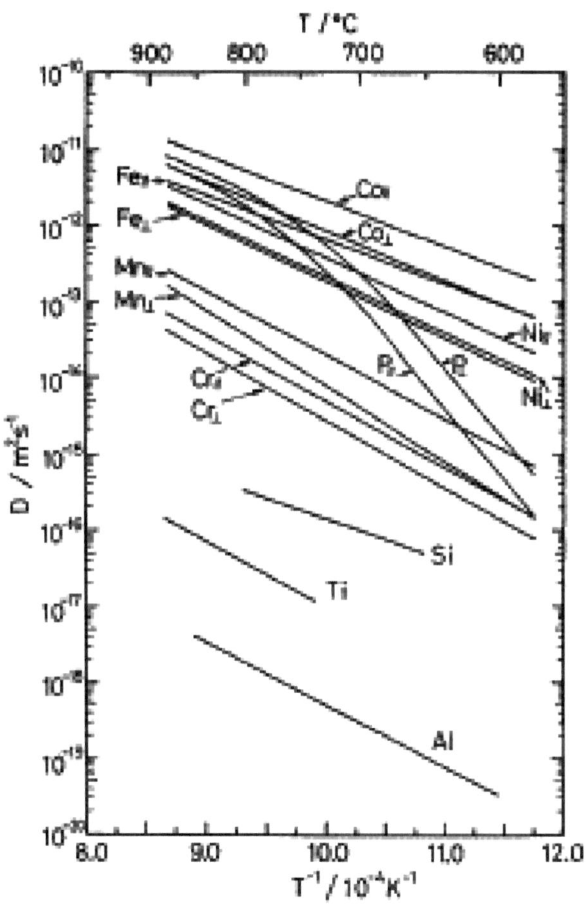

Figure 6.
Temperature dependence of self- and temperature diffusivity in β type titanium alloy (as called from [33]).

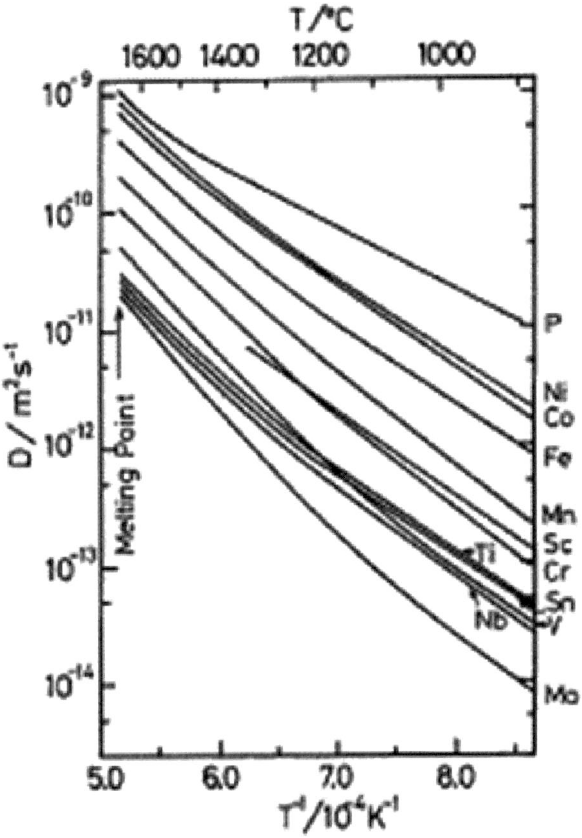

Figure 7.
Temperature dependence of impurity diffusion coefficient in α titanium: Co, Fe, Ni, Mn, Cr and P in single crystal and Si, Al and Ti in polycrystal, culled from [33].

Al and Mo are shown as examples of slow diffusing elements from the group of slow diffusing elements, Others includes the other alloying elements, such as, V and Sn, which are close to Al, and Nb lies in between Al and Mo. Element Fe is shown as an example in the group of fast diffusing elements in the figure. However, Ni is slightly faster, whereas Cr and Mn fall in between Fe and the βTi-self-diffusion line.

Subsequently systematic measurements were hitherto made for the diffusion of Fe [51] Ni, [52] Mn, [53]Cr [54] and P [55]. On the other hand, Raiszinen and Keinonen measured diffusivities of Al [56] and Si [56] in polycrystalline Ti by a nuclear-reaction method. A detailed analysis of the data is compiled in the form of Arrhenius plots in the review by [57] and presented in **Figure 7**. The findings has shown that transition metal elements and phosphorus exhibit fast diffusion, which are three to five orders of magnitude faster than the self-diffusion. While, measurements done on ultrahigh purity α titanium with respect to Fe, Ni, and Co impurities resulted in very low diffusivity rates for self-diffusion in titanium and about two orders of magnitude slower than Fe, Co, and Ni [58].

5. Self-healing concepts in titanium based materials

It is well known that research in the field metallic self-healing is still in its infancy stage. Self-healing metallic materials has received attention only in the past decade [13, 18]. While previous reviews on self-healing materials [59] have focused on describing the various routes to obtain self-healing mostly in polymeric materials, the present chapter is directed toward physical or chemical mechanism can be used to engineer self-healing in Ti-metals and its alloys.

5.1 Electrochemically induced self-healing

Self-healing coatings inspired by biological systems possess the ability to repair physical damage or recover functional performance with minimal or no intervention. When the kinetics are extremely fast, the phenomenon is controlled by the diffusion (mass transport) of the species that enters or leaves the surface of the material under consideration. Consequently, the composition of the system will also be changing. Analogous effects have been found by other workers in systems of biological interest, e.g., with processes involving membranes and enzymes. It is well known that the basic diffusion controlled modes, such as surface diffusion, Ds; grain boundary diffusion, Dgb; vacancy diffusion, Dv and pipe diffusion, Dp, are fundamental to determine the rate of atomic diffusion in polycrystalline metals. In general, surface diffusion occurs much faster than grain boundary diffusion, and grain boundary diffusion occurs much faster than lattice diffusion. Atomic diffusion and indeed electrochemical-induced self-healing in polycrystalline materials is therefore often modelled using a combination of diffusion kinetics (see previous section). More details of the transformation modes in titanium have been discussed elsewhere [31–33]. For this, electrochemically induced self-healing are said to be a good strategy to be exploited in metals. For instance, a damage to oxide films, which normally protect the surfaces of Ti materials from corrosion, can be repaired by reoxidation in air. Recently, Gang Lu et al. [60] studied the oxidation of a polycrystalline titanium surface by oxygen and water and found that at 150 K O_2 can oxidize Ti to Ti_5, Ti_3 and Ti_2, while exposure of Ti to H_2O at this temperatures only produces Ti_2 species. At temperatures above 300 K, H_2O can by both O_2 and H_2O slightly increases a further oxidize Ti_2 to higher oxidation states. They observed rising temperature promotes the diffusion of oxygen into the bulk of the sample, which increasing overall oxidation. This is because the thickness of the oxide coating on Ti surface depends on both the duration of O_2 exposure and on the sample temperature. At a given temperature, Ti oxidation by both O_2 and H_2O slightly increases as exposure increases.

Additionally, a crack on the surface of a titanium component can also be healed, when the oxidation reaction products fill up the crack cap. Therefore, cracks developed due to operational related stress can be autonomously self-healed or repaired by re-oxidative reaction that occur in Ti-based materials. Although self-healing coatings are considered as an alternative route for efficient anti-corrosion protection, intense research and development effort are been done in the area of corrosion protection coatings of metals and alloys. However, in order to improve the equipment service prediction capabilities of infrastructure, the use of Ti-based materials in infrastructures are beneficial as it can act as a second line of safety assurance even after the coating has failed. In this context, autonomic healing materials respond without external intervention to environmental stimuli, and have great potential for advanced engineering systems [61]. However, the limitation of this self-healing approach is that the extent of oxidation depends on sample temperature. A recent study identified 550–600 K as maximum oxidation in Ti based alloy. Upon heating the oxidized Ti above 850 K the titanium oxide layer is completely reduced to Ti^0, which is effective.

5.2 Thermally activated solid phase healing in titanium

This mechanism is based on a thermoelastic displacive phase transformation design methodology. Certain strongly ordered intermetallic systems exhibit shear-dominated thermoelastic displacive transformations that involve minimal volume dilatation, a high degree of crystallographic reversibility, and a low-temperature allotrope that readily twins during plastic deformation. This combination gives rise to the well-known shape-memory effect in which plastic deformation imparted to the low-temperature martensitic phase can be reversed almost completely during transformation to the high-temperature austenitic phase.

In a recently study, by Elena et al. [62] thermally engineered self-healing was proven by observation of crack healing after annealing. Microscopic deformation and recovery of a shape-memory nickel-titanium alloy were studied. The deformation was induced by microindentation using spherical and Vickers diamond indenters. The recovery of the indents, caused by thermal annealing, was measured quantitatively using an optical surface profiler. Microindents formed by a spherical diamond in an equiatomic nickel-titanium martensite almost completely recover at low indentation load when moderately heated [63]. A smaller recovery ratio was observed for microindents formed by Vickers indenters. These observations suggest that the shape-memory effect exists at the microscopic level and under complex loading conditions. The observations were rationalized using the concept of representative strain and maximum stress under the spherical and pyramidal indenters. A representative surface profiles measured before and after heating for both spherical and Vickers indents are shown in **Figure 8(a)** and **(b)**, respectively. The degree of indent recovery was determined quantitatively from the surface profiles by defining a recovery ratio, d, as.

5.3 Precipitation in under-aged alloys

This is an approach, of self "healing" is likened to an established metallurgical process of aging. In this mechanism, a defect sites (primarily microscopic voids) serve as nucleation centers for diffusion driven precipitation of oversaturated

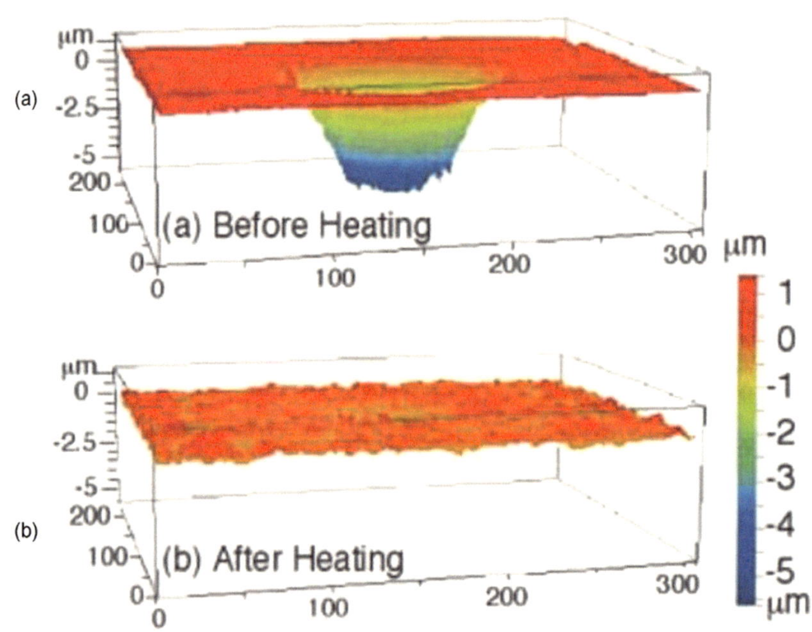

Figure 8.
A representative three-dimensional profile of a spherical indent at load of 15 N: (a) fresh indent and (b) after heating above the austenite finish temperature.

solute in the alloy and are thereby immobilized from further growth until failure. Consequently, the newly formed voids are sealed before they grow and this results in improvement of creep and fatigue properties of the alloy. This form of 'preventive' healing has been used Al-Cu alloys, known for decreasing solute solubility with decreasing temperatures. The process involves a high-temperature solution treatment, accompanied by quenching and annealing for relatively short periods of time, results in an underage microstructure that maintains substantial amounts of solute and serves as the healing agent. The processes of "secondary precipitation" in Al-Cu alloys that results into much finer precipitates from low-temperature aging [64] and investigations into dynamic precipitation in Al-Cu-Mg-Ag alloys occurring in response to moving dislocation generation under load [65], have been identified as potential healing mechanisms during fatigue and creep. This can be extended to other metallic materials. The limitation to this approach is that not all Ti alloy is heat treatable like Al-Cu alloys.

5.4 Micro-encapsulated low-melting healing agent reinforced metal matrix composites (MMCs)

This technique derives its inspiration from polymer healing and was recently conceptualized by Rohatgi et al. [63] includes the embedding of a hollow reinforcement (micro-sphere, micro-tube) containing a low-melting alloy in a higher melting metal matrix. The encapsulation of a metallic healing agent, however, allows the microcapsule to serve as a diffusion barrier and the interface should be sufficiently fragile to break upon an advancing crack and not deflect.

In the line of work done on hollow fibers reinforced polymers [66], attempts were also made to integrate hollow microfibers containing low-melting healing agents into metallic systems [67]. This attempted healing was implement by incorporating indium as a healing agent in carbon tubes embedded in a higher melting solder matrix. Upon heating beyond the Indium melting point, a macroscopic crack that was directed downward to gravity was repaired. Computational fluid dynamics studies were conducted by Lucci et al. [67] on this healing method and interface wettability and gravity-related crack orientation were highlighted as major factors affecting the flow of healing liquid, with more wetting systems and gravity-oriented cracks being more desirable to fill.

5.5 Shape memory alloy (SMA) reinforced metal matrix composites (MMCs)

In one of the earliest trial, Manuel and co-workers [68] used this approach to heal Sn and Mg based metallic materials. The method involves reinforcing an alloy matrix with wires made of a shape-memory alloy (SMA), such as nitinol (NiTi). SMA wires have the ability to recover their original shape when heated above a critical temperature [69]. Thus, when the metal matrix composite cracks, the resulting plastic strain stretches the SMA that bridges the crack. On heating above the shape transformation temperature of the SMA, the wire shrinks back to its original shape applying compressive force to the matrix and clamping the crack. This is accompanied by welding of the crack in the matrix alloy which is so-designed as to partially liquefy at the healing temperature [70].

6. Summary

Although the production of autonomous self-healing in metallic materials has been the subject of numerous studies and will continue to be in the near future.

It is still at the infancy stage. Thus far, the field of autonomous self-healing engineering materials has been dominated by polymeric based substances for several years with only one route and mechanism, that of the encapsulated liquid adhesive route. The current chapter discusses the framework for engineered self-healing Ti-alloys has been presented with some practical examples revealing how self-healing functionality can be incorporated in Ti alloys. Nonetheless, the functionality of the mechanism comes down to the turning these experimental approaches under laboratory conditions into commercially available materials and products. Albeit, the studies conducted thus far in Ti and indeed metals have been relatively limited, the information contained within this chapter should not be considered exhaustive, and should be open to further improvement, modification and discussion. There are obvious indications that, the once impossible task of achieving autonomously engineered self-healing metallic materials in service will be feasible in no distant time.

Author details

Paul Sunday Nnamchi[1,2,3*] and Camillus Sunday Obayi[1]

1 Department of Metallurgical and Materials Engineering, University of Nigeria, Nsukka, Enugu State, Nigeria

2 Department of Mechanical and Construction Engineering, Wynne Jones Building, Northumbria University, Newcastle upon Tyne, United Kingdom

3 Department of Metallurgical and Materials Engineering, Enugu State University of Science and Technology, Agbani, Enugu State, Nigeria

*Address all correspondence to: paul.nnamchi@esut.edu.ng; paul.nnamchi@northumberia.ac.uk

References

[1] Alaneme KK, Bodunrin MO. Self-healing using metallic material systems—A review. Applied Materials Today. 2017;**6**:9-15

[2] Martin P. Wound healing—Aiming for perfect skin regeneration. Science. 4 Apr 1997;**276**(5309):75-81

[3] Trask RS, Williams HR, Bond IP. Self-healing polymer composites: Mimicking nature to enhance performance. Bioinspiration & Biomimetics. 2007;**2**(1):1

[4] Van Tittelboom K, De Belie N. Self-healing in cementitious materials—A review. Materials. 2013;**6**:2182-2217

[5] Shi S-C, Huang T-F. Self-healing materials for ecotribology. Materials. 2017;**10**:91

[6] Gupta RK, Mirza F, Khan MUF, Esquivel J. Aluminum containing Na_2CrO_4: Inhibitor release on demand. Materials Letters. 2017;**205**:194-197

[7] Fan Z, Wang K, Dong X, Wang R, Duan W, Mei X, et al. Enhanced cyclic oxidation resistance through the self-healing of segmented cracks using nano-Al_2O_3/Ni-20 wt% Al particles in laser re-melted thermal barrier coatings. Materials Letters. 2017;**201**:156-160

[8] Du K, Guo X, Guo Q, Wang F, Tian Y. A monolayer PEO coating on 2024 Al alloy by transient self-feedback control mode. Materials Letters. 2013;(91):45-49

[9] Zhang Y, Li Y, Ren Y, Wang H, Chen F. Double-doped LDH films on aluminum alloys for active protection. Materials Letters. 2017;**192**:33-35

[10] Rohatgi PK. Al-shape memory alloy self-healing metal matrix composite. Materials Science and Engineering A. 2014;**619**:73-76

[11] Grabowski B, Cem T. Self-healing metals. In: Advances in Polymer Science. New York, NY, USA: Springer; 2015

[12] Li W, Jiang Z, Yang Z. Extension and possibility of debonding in encapsulation-based self-healing materials. Materials. 2017;**10**:589

[13] Kim D-M, Cho Y-J, Choi J-Y, Kim B-J, Jin S-W, Chung C-M. Low-temperature self-healing of a microcapsule-type protective coating. Materials. 2017;**10**:1079-1093

[14] Hautakangas S, Schut H, van Dijk NH, Rivera Diaz del Castillo PEJ, van der Zwaak S. Self-healing of deformation damage in underaged Al–Cu–Mg alloys. Scripta Materialia. 2008;**58**:719-722

[15] Sheng X, Lei Q, Xiao Z, Wang M. Precipitation behavior and quenching sensitivity of a spray deposited Al-Zn-Mg-Cu-Zr alloy. Materials. 2017;**10**:1100

[16] Hastings GW, Mahmud FA. Intelligent orthopaedic materials. Journal of Intelligent Material Systems and Structures. 1 Oct 1993. pp. 452-456

[17] Varley RJ, van der Zwaag S. Towards an understanding of thermally activated self-healing of an ionomer system during ballistic penetration. Acta Materialia. 2008;**56**(19):5737

[18] Hanson BH. Present and future uses of titanium uses in engineering. Materials and Design. 1986;**7**:301-307

[19] Klinger MM, Rahemtulla F, Prince CW, Lucas LC, Lemons JE. Proteoglycans at the bone-implant interface. Critical Reviews in Oral Biology and Medicine. 1998;**9**:449

[20] Collings EW, Gegel HL. A physical basis for solid solution strengthening and phase stability in alloys of titanium. Scripta Metallurgica. 1973;**7**:437-443

[21] Welsch G, Boyer RR, Collings EW. Materials Properties Handbook: Titanium Alloys, ASM Handbook. USA: Institute of Metals; 1994

[22] Newkirk JB, Geyser AH. Crystallographic aspects of beta to alpha transformation in titanium. Acta Metallurgica. 1953;**1**:370

[23] Oka M, Taniguchi Y. {332} diffusion twins in a Ti-15.5 pct V alloy. Metallurgical and Materials Transactions. 1979;**10A**:651

[24] Collings EW, Ho JC, Jaffee RI. Superconducting transition temperature, lattice instability and electron-to-atom ratio of transition-metal binary solid solutions. Physics Review. 1972;**5**:4435-4449

[25] Burgers WG. On the process of transition of the cubic-body-centered modification into the hexagonal-close-packed modification of zirconium. Physica. 1934;**1**:561-586

[26] Blanco MA, Francisco E, Luaña V. GIBBS: Isothermal-isobaric thermodynamics of solids from energy curves using a quasi-harmonic Debye model. Computer Physics Communications. 2004;**158**:57-72

[27] Reedhill RE, Abbaschian R. Physical metallurgy principles. In: The Pws-Kent Series in Engineering. United Kingdom: Cengage Learning; 1991

[28] Hirao M, Ogi H, Kai S, Ledbetter H, Tarumi R, Takashima K. Titanium's high temperature elastic constants through the hcp-bcc phase transformation. Acta Metallurgica. 2004;**52**:2075-2080

[29] Deurig TW, Williams JC. Beta Titanium Alloys in the 1980s. In: Boyer RR, Rosenberg HW, editors. New York, USA: AIME; 1984. pp. 19-67

[30] Burgers WG. On the process of transformation of the cubic body centered modification into hexagonal close packed modification in zirconium. Physica. 1934:561-586

[31] Banerjee D, Muraleedharan K. Substructure in titanium alloy martensite. Philosophical Magazine A. 1998;**77**(2):299-323

[32] Paul S. Nnamchi. Published thesis presented for the award of PhD. United Kingdom: University of Sheffield; 2014; p. 127

[33] Liu Z, Welsch G. Materials Transactions. 1988;**19A**:121

[34] Jepson KS, Brown ARG, Gray JA. In: Jaffee RI, Promisel WE, editors. Proceedings of the First International Conference on Titanium. London: Pergamon; 1970. p. 677

[35] Furuya Y, Shimada H. Shape memory actuators for robotic applications. Materials and Design. 1991;**1**(12):21-28

[36] Petrini L, Migliavacca F. Biomedical applications of shape memory alloys. Journal of Metallurgy. 2011;**2011**:1-15. DOI: 10.1155/2011/501483

[37] Lieva VL, Carla H. Smart clothing: A new life. International Journal of Clothing Science and Technology. 2004;**16**:63-72. DOI: 10.1108/09556220410520360

[38] Cayton C. Shifting the shear paradigm in the crystallographic models of displacive transformations in metals and alloys. Crystals. 2018;**8**(4):181. DOI: 10.3390/cryst8040181

[39] Nnamchi P, Younes A, González S. A review on shape memory metallic alloys and their critical stress for twinning. Intermetallics. 2019;**105**:61-78

[40] O'Brien B, Weafer FM, Bruzzi MS. Shape memory alloys for use in medicine. In: Reference Module

in Materials Science and Materials Engineering. Elsevier; 2017

[41] Buehler WJ, Gilfrich J, Wiley R. Effect of low-temperature phase changes on the mechanical properties of alloys near composition TiNi. Journal of Applied Physics. 1963;**34**:1475-1477

[42] Stoeckel D. Shape memory actuators for automotive applications. Materials and Design. 1990;**11**:302-307. DOI: 10.1016/0261-3069(90)90013-A

[43] Bellini A, Colli M, Dragoni E. Mechatronic design of a shape memory alloy actuator for automotive tumble flaps: A case study. IEEE Transactions on Industrial Electronics. 2009;**56**:2644-2656

[44] Hartl DJ, Lagoudas DC. Aerospace applications of shape memory alloys. Proceedings of the Institution of Mechanical Engineers, Part G: Journal of Aerospace Engineering. 2007;**221**:535-552. DOI: 10.1243/09544100JAERO211

[45] Fujita H, Toshiyoshi H. Micro actuators and their applications. Microelectronics Journal. 1998;**29**:637-640. DOI: 10.1016/S0026-2692(98)00027-5

[46] Valiev RZ, Islamgaliev RK, Alexandrov IV. Bulk nanostructured materials from severe plastic deformation. Progress in Materials Science. 2000;**45**:103-108

[47] Rogers HC, Hirth JP, Reed-Hill RE. Deformation Twinning; Proceedings. New York: Gordon and Breach Science Publishers; 1964

[48] Zhu Y, Liao X, Wu X. Deformation twinning in nanocrystalline materials. Progress in Materials Science. 2012;**57**:1-62

[49] Wang J, Zeng Z, Weinberger CR, Zhang Z, Zhu T, Mao SX. In situ atomic-scale observation of twinning-dominated deformation in nanoscale body-centred cubic tungsten. Nature Materials. 2015;**14**:594-600

[50] Zwicker U. Titan and Titanlegierungen. Berlin, Germany: Springer-Verlag; 1974. p. 102

[51] Perez RA, Nakajima H, Dyment F. Diffusion in α-Ti and Zr. Materials Transactions. 2003;**44**(1):2-13

[52] Nakamura Y, Nakajima H, Ishioka S, Koiwa M. Effect of oxygen on diffusion of manganese in α-titanium. Acta Meterialia. 1988;**36**:2787

[53] Nakamura Y, Nakajima H, Ishioka S, Koiwa M. Effect of oxygen on diffusion of manganese in α-titanium. Acta Metallurgica. 1988;**36**:2787

[54] Nakajima H, Ogasawara K, Yamaguchi S, Koiwa M. Diffusion of chromium in α-titanium and its alloys. Materials Transactions, JIM. 1990;**31**:249

[55] Nakajima H, Nakazawa J, Minonishi Y, Koiwa M. Diffusion in titanium. Philosophical Magazine. 1986;**A53**:427

[56] Raisnen J, Anttila A, Keinonen J. Diffusion of aluminum in ion-implanted α-Ti. Journal of Applied Physics. 1985;**57**:613

[57] Nakajima H, Koiwa M. Diffusion in titanium. ISIJ International. 1991;**31**:757-766

[58] Mishin Y, Hertig C. Diffusion in the Ti–Al system. Acta Materialia. 2000;**48**:589

[59] Sharma S et al. Recent advances in self-healing. Materials Today: Proceedings. 2019;**18**:4729-4737

[60] Lu G, Bernasek SL, Schwartz J. Oxidation of a polycrystalline titanium

surface by oxygen and water. Surface Science. 2000;**458**:80-90

[61] Kamat S, Su X, Ballarini R, Heuer AH. Structural basis for the fracture toughness of the shell of the conch *Strombus gigas*. Nature. 2000;**405**(6790):1036

[62] Lumley RN, Polmear LJ, Morton AJ. Interrupted aging and secondary precipitation in aluminum alloys. Materials Science and Technology. 2003;**19**:1483-1490

[63] Rohatgi PK. Self-Healing Metals and Alloys-Including Structural Alloys and Self-healing Solders. U.S. Patent Application 12/537, 675; 2009

[64] Lumley RN, Polmear LJ, Morton AJ. Interrupted aging and secondary precipitation in aluminum alloys. Materials Science and Technology. 2003;**19**:1483-1490

[65] Hautakangas S, Schut H, van der Zwaag S, del Castillo PEJRD, van Dijk NH. Positron annihilation spectroscopy as a tool to develop self-healing in aluminum alloys. Physica Status Solidi. 2007;**4**:3469-3472

[66] Trask RS, Bond IP. Biomimetic self-healing of advanced composite structures using hollow glass fibers. Smart Materials and Structures. 2006;**15**(3):704-710

[67] Lucci JM, Ruzek A, Misra SK, Rohatgi PK, Amano R. Self-healing in metal castings. In: Proceedings of American Foundry Society. Illinois, USA: Schaumburg; 2011

[68] Manuel M, Olson GB. Biologically inspired self-healing metals. In: Proceedings of the 2nd International Conference on Self-Healing Materials. Chicago; 2009

[69] Feng QL, Cui FZ, Pu G, Wang RZ, Li HD. Crystal orientation, toughening mechanisms and a mimic of nacre. Materials Science and Engineering: C. 2000;**11**:19-25

[70] Lumley RN, Morton AJ, Polmear LJ. Enhanced creep performance in an Al-Cu-Mg-Ag alloy through underageing. Acta Materialia. 2002;**50**:3597-3608

Chapter 10

Biomaterials for Drug Delivery: Sources, Classification, Synthesis, Processing, and Applications

Samson O. Adeosun, Margaret O. Ilomuanya, Oluwashina P. Gbenebor, Modupeola O. Dada and Cletus C. Odili

Abstract

A way to avoid or minimize the side effect that could result in drug delivery to cells with increased efficiency and performance in the health rehabilitation process is to use biocompatible and biodegradable drug carriers. These are essentially biomaterials that are metallic, ceramic, or polymeric in nature. The sources of these materials must be biological in its entire ramification. The classification, synthesis, processing, and the applications to which these materials are put are the essential components of having suitable target cell drug carriers. This chapter will be devoted to discussing biomaterials suitable as drug carrier for use in the health-related matters of rehabilitation.

Keywords: biomaterials, drug carriers, cell, drug delivery, health, synthesis, target

1. Introduction

The quest for controlled drug release emanating from side effects associated with the application and delivery of conventional drugs has necessitated the need for materials that can transport drugs to target site without difficulty or problem during and after delivery. Normally, drugs are delivered repeatedly on prescription to the body in measures that will bring about remediation and quick recovery to the patient during the treatment period. In this wise, drug concentration levels will increase and when above the body's tolerance level, the problems associated with over therapeutic concentrations could occur that could result into toxic side [1]. It is also possible that the drug release rate is so fast that therapeutic actions are no longer effective owing to low drug concentrations at the delivery site, which may occur through drug metabolism, degradation, and transport out of the target [1]. Consequently, this phenomenon would result in drug wastage and transport medium loss with high risk offside effects on surrounding body cells, tissues, and organs. The solution to these problems is to have drug carriers that can provide controlled release rate to the target and would allow for complete therapeutic rehabilitation before degradation and transport of excess concentration of drug and carrier medium [2]. The drug and its carrier in form of capsules are orally administered and may be formulated for parenteral administration [3]. The drug release rate of the capsule can be controlled via the use of cellulose coatings exhibiting slow dissolution, incorporation of drug-complexing elements or compounds which hinder fast dissolution of drug, use of compressed tablets, and the inclusion of emulsion and suspensions. Materials that can permit drug release without changing or decaying over time with longer therapeutic windows (days to years) are required. These carries are such that they can be injected and/or implanted directly to target diseased tissues/cells for enhancing delivery efficiency [4]. To achieve target drug delivery, the use of affinity ligands deposited on biomaterial surfaces to allow for a set retention and usage by infirm tissues and cells have been employed [5]. The design of biomaterials for drug carriers aside permitting surface modification using ligands should also shield drugs from speedy break down and/or degeneracy within the target site.

Thus, the design parameters include: (i) the encapsulation of the sufficient drug of the biomaterial for lengthened release pattern to achieve efficient healing, (ii) sustaining drug stability for effective therapeutics through body transport and at the target site while preserving biological activity, (iii) predictable release rate in the therapeutic period from days to years, (iv) biomaterials and its degradation products must be biocompatible and nontoxic within the body, and (v) the cost of biomaterial synthesis and/or fabrication.

2. Health implications of materials used for drug delivery

Lupron Depot, a poly (lactic-co-glycolic) acid (PLGA) microsphere encapsulating the hormone leuprolide, for the treatment of advanced prostate cancer, and endometriosis [6], PLGA, poly (lactic acid) (PLA), and polyglycolic acid (PGA) materials have FDA approval as micro-particle depot systems as they versatile in controlling material biodegradation time, are biocompatible with nontoxic natural degradation products (lactic acid and glycolic acid). Clinical nanoparticles with FDA approval for cancer nanomedicine treatment of Kaposi's sarcoma (approved 1995) and for recurrent ovarian cancer (approved 1998) is Doxil [7], a poly (ethylene glycol) (PEG) coated (i.e., PEGylated) liposomal encapsulating the chemotherapeutic doxorubicin [8]. This enhances circulation half-life and tumor uptake of the drug, and also reduces its toxicological activity in patients in comparison to the use of free drug [9]. Other approved nanoparticle drug carriers include Marqibo, a liposomal encapsulating vincristine for rare leukemia treatment [10] and Abraxane an albumin-bound paclitaxel nanoparticle for the treatment of breast cancer [11]; Duragesic-transdermal drug delivery system patch containing the opioid fentanyl embedded within an acrylate polymer matrix, in the treatment of chronic pain [12]; and OROS, an osmotically controlled oral drug delivery technology, incorporated into several oral delivery products including Concerta [13]. Implantable biomaterials used include the Gliadel wafer, which consists of dime-sized wafers comprised of the chemotherapeutic agent carmustine and a polymer matrix made of poly (carboxyphenoxy-propane/sebacic acid), which are surgically inserted into the brain post-tumor resection [14–16] use as an adjunct to surgery in patients with recurrent glioblastoma multiforme.

3. Bioresponsive polymers: from design to implementation

An ideal therapeutic drug is expected to treat or cure a disease without resulting to any side effects [17–19]. However, this goal has not been achieved. Many chemotherapeutics are found to destroy both cancerous and healthy cells within the vicinity of the target site [20]. An efficient chemotherapeutics would administer drug, directly to diseased cell populations. Polymers have been found to permit the creation of "responsive" materials within the host environment and can be formulated with drugs to control release [21]. This polymer attribute is due to tuning propensity of the molecular weight of polymers that can be controlled via monomer stoichiometry using controlled polymerization strategies like ATRP [22], RAFT [23], NMO [24], and ROMP [25]. A bioresponsive material is one that can respond to a specific "trigger" inside or outside of the human body. Because the body have unique pathological parameters as pH gradients, temperatures, enzymes, small molecules, etc., the creation of materials that will respond to physiological alterations in both space and time are required.

Triggers include chemical, biological, and physical stimuli [26, 27], the chemical and biological ones are intrinsic to the body, while the physical stimuli are extrinsic to the body can thus be used to quicken sole drug delivery.

4. Redox-sensitive polymers

Bioresponsive materials are initiated by redox potential difference tissue environment and its surrounding [28]. There are materials that can respond to both oxidation and reduction triggers, which are incorporated into responsive polymers, e.g., diselenides with chemical structure like those of disulfides [29]. Diselenides allows for alternative triggers within nano-biotechnology applications [30].

5. pH-responsive polymers

The constituents of the human body such as tissues, fluids, and organelles have varied pH values. Areas like stomach, vagina, and lysosomes display acidic pHs (<7); ocular surface (7.1), the blood (≈7.4), and bile (7.8) [21]. Owing to these varied

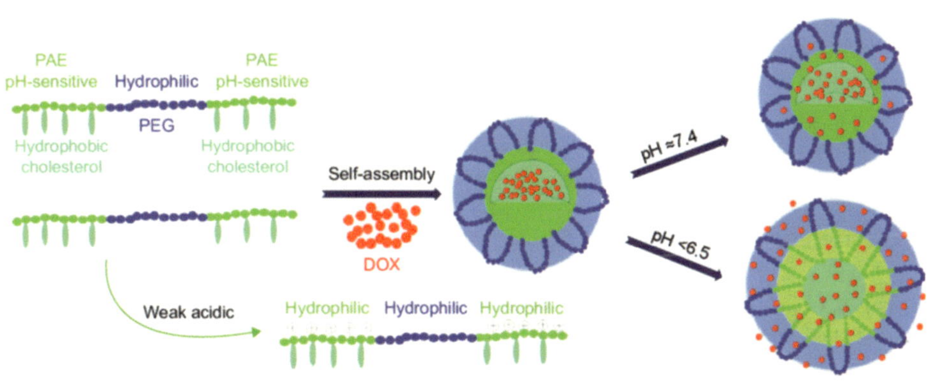

Figure 1.
Schematic illustration of drug loading and controlled release of poly (ethylene glycol) [34]. DOX, doxorubicin; PAE, poly (β-amino esters); PEG, poly (ethylene glycol).

pHs of systems and organs in the body improvement in the efficacy and precision of therapeutic molecules will necessitate the design of polymeric drug delivery systems that are pH specific. pH-responsive materials have been useful in nucleic acid delivery, doxorubicin delivery, and taste masking [31, 32]. The target treatment of tumors has been enhanced using the pH-responsive materials. Such known target delivery includes multifunctional acid sensitive nanocomposites for anticancer drugs and acid-responsive poly(ethylene glycol) derivatives [33] for the controlled release of therapeutics in tumor target treatment (**Figure 1**).

6. Hydrolysis and enzymatically responsive polymers

Hydrolysis prone materials can be degraded by body fluid via nucleophilic addition of water into an electrophilic functional group on a polymer. The electrophilic functional groups often used on polymers include esters and anhydrides [35]. The Gliadel wafer consisting of chemotherapeutic Carmustine impregnated within a polyanhydride material has been demonstrated as hydrolysis-sensitive materials for drug delivery [36] in the treatment of brain tumors. Enzyme-responsive polymers such as matrix metallo-proteins, hyaluronidases, phospholipases, and prostate-specific antigen [21] have been incorporated into polymers for target drug delivery in areas like tumor imaging, doxorubicin delivery, and minimization of inflammation in the colon [37].

7. Temperature-responsive polymers

Another drug delivery vehicle is the temperature-sensitive polymers that can operate at both human body temperature of 37°C and at ambient temperature such as 25°C [38]. These polymers include poloxamers, poly(N-alkyl acryl amides), poly(N vinyl caprolactams), cellulose, xyloglucan, and chitosan. These thermo-responsive polymers can be modified via [39] varying the ratio of monomers, end-group modifications, and post-polymerization modifications to make them suitable for varying applications [40].

8. Magnetic-responsive polymers

Magnetic-responsive polymers are therapeutic drug-loaded polymers that work under the influence of magnetic resonance imaging (MRI) to delivery its drug to the target [41]. These include the following: systematic release of dopamine from alginates impregnated with magnetic beads; targeted plasmid delivery to the lung via chitosan nanoparticles; and insulin delivery [42].

9. Light-responsive polymers

Light-responsive polymers are used as external drug delivery systems that use noninvasive and painless techniques [26, 43–48] as drugs are delivered by light UV- and visible-wavelength irradiation stimulation. In this technique, a remote-activated approach without direct patient contact is used [49]; this includes the release of drugs from a light-responsive azobenzene modified amphiphilic block copolymer to target melanoma cells [50].

10. Swelling and contracting polymers

There are polymers that can swell or shrink in response to external stimuli [51]. This phenomenon can have stemmed from changes in porosity occasioned as ionic cross-linking molecules are leached, resulting in alteration of the diffusion pathways for sensing molecules. Alginate is a commonly employed polymer that is isolated from seaweed, is relatively biocompatible, and has been used for sustained delivery of vascular endothelial growth factor (VEGF) to a target within the body.

11. Biomaterial-based drug delivery systems

While a limited number of affinity-based delivery systems have been developed for the delivery of neurotrophic factors, we also examine the broad spectrum of reservoir-based delivery systems, including microspheres, electrospun nanofibers, hydrogels, and combinations of these systems.

Drug delivery systems transport biological active agents, such as growth factors and genetic material, into the desired location to promote beneficial effects for the treatment of diseases and disorders [52], osmotic pumps for the delivery of neurotrophic factors [53] to target site, affinity-based delivery systems (ABDS) in which drug loading and controlled release are achieved through the interactions of therapeutic drug and the delivery system, and reservoir-based delivery systems, where a polymer structure encapsulates the drug while its release is controlled via the material properties.

12. Affinity-based delivery systems (ABDS)

ABDS operate through the noncovalent interactions between device material and target drug [54] in a similar pattern to the interactions that occur in the extracellular matrix where the delivery of proteins and other biomolecules are controlled [55]. ABDS include molecular imprinting, cyclodextrin-based delivery, and heparin-based delivery [56]. Molecular imprinting uses polymer networks synthesized via a precursor molecule that is removed to reveal an imprint that acts as an affinity binding zone. In cyclodextrin-based delivery systems, small hydrophobic drugs are attracted to the hydrophobic center of an oligosaccharide cyclodextrin torus, which permits the complexes formation with enhanced solubility when compared to the drug itself. ABDS is observed to be superior to traditional reservoir-based systems as the release characteristics are dependent on the activities occurring between the drug and the matrix in a way not affected by the matrix properties [57].

13. Reservoir-based delivery systems (RBDS)

Reservoir-based delivery systems (RBDS) are porous with drug release rate controlled by diffusion [58]. In RBDS, the drug is immersed or dissolved in a polymer solvent/reservoir. The drug penetrates via the biodegradable polymer structure to control the initial release followed by another release as surface and bulk erosion occurs in polymer reservoir. RBDS include nanogels, nanoparticles, micelles, hydrogels, microspheres, and electrospun nanofibers.

14. Microspheres

Microspheres are usually used as controlled drug release systems for stereotactic injections to isolated disease or injury sites in medicine and pharmacology [59]. Drugs like neurotransmitters, hormones, and neurotrophic factors have been encapsulated using microspheres obtained from biodegradable polymers [60]. These polymers include poly(lactic acid) (PLA), poly(glycolic acid) (PGA), and poly(ε-caprolactone) (PCL). Microsphere-based drug delivery uses localized surgical injection to circumvent the blood-brain barrier; this is better in performance to orthodox methods like intravenous injection and oral drug delivery. The parameters of the microsphere such as the particle size, polymer degradation rate, and method of erosion (bulk versus surface degradation) can be utilized to control the rate of drug delivery rates [61]. PCL has been found useful as a microsphere for the carrier of sustained long period drug delivery as it demonstrates the slowest degradation rate [62]. The double emulsion method is often used in synthesizing of microspheres. The method involves dissolving the desired polymer in a nonpolar solvent to form an oil emulsion. The hydrophilic compound that is to be encapsulated is dissolved in an aqueous solution and then emulsified with the dissolved polymer-solvent solution to give a water-in-oil emulsion. After this the solvent evaporates, the polymer solidifies as it forms microspheres that encapsulated the inner aqueous solution [63].

15. Electrospun nanofibers

Electrospinning process involves the application of an electric potential to draw out thin nanometer to micrometer diameter polymer fibers (natural or synthetic). A viscous solution of the polymer is prepared (at room or elevated temperature), then pumped via a spinneret nozzle (positive terminal) into an electric field such that the applied force due to the high voltage counters the surface tension leading to the formation of fiber droplets onto a collector plate that serves as negative terminal. The nanofibers produced are often used as drug based-reservoir delivery systems as the pores in the matrix serves as receptive sites for bioactive agents [64]. This fiber production process advantages include surface flexibility with respect to function or application, reduced initial burst release, and the possibility of producing different fiber configuration depending on usage [65]. Drugs are embedded in the pores of electrospun nanofibers by emulsion electrospinning; the target drug is dissolved in a desired polymer solution [64] such as in diclofenac sodium (DS) and human serum albumin (HSA) [66]. Electrospun nanofibers show some draw backs that include formation of drug aggregates during encapsulation along nonsmooth fibers, maintaining uniform fiber size distribution, the use of toxic solvents to form polymer-drug emulsion in drug delivery and its attendant health concerns. Despite these drawbacks, advances in the development of less toxic electrospun fibers, which contain extracellular matrix components such as keratin and collagen, have been developed for wound healing application. The biocompatibility potential of PVA with the bioactive nature of keratin, CoQ10, and antimicrobial mupirocin has been evaluated for wound care due to its ability to support the growth of keratinocytes and hasten skin regeneration [67].

16. Hydrogels

Hydrogel is a hydrophilic network of cross-linked polymer chains with swelling capability but does not dissolve in aqueous solution in the presence of water to create a three-dimensional gel-like structure. The synthesis of hydrogels is through polymerization [68], its properties, and drug release mechanism that depend on the polymer type used. The mechanisms involved in the drug delivery of hydrogel may be diffusion controlled, chemical controlled, swelling controlled, and modulated release systems. The use of acetyl-(Arg-Ala-Asp-Ala)4-$CONH_2$ self-assembling peptide hydrogel to carry model factors such as lysozyme, trypsin inhibitor, BSA, and IgG [69] reveals the potential of these hydrogels carriers of therapeutic agents with the preservation of protein activity. An agarose hydrogel has been found capable of delivering sustained bioactive lysozyme release [70] and was used for the local delivery of BDNF in adult rat models.

17. Surface-modified biomaterials

The biomaterial surface chemistry and topography impact protein adsorption, cell interaction, and host site response. Monocyte adhesion in vitro [71] have been shown to be altered by its surface chemistry, while in vivo surface chemistry does not significantly influence the foreign body reaction. Polymeric, ceramic, or metallic-based biomaterials exhibit variability in surface properties such as hydrophilic to hydrophobic; hard to soft in vivo [72].

18. Surface specificity

Cell adhesion to adsorbed proteins is achieved via integrin and other receptors in the cell membrane and the occurrence of this triggered intracellular signaling events. Thus, the control of protein adsorption on biomaterials surfaces is crucial to controlling and directing cell responses. Oligopeptides with specific binding sites have been incorporated to control cell adsorption to the protein surface and these include short oligopeptide, e.g., adhesive oligopeptide is an arginine-glycine-aspartic acid (or RGD) [73] that is found in a number of different extracellular matrix proteins, such as fibronectin [74], laminin [75], collagen [76], and vitronectin [77]. Short oligopeptides are less expensive, easy to synthesize, and has greater flexibility for surface modification compared to bulky and labile intact proteins. To a surface modified using nonfouling PEG (99%) and RGD (1%), the protein adsorption was minimal (2 ng/cm2) leaving the sufficient RGD sites for fibroblast cell adhesion [78]. Structure and conformation of oligopeptides influence modulating cell adhesion as demonstrated with the use of immobilized cyclic RGD peptide which increased human bone marrow stromal cell adhesion to that of linear RGD peptides [79] (**Table 1**).

S/No.	Drug delivery systems	Biomaterial	API	Significance of the study	Reference
	ORAL DRUG DELIVERY SYSTEMS				
	Silk Nanoparticles	Silk and fibrin	Celecoxib and curcumin	Silk fibroin nanoparticles were seen to promote anti-inflammatory properties of celecoxib or curcumin and could be exploited for oral osteoarthritis management since a controlled drug release was achieved by varying the drug loading	[80]
	Electrospun fibers	Polylactic acid	Metronidazole	PLA nanofibers associated with metronidazole (MNZ) were used to control microbiological proliferation during periodontitis treatment, inhibiting bacteria growth during the treatment	[81]
	OCULAR DRUG DELIVERY SYSTEMS				
	Nanocomposite hydrogel	Hyaluronic acid	Latanoprost	The hyaluronic acid nanocomposite hydrogels, with controlled degradation properties and sustained release, could serve as potential drug delivery systems for many ocular diseases as they controlled the release of latanoprost in vitro	[82]
	Hydrogel contact lens	Silicone	Ofloxacin and Chloramphenicol	The drug release from the lenses was directly proportional to the amount of drug loaded and the lenses at the different loading concentrations showed transmittance of 95–97%. The silicone hydrogel contact lenses can be used to control drug delivery to the eye and is an alternative ocular delivery technique in the treatment or prevention of corneal infections	[83]
	PULMONARY DRUG DELIVERY SYSTEMS				
	Porous particles	Poly(lactide-co-glycolide) (PLGA)	Celecoxib	Large porous celecoxib-PLGA microparticles prepared using supercritical fluid technology exhibited sustained drug delivery and antitumor efficacy, without causing any significant toxicity	[84]
	Nanoparticles	Nanopolymeric particles consisting of hydroxyl propyl methylcellulose (HPMC), poly-vinylpyrrolidone (PVP)	Fluticasone	The in vitro antibacterial studies showed that HPMC-PVP-FLU nanoparticles displayed superior effect against Gram-positive bacteria compared to the unprocessed FLU and positive control	[85]

S/No.	Drug delivery systems	Biomaterial	API	Significance of the study	Reference
IV	**IMPLANT DRUG DELIVERY SYSTEMS**				
	Silk disc implants	Silk fibrin	IgG antibody or human immunodeficiency virus (HIV) inhibitor 5P12-RANTES	SF was formulated into insertable discs that can encapsulate either IgG antibody or human immunodeficiency virus (HIV) inhibitor 5P12-RANTES. The water vapor annealing showed a sustained release for 31 days and this released protein could inhibit HIV infection in both blood and human colorectal tissue	[86]
	Bone biomaterials implant	Hydroxyapatite	Doxorubicin-loaded cyclodextrin	Hydroxyapatite-cyclodextrin-doxorubicin chemotherapeutic strategy enhanced the drug-targeting effect on tumor cells while protecting the more sensitive healthy cells after implantation. A successful integration of such a drug delivery system might allow healthy cells to initially survive during the doxorubicin exposure period	[87]
V	**SYSTEMIC DRUG DELIVERY**				
	Polylactide scaffold hydrogel injections	Cholesterol-modified poly(ethylene glycol)–polylactide	Chondrocytes	The formulation shows lower critical gelation temperature, higher mechanical strength, larger pore size, better chondrocyte adhesion, and slower degradation compared to plain polylactide scaffold gels. The hydrogel serves as a promising chondrocyte carrier for cartilage tissue engineering and gives an alternative solution to surgical cartilage repair	[88]
	ANG-(1–7) functionalized plant chloroplast	Lyophilized lettuce cells (ACE2/ANG-(1–7))	Lyophilized lettuce cells (ACE2/ANG-(1–7))	Toxicology studies showed that both male and female rats tolerated ~10-fold ACE2/ANG-(1–7) higher than efficacy dose. The efficient attenuation of pulmonary arterial hypertension with no toxicity augurs well for the clinical advancement of the first oral protein therapy to prevent/treat underlying pathology for this disease.	[89]
VI	**VAGINAL DRUG DELIVERY SYSTEMS**				
	Organogel	Palm oil and hyaluronic acid	Maraviroc	There was a 2.5-fold increase in the percentage of maraviroc release in the presence of hyaluronidase, hence the effectiveness of hyaluronidase enzyme acting as a trigger. This shows the potential use of palm oil/hyaluronic acid-based organogel for the vaginal delivery of anti-HIV microbicide for HIV prevention	[90]
	Vaginal rings	Silicone matrix polymer	Dapivirine	A monthly vaginal ring containing dapivirine reduced the risk of HIV-1 infection among African women, with increased efficacy in subgroups with evidence of increased adherence	[91]

S/No.	Drug delivery systems	Biomaterial	API	Significance of the study	Reference
VII	TOPICAL DRUG DELIVERY SYSTEMS				
	Electrospun fibers	Polylactic acid and collagen	Collagen and silver sulfadiazine	The electrospun fibers were nontoxic to the cells and provided favorable substrates for the neonatal epidermal keratinocytes cells to undergo cell attachment and proliferation, hence its potential for use in chronic wound management	[92]
	Hydrogel	Polyvinyl alcohol and carbopol	Diclofenac diethylamine	In vitro skin permeation for 10 h showed that the enhancement ratios of the flux of diclofenac was higher compared to the marketed formulations. The study highlighted the advantage of the experimental transdermal hydrogel over the hydrogel with microsized drug particles	[93]

Table 1.
Drug delivery systems showing the significance of the biomaterials utilized in delivering active pharmaceutical ingredients at their biological target site.

19. Nonfouling surfaces

Poly (ethylene glycol) (PEG), or poly(ethylene oxide) (PEO) having nonfouling surfaces demonstrates protein and cell resistance capabilities. PEG have been attached to materials in such a manner to render them nonfouling through processes like covalent immobilization, adsorption, or interpenetration. PEG has been covalently attached to mussel adhesive protein to form a nonfouling and a sticky segment copolymer [94] with gold and titanium surfaces attached to the sticky segment, while the PEG chains occur at the new interface. It should be noted that the nonfouling ability/attribute of PEG is dependent on the surface chain density that is prone to oxidants damaged. However, the use of plasma deposition of tetra ethylene glycol dimethyl ether (tetraglyme) on PEG will reduce protein surface adsorption [95]. Other materials with nonfouling surfaces include phospholipid surfaces [96] and saccharide surfaces [97], and these biomaterials ensure increased compatibility issues between the drug carrier systems and biological systems to which they are introduced to elicit a pharmacological activity.

20. Smart biomaterials

Materials which respond to environmental changes are attractive particularly in vivo as these can be utilized to control drug release, cell adhesiveness, mechanical properties, or permeability. These environmental changes can be brought about by stimulants like pH [98], temperature [99], and light [100]. The body employs changes in pH to facilitate a range of different processes. For example, along the gastrointestinal track, food is broken down into nutritive substances in the stomach under acidic pH ~ 2 and subsequently absorbed in the small intestine (pH ~ 7). Patient often prefers the oral drug delivery requiring routine, periodic delivery of drugs and for effectiveness, the drug must resist the stomach acidic pH. The pH-sensitive materials that are mindful of gastrointestinal tract pH variation have been developed to transport drugs successfully through the stomach to the small intestine. Such successful materials include pH responsive hydrogels prepared from poly(methacrylic acid) grafted with poly(ethylene glycol) (PMAA-g-PEG) that swells in response to pH. For instance, the gel shrinks by trapping the drug cargo pH ~ 2 as interpolymer complexes are formed, but at physiological pH ~ 7, the gel can swell 3–25 times based on its composition as it releases its cargo in the target site [101]. Insulin-loaded PMAA-g-PEG gels have been orally delivered to diabetic mice with a significant decrease in glucose levels as protein function is protected in acidic and digestive enzymes environments [102].

20.1 Self-assembled biomaterials

Self-organization or self-assembly is based on the formation of weak noncovalent bonds, like hydrogen, ionic, or Van der Waals bonds or hydrophobic interactions [103]. In amphiphilic molecules, there are hydrophobic and hydrophilic segments that self-assemble to form nanometer 3D structures like micelles, vesicles, and tubules, which depend on the molecule's length and composition [104–107]. When any of these are dispersed in aqueous solvent, the hydrophobic segments agglomerate and water is expelled to produce a well-ordered structure useful in biomedical applications. Phospholipid a naturally occurring amphiphilic molecule that is largely compose of cell membrane is one such amphiphilic molecules while an oligomer, a polymer of amino acids, can be synthesized to have hydrophobic, hydrophilic, charged, etc., regions that can self-assemble into a macroscopic hydrogel [108]. The self-assembled biomaterials can be engineered for use in nanotechnology, tissue engineering for drug and cell carriers.

21. Controlled drug delivery

Polymers are large molecules formed from simple monomers and may be synthetic or biopolymers that are the constituents of living organisms like proteins, nucleic acids, and sugars. Biopolymers are active in controlling and regulating many biochemical and biophysical functions of living cells, and thus can participate in cooperative interactions, resulting in nonlinear response to external stimuli. The cooperative interaction mechanism of biopolymers is utilized in producing synthetic polymers that are similar in behavior to biopolymers, which are used as biomaterials with ability to interface with biological systems for a variety of living cells functions.

Polymeric, biodegradable materials are often useful in biomedical applications, as the polymers degrade into normal metabolites of the body or eliminated from the body with or without further metabolic transformation [109, 110]. Developed polymeric biomaterials have physical and chemical properties that are maintained and are not tampered with during synthesis. The use of synthetic polymeric biomaterials includes artificial corneal substitute, blood contacting devices, hip joint replacements, and formation of intraocular lenses [111, 112]. Biodegradable polymers are either natural or synthetic. Natural polymers are derived from natural resources and have potential to be considered for biomedical and pharmaceutical applications owing to biocompatibility, biomimicking environments, unique mechanical properties, and biodegradability. Natural polymers are prone to viral infection, antigenicity, and unstable material supply, which limit biomedical application. On the other hand, synthetic polymers are flexible in synthesis procedure technique with excellent reproducibility which made them useful for surgical and short-term medical application, orthopedic applications that may slowly transfer the load as it degrades [113].

The drug administration into the body is either via an oral or intravenous route with repeated administration done to increase concentration and performance. But this may reach an extreme level before it declines rapidly especially when the elimination rate from the body is high. A too low or too high drug concentration in the body will not benefit the patient because of the side effects. This phenomenon then becomes a concern requiring the use of controlled drug release mechanism which can only be offered by biomaterials [114]. For controlled drug release, the therapeutic and bioactive agents are enveloped or encapsulated in an insoluble biodegradable subnano, nano, micropolymer matrix cavity where the therapeutic agents are released in a controlled fashion.

22. Various drug delivery systems

Widely used drug delivery systems include a liposomal drug delivery system [115, 116] that consists of phospholipids, i.e., fatty acid esters and fat alcohol ethers of glycerol phosphatides; they are negatively charged at physiological pH due to their phosphate groups. Cationic liposomes are prepared using lipid molecules having a quaternary ammonium head group. Because cellular membranes carry negative charges, cationic liposomes interact with these cellular membranes [117]. The stability of liposomes in biological environment is improved with steric stability that can extend its blood circulation time after being administered [118]. Biodegradable polymers are usually used to enhance the steric stability of the liposomes. Natural biodegradable polymers that are suitable for drug delivery systems include proteins (collagen, gelatin, albumin, etc.) and polysaccharides (starch, dextran, chitosan, etc.) [119].

22.1 Polysaccharides

Polysaccharides are many monosaccharide repeating units with high molecular weight. It is biodegradable, biocompatible, and water soluble which make suitable for drug delivery. There are several different types of polysaccharides having different functional groups, which are as follows:

22.2 Alginic acid or alginates

Alginic acid is a linear hetero polysaccharide, nonbranched, high-molecular-weight binary copolymer of (1–4) glycosidic linkage with β-D-mannuronic acid and α-L guluronic acid monomers [120, 121]. Natural alginic acid can be obtained from the cell walls of brown algae. Its acidic nature helps in its spontaneous formation of salts and later gels in the presence of divalent cations like calcium ions. This occurs by the interaction of divalent cations with guluronic acid blocks present on other polysaccharide chains. The gel property paves way for the encapsulation of molecules that can act as drugs within alginate gels with negligible side effects. The drug delivery mechanism of alginates is hinged on the drug polymer interaction and chemical immobilization of the drug on the polymer backbone via reactive carboxylate groups [122–124].

22.3 Starch

Starch, which is a carbohydrate source can be isolated from corn, wheat, potato, tapioca, rice, etc., and consists of two glucosidic macromolecules: 20–30% of linear molecule—amylase and 70–80% of branched molecule—amylopectin. The products of starch processing include thin films, fibers, and porous matrices. It is an important polymer for thermoplastic biodegradable materials due to its low cost, availability, biocompatibility, biodegradability, and having renewable resources [125]. The products of starch degradation include fructose and maltose that are low molecular weight sugar [126]. Microspheres from starch have bioadhesive drug delivery system potential for nasal delivery of proteins [127].

22.4 Dextran

Dextran is a natural polysaccharide of large glucose molecules with long and branched chains of varying lengths from 3 to 2000 Kd at 1,6- and partly at 1,3-glucosidic linkages. It is synthesized from sucrose via lactic-acid bacteria like *Leuconostoc mesenteroides*, *Streptococcus mutans*, and lactic acid bacterium *Lactobacillus brevis*. It is colloidal and hydrophilic in nature; it is inert to the in vivo environment with no effect on cell viability [128]. Dextran is used as an antithrombotic (antiplatelet), to reduce blood viscosity, and as a volume expander in anemia [129]. Dextran can be degraded by enzyme dextranase in the colon and thus can serve as a colonic drug delivery system.

22.5 Pullulan

Pullulan occurs naturally as linear homopolysaccharide polymer with maltotriose units of 3-glucose or D-glucopyranose units which are linked by α-(1 → 4) glycosidic linkages. It is edible, bland, and tasteless and thus is added to food and beverages. It serves as a coating agent in pharmaceutics, breath fresheners, or oral hygiene products [130]. Consecutive maltotriose units are linked to one another via α-(1 → 6) glycosidic bond. The pullulan backbone structure is similar to dextran, as both are plasma expanders. Pullulan is commercially synthesized via fermentation process involving the growth of fungus *Aureobasidium pullulans* on a carbohydrate substrate, which is then harvested. This process is followed by the rupture of cell using either an enzyme or a physical force, and pullulan is then extracted via simple water extraction method [130]. This method does not constitute any threat to the environment and therefore it is ecofriendly. This then makes pullulan suitable as a drug delivery vehicle. Pullulan hydrogel micro and nanoparticles are employed in oral administration of gastro-sensitive drugs.

22.6 Hyaluronic acid

Hyaluronic acid also a natural occurring negatively charged linear polysaccharide made of repeating disaccharide units of D-glucuronic acid and 2-acetamido-2-deoxy-D-glucose monosaccharide units. It exists majorly in articular cartilage, connective tissues, synovial fluids of mammals and the mesenchyme of developing embryos. It is water soluble and forms highly viscous solutions and therefore suitable for use as wound dresser as it can act as scavenger for free radicals in wound sites to modulate inflammation [131]. Its use in tissue repair application include to protect delicate tissue in the eye in removal of cataract, corneal transplantation, and glaucoma surgery, as vitreous substitute in retina re-attachment surgery, to relieve pain and improve joint mobility in osteoarthritis (knee) patients suffering and accelerate bone fracture healing [132].

22.7 Chitin and chitosan

Chitin a natural occurring polysaccharide of 1 → 4 β-linked glycan containing 2-acetamido-2-deoxy-D-glucose is a component of shells of crustaceans, cell walls of fungi, etc. When chitin is deacetylated chitosan a semi-crystalline linear copolymer polysaccharide is produced with (1 → 4) β-linked D-glucosamine and some N-acetyl glucosamine groups. The degree of deacetylation (DD) of chitosan may be from 70% and 90% and the MW is in between 10 and 1000 k [133]. While chitin is insoluble in regular solvents, chitosan is fully soluble in aqueous solutions with pH <5.0 [134]. Chitosan degrades in vivo enzymatically via lysozyme to nontoxic products [134]. Chitosan is easy to process and applied, oxygen permeability, water absorptivity, hemostatic property, and ability to induce interleukin-8 from fibroblasts. It uses include wound and burn dressing material, drug delivery and controlled drug release.

22.8 Polyurethane

Polyurethane is a polymer with a chain of organic units linked by carbamate (urethane), which is formed from two or several bi- or higher-functional monomers, one having two or more isocyanate functional groups (–N=C=O) and the other with two or more hydroxyl groups (–OH) [135]. It is a material with similar elasticity to rubber, possess toughness and durability comparable to metal, and is chemically inert. Polyurethane micelles are suitable drug delivery systems.

23. Conclusion

Advances in medical research have led to the exploration of various materials as drug carriers for suitable delivery. Biomaterials are currently well explored in recent years as a result of their ubiquitous nature, ease of accessibility, biodegradability, and biocompatibility with living tissues. They have been singly used or blended with other materials as composites. This chapter has thus discussed the different biomaterials with their functionalities in the area of drug release. More biomaterials can be explored by processing and characterizations from natural origin to ensure effective performance and limit health complications associated with drug release.

Author details

Samson O. Adeosun[1]*, Margaret O. Ilomuanya[2], Oluwashina P. Gbenebor[1], Modupeola O. Dada[3] and Cletus C. Odili[1]

1 Department of Metallurgical and Materials Engineering, University of Lagos, Lagos, Nigeria

2 Department of Pharmaceutics and Pharmaceutical Technology, Faculty of Pharmacy, University of Lagos, Lagos, Nigeria

3 Department of Chemical, Metallurgical and Materials Engineering, Tshwane University of Technology, Pretoria, South Africa

*Address all correspondence to: sadeosun@unilag.edu.ng

References

[1] Schneider C, Langer R, Loveday D, Hair D. Applications of ethylene vinyl acetate copolymers (EVA) in drug delivery systems. Journal of Controlled Release. 2017;**262**:284-295. DOI: 10.1016/j.jconrel.2017.08.004

[2] Park K. Controlled drug delivery systems: Past forward and future back. Journal of Controlled Release. 2014;**190**:3-8. DOI: 10.1016/j.jconrel.2014.03.054

[3] Yun YH, Lee BK, Park K. Controlled drug delivery: Historical perspective for the next generation. Journal of Controlled Release. 2015;**219**:2-7. DOI: 10.1016/j.jconrel.2015.10.005

[4] Bertrand N, Wu J, Xu X, Kamaly N, Farokhzad OC. Cancer nanotechnology: The impact of passive and active targeting in the era of modern cancer biology. Advanced Drug Delivery Reviews. 2014;**66**:2-25. DOI: 10.1016/j.addr.2013.11.009

[5] Kamaly N, Xiao Z, Valencia PM, Radovic-Moreno AF, Farokhzad OC. Targeted polymeric therapeutic nanoparticles: Design, development and clinical translation. Chemical Society Reviews. 2012;**41**(7):2971-3010. DOI: 10.1039/c2cs15344k

[6] Shi N-Q, Zhou J, Walker J, Li L, Hong JKY, Olsen KF, et al. Microencapsulation of luteinizing hormone-releasing hormone agonist in poly (lactic-co-glycolic acid) microspheres by spray-drying. Journal of Controlled Release. 2020;**321**:756-772. DOI: 10.1016/j.jconrel.2020.01.023

[7] Barenholz Y. Doxil®—the first FDA-approved nano-drug: Lessons learned. Journal of Controlled Release. 2012;**160**:117-134. DOI: 10.1016/j.jconrel.2012.03.020

[8] Gabizon A, Catane R, Uziely B, Kaufman B, Safra T, Cohen R, et al. Prolonged circulation time and enhanced accumulation in malignant exudates of doxorubicin encapsulated in polyethylene-glycol coated liposomes. Cancer Research. 1994;**54**(4):987-992

[9] Miller JB, Zhang S, Kos P, Xiong H, Zhou K, Perelman SS, et al. Non-viral CRISPR/Cas gene editing in vitro and In vivo enabled by synthetic nanoparticle co-delivery of Cas9 mRNA and sgRNA. Angewandte Chemie, International Edition. 2017;**56**(4):1059-1063. DOI: 10.1002/anie.201610209

[10] Silverman JA, Deitcher SR. Marqibo® (vincristine sulfate liposome injection) improves the pharmacokinetics and pharmacodynamics of vincristine. Cancer Chemotherapy and Pharmacology. 2013;**71**(3):555-564. DOI: 10.1007/s00280-012-2042-4

[11] Ferrari M. Cancer nanotechnology: Opportunities and challenges. Nature Reviews Cancer. 2005;**5**(3):161-171. DOI: 10.1038/nrc1566

[12] Prausnitz MR, Mitragotri S, Langer R. Current status and future potential of transdermal drug delivery. Nature Reviews Drug Discovery. 2004;**3**(2):115-124. DOI: 10.1038/nrd1304

[13] DeRuiter J, Holston PL. Review of selected NMEs. U.S. Pharmacist. 2012;**37**(10):HS2-HS8

[14] Brem H, Mahaley MS, Vick NA, Black KL, Schold SC, Burger PC, et al. Interstitial chemotherapy with drug polymer implants for the treatment of recurrent gliomas. Journal of Neurosurgery. 1991;**74**(3):441-446. DOI: 10.3171/jns.1991.74.3.0441

[15] Brem H, Piantadosi S, Burger PC, Walker M, Selker R, Vick NA, et al. Placebo-controlled trial of safety and efficacy of intraoperative controlled

delivery by biodegradable polymers of chemotherapy for recurrent gliomas. The Lancet. 1995;**345**(8956):1008-1012. DOI: 10.1016/S0140-6736(95)90755-6

[16] Giese A, Kucinski T, Knopp U, Goldbrunner R, Hamel W, Mehdorn HM, et al. Pattern of recurrence following local chemotherapy with biodegradable carmustine (BCNU) implants in patients with glioblastoma. Journal of Neuro-Oncology. 2004;**66**(3):351-360. DOI: 10.1023/B:NEON.0000014539.90077.db

[17] Langer R, Folkman J. Polymers for the sustained release of proteins and other macromolecules. Nature. 1976;**263**(5580):797-800. DOI: 10.1038/263797a0

[18] Langer R, Peppas NA. Advances in biomaterials, drug delivery, and bionanotechnology. AICHE Journal. 2003;**49**(12):2990-3006. DOI: 10.1002/aic.690491202

[19] Langer R. New methods of drug delivery. Science. 1990;**249**(4976):1527-1533. DOI: 10.1126/science.2218494

[20] Iwamoto T. Clinical application of drug delivery systems in cancer chemotherapy: Review of the efficacy and side effects of approved drugs. Biological and Pharmaceutical Bulletin. 2013;**36**(5):715-718. DOI: 10.1248/bpb.b12-01102

[21] Lu Y, Aimetti AA, Langer R, Gu Z. Bioresponsive materials. Nature Reviews Materials. 2016;**2**(1):16075. DOI: 10.1038/natrevmats.2016.75

[22] Matyjaszewski K. Atom transfer radical polymerization (ATRP): Current status and future perspectives. Macromolecules. 2012;**45**(10):4015-4039. DOI: 10.1021/ma3001719

[23] Boyer C, Bulmus V, Davis TP, Ladmiral V, Liu J, Perrier S. Bioapplicationsof RAFT polymerization. Chemical Reviews. 2009;**109**(11):5402-5436. DOI: 10.1021/cr9001403

[24] Nicolas J, Guillaneuf Y, Lefay C, Bertin D, Gigmes D, Charleux B. Nitroxide-mediated polymerization. Progress in Polymer Science. 2013;**38**(1):63-235. DOI: 10.1016/j.progpolymsci.2012.06.002

[25] Bielawski CW, Grubbs RH. Living ring-opening metathesis polymerization. Progress in Polymer Science. 2007;**32**(1):1-29. DOI: 10.1016/j.progpolymsci.2006.08.006

[26] Lee TT, García JR, Paez JI, Singh A, Phelps EA, Weis S, et al. Light-triggered in vivo activation of adhesive peptides regulates cell adhesion, inflammation and vascularization of biomaterials. Nature Materials. 2015;**14**(3):352-360. DOI: 10.1038/nmat4157

[27] Lei P, Padmashali RM, Andreadis ST. Cell-controlled and spatially arrayed gene delivery from fibrin hydrogels. Biomaterials. 2009;**30**(22):3790-3799. DOI: 10.1016/j.biomaterials.2009.03.049

[28] Meng F, Hennink WE, Zhong Z. Reduction-sensitive polymers and bioconjugates for biomedical applications. Biomaterials. 2009;**30**(12):2180-2198. DOI: 10.1016/j.biomaterials.2009.01.026

[29] Cao W, Wang L, Xu H. Selenium/tellurium containing polymer materials in nanobiotechnology. Nano Today. 2015;**10**(6):717-736. DOI: 10.1016/j.nantod.2015.11.004

[30] Cheng G, He Y, Xie L, Nie Y, He B, Zhang Z, et al. Development of a reduction-sensitive diselenide-conjugated oligoethylenimine nanoparticulate system as a gene carrier. International Journal of Nanomedicine. 2012;7:3991-4006. DOI: 10.2147/IJN.S32961

[31] Boussif O, LezoualC'H F, Zanta MA, Mergny MD, Scherman D, Demeneix B,

et al. A versatile vector for gene and oligonucleotide transfer into cells in culture and in vivo: Polyethylenimine. Proceedings of the National Academy of Sciences of the United States of America. 1995;**92**(16):7297-7301. DOI: 10.1073/pnas.92.16.7297

[32] Sun W, Jiang T, Lu Y, Reiff M, Mo R, Gu Z. Cocoon-like self-degradable DNA nanoclew for anticancer drug delivery. Journal of the American Chemical Society. 2014;**136**(42):14722-14725. DOI: 10.1021/ja5088024

[33] Wang S, Wang H, Liu Z, Wang L, Wang X, Su L, et al. Smart pH- and reduction-dual-responsive folate-PEG-coated polymeric lipid vesicles for tumor-triggered targeted drug delivery. Nanoscale. 2014;**6**(13):7635-7642. DOI: 10.1039/c4nr00843j

[34] Huang X, Liao W, Zhang G, Kang S, Zhang CY. pH Ssensitive micelles self-assembled from polymer brush (PAE-g-cholesterol)-b-PEG-b- (PAE-g-cholesterol) for anticancer drug delivery and controlled release. International Journal of Nanomedicine. 2017;**12**:2215-2226

[35] Ulery BD, Nair LS, Laurencin CT. Biomedical applications of biodegradable polymers. Journal of Polymer Science, Part B: Polymer Physics. 2011;**49**(12):832-864. DOI: 10.1002/polb.22259

[36] Attenello FJ, Mukherjee D, Datoo G, McGirt MJ, Bohan E, Weingart JD, et al. Use of gliadel (BCNU) wafer in the surgical treatment of malignant glioma: A 10-year institutional experience. Annals of Surgical Oncology. 2008;**15**(10):2887-2893. DOI: 10.1245/s10434-008-0048-2

[37] Gajanayake T, Olariu R, Leclère FM, Dhayani A, Yang Z, Bongoni AK, et al. A single localized dose of enzyme-responsive hydrogel improves long-term survival of a vascularized composite allograft. Science Translational Medicine. 2014;**6**(249):249ra110. DOI: 10.1126/scitranslmed.3008778

[38] Wiltsey C, Christiani T, Williams J, Scaramazza J, Van Sciver CV, Toomer K, et al. Thermogelling bioadhesive scaffolds for intervertebral disk tissue engineering: Preliminary in vitro comparison of aldehyde-based versus alginate microparticle-mediated adhesion. Acta Biomaterialia. 2015;**16**(1):71-80. DOI: 10.1016/j.actbio.2015.01.025

[39] Mura S, Nicolas J, Couvreur P. Stimuli-responsive nanocarriers for drug delivery. Nature Materials. 2013;**12**(11):991-1003. DOI: 10.1038/nmat3776

[40] Chen CY, Kim TH, Wu WC, Huang CM, Wei H, Mount CW, et al. PH-dependent, thermosensitive polymeric nanocarriers for drug delivery to solid tumors. Biomaterials. 2013;**34**(18):4501-4509. DOI: 10.1016/j.biomaterials.2013.02.049

[41] Hsieh DST, Langer R, Folkman J. Magnetic modulation of release of macromolecules from polymers. Proceedings of the National Academy of Sciences of the United States of America. 1981;**78**(3 I):1863-1867. DOI: 10.1073/pnas.78.3.1863

[42] Mir M, Ishtiaq S, Rabia S, Khatoon M, Zeb A, Khan GM, et al. Nanotechnology: From in vivo imaging system to controlled drug delivery. Nanoscale Research Letters. 2017;**12**(1):500. DOI: 10.1186/s11671-017-2249-8

[43] Iqbal D, Samiullah M. Photo-responsive shape-memory and shape-changing liquid-crystal polymer networks. Materials. 2013;**6**(1):116-142. DOI: 10.3390/ma6010116

[44] Jochum FD, Theato P. Temperature- and light-responsive smart polymer

materials. Chemical Society Reviews. 2013;**42**(17):7468-7483. DOI: 10.1039/c2cs35191a

[45] Kim MS, Diamond SL. Photocleavage of o-nitrobenzyl ether derivatives for rapid biomedical release applications. Bioorganic & Medicinal Chemistry Letters. 2006;**16**(15):4007-4010. DOI: 10.1016/j.bmcl.2006.05.013

[46] Kloxin AM, Kasko AM, Salinas CN, Anseth KS. Photodegradable hydrogels for dynamic tuning of physical and chemical properties. Science. 2009;**324**(5923):59-63. DOI: 10.1126/science.1169494

[47] Lin CC, Anseth KS. PEG hydrogels for the controlled release of biomolecules in regenerative medicine. Pharmaceutical Research. 2009;**26**(3):631-643. DOI: 10.1007/s11095-008-9801-2

[48] Timko BP, Arruebo M, Shankarappa SA, McAlvin JB, Okonkwo OS, Mizrahi B, et al. Near-infrared-actuated devices for remotely controlled drug delivery. Proceedings of the National Academy of Sciences of the United States of America. 2014;**111**(4):1349-1354. DOI: 10.1073/pnas.1322651111

[49] Marturano V, Cerruti P, Giamberini M, Tylkowski B, Ambrogi V. Light-responsive polymer micro-and nano-capsules. Polymers (Basel). 2017;**9**(1):1-19. DOI: 10.3390/polym9010008

[50] Pearson S, Vitucci D, Khine YY, Dag A, Lu H, Save M, et al. Light-responsive azobenzene-based glycopolymer micelles for targeted drug delivery to melanoma cells. European Polymer Journal. 2015;**69**:616-627. DOI: 10.1016/j.eurpolymj.2015.04.001

[51] Yu J, Zhang Y, Ye Y, DiSanto R, Sun W, Ranson D, et al. Microneedle-array patches loaded with hypoxia-sensitive vesicles provide fast glucose-responsive insulin delivery. Proceedings of the National Academy of Sciences of the United States of America. 2015;**112**(27):8260-8265. DOI: 10.1073/pnas.1505405112

[52] Hosseinkhani H, Hosseinkhani M. Biodegradable polymer-metal complexes for gene and Drug delivery. Current Drug Safety. 2009;**4**(1):79-83. DOI: 10.2174/157488609787354477

[53] Agterberg MJH, Versnel H, Van Dijk LM, De Groot JCMJ, Klis SFL. Enhanced survival of spiral ganglion cells after cessation of treatment with brain-derived neurotrophic factor in deafened guinea pigs. JARO-Journal of the Association for Research in Otolaryngology. 2009;**10**(3):355-367. DOI: 10.1007/s10162-009-0170-2

[54] Willerth SM, Sakiyama-Elbert SE. Approaches to neural tissue engineering using scaffolds for drug delivery. Advanced Drug Delivery Reviews. 2007;**59**(4-5):325-338. DOI: 10.1016/j.addr.2007.03.014

[55] Maxwell DJ, Hicks BC, Parsons S, Sakiyama-Elbert SE. Development of rationally designed affinity-based drug delivery systems. Acta Biomaterialia. 2005;**1**(1):101-113. DOI: 10.1016/j.actbio.2004.09.002

[56] Wang NX, Von Recum HA. Affinity-based drug delivery. Macromolecular Bioscience. 2011;**11**(3):321-332. DOI: 10.1002/mabi.201000206

[57] Fu AS, Thatiparti TR, Saidel GM, Von Recum HA. Experimental studies and modeling of drug release from a tunable affinity-based drug delivery platform. Annals of Biomedical Engineering. 2011;**39**(9):2466-2475. DOI: 10.1007/s10439-011-0336-z

[58] Yang S, Chen D, Li N, Mei X, Qi X, Li H, et al. A facile preparation of targetable pH-sensitive polymeric nanocarriers with encapsulated

magnetic nanoparticles for controlled drug release. Journal of Materials Chemistry. 2012;**22**(48):25354-25361. DOI: 10.1039/c2jm34817a

[59] Checa-Casalengua P, Jiang C, Bravo-Osuna I, Tucker BA, Molina-Martínez IT, Young MJ, et al. Retinal ganglion cells survival in a glaucoma model by GDNF/Vit e PLGA microspheres prepared according to a novel microencapsulation procedure. Journal of Controlled Release. 2011;**156**(1):92-100. DOI: 10.1016/j.jconrel.2011.06.023

[60] Benoit JP, Faisant N, Venier-Julienne MC, Menei P. Development of microspheres for neurological disorders: From basics to clinical applications. Journal of Controlled Release. 2000;**65**(1-2):285-296. DOI: 10.1016/S0168-3659(99)00250-3

[61] Xu X, Yu H, Gao S, Mao HQ, Leong KW, Wang S. Polyphosphoester microspheres for sustained release of biologically active nerve growth factor. Biomaterials. 2002;**23**(17):3765-3772. DOI: 10.1016/S0142-9612(02)00116-3

[62] Sinha VR, Bansal K, Kaushik R, Kumria R, Trehan A. Poly-ε-caprolactone microspheres and nanospheres: An overview. International Journal of Pharmaceutics. 2004;**278**(1):1-23. DOI: 10.1016/j.ijpharm.2004.01.044

[63] Sinha VR, Trehan A. Biodegradable microspheres for protein delivery. Journal of Controlled Release. 2003;**90**(3):261-280. DOI: 10.1016/S0168-3659(03)00194-9

[64] Meinel AJ, Germershaus O, Luhmann T, Merkle HP, Meinel L. Electrospun matrices for localized drug delivery: Current technologies and selected biomedical applications. European Journal of Pharmaceutics and Biopharmaceutics. 2012;**81**(1):1-13. DOI: 10.1016/j.ejpb.2012.01.016

[65] Han D, Cheung KC. Biodegradable cell-seeded nanofiber scaffolds for neural repair. Polymers. 2011;**3**(4):1684-1733. DOI: 10.3390/polym3041684

[66] Piras AM, Chiellini F, Chiellini E, Nikkola L, Ashammakhi N. New multicomponent bioerodible electrospun nanofibers for dual-controlled drug release. Journal of Bioactive and Compatible Polymers. 2008;**23**(5):423-443. DOI: 10.1177/0883911508093357

[67] Amajuoyi JN, Ilomuanya MO, Asantewaa-Osei Y, Azubuike CP, Adeosun SO, Igwilo CI. Development of electrospun keratin/coenzyme Q10/poly vinyl alcohol nanofibrous scaffold containing mupirocin as potential dressing for infected wounds. Future Journal of Pharmaceutical Sciences. 2020;**6**:25. DOI: 10.1186/s43094-020-00043-z

[68] Aurand ER, Lampe KJ, Bjugstad KB. Defining and designing polymers and hydrogels for neural tissue engineering. Neuroscience Research. 2012;**72**(3):199-213. DOI: 10.1016/j.neures.2011.12.005

[69] Koutsopoulos S, Unsworth LD, Nagai Y, Zhang S. Controlled release of functional proteins through designer self-assembling peptide nanofiber hydrogel scaffold. Proceedings of the National Academy of Sciences of the United States of America. 2009;**106**(12):4623-4628. DOI: 10.1073/pnas.0807506106

[70] Mehrotra S, Lynam D, Maloney R, Pawelec KM, Tuszynski MH, Lee I, et al. Time controlled protein release from layer-by-layer assembled multilayer functionalized agarose hydrogels. Advanced Functional Materials. 2010;**20**(2):247-258. DOI: 10.1002/adfm.200901172

[71] Anderson JM, Rodriguez A, Chang DT. Foreign body reaction to biomaterials. Seminars in Immunology.

2008;**20**(2):86-100. DOI: 10.1016/j.smim.2007.11.004

[72] Pigłowski J, Gancarz I, Staniszewska-Kuś J, Paluch D, Szymonowicz M, Konieczny A. Influence of plasma modification on biological properties of poly(ethylene terephthalate). Biomaterials. 1994;**15**(11):909-916. DOI: 10.1016/0142-9612(94)90116-3

[73] Shin H, Jo S, Mikos AG. Biomimetic materials for tissue engineering. Biomaterials. 2003;**24**(24):4353-4364. DOI: 10.1016/S0142-9612(03)00339-9

[74] Pierschbacher MD, Ruoslahti E. Cell attachment activity of fibronectin can be duplicated by small synthetic fragments of the molecule. Nature. 1984;**309**(5963):30-33. DOI: 10.1038/309030a0

[75] Aumailley M, Gerl M, Sonnenberg A, Deutzmann R, Timpl R. Identification of the Arg-Gly-asp sequence in laminin A chain as a latent cell-binding site being exposed in fragment P1. FEBS Letters. 1990;**262**(1):82-86. DOI: 10.1016/0014-5793(90)80159-G

[76] Staatz WD, Fok KF, Zutter MM, Adams SP, Rodriguez BA, Santoro SA. Identification of a tetrapeptide recognition sequence for the alpha 2 beta 1 integrin in collagen. Journal of Biological Chemistry. 1991;**266**:7363-7367

[77] Smith JW, Cheresh DA. The Arg-Gly-asp binding domain of the Vitronectin receptor. Photoaffinity cross-linking implicates amino acid residues 61-203 of the beta subunit. The Journal of Biological Chemistry. 1988;**263**(35):18726-18731

[78] VandeVondele S, Vörös J, Hubbell JA. RGD-grafted poly-L-lysine-graft-(polyethylene glycol) copolymers block non-specific protein adsorption while promoting cell adhesion. Biotechnology and Bioengineering. 2003;**82**(7):784-790. DOI: 10.1002/bit.10625

[79] Verrier S, Pallu S, Bareille R, Jonczyk A, Meyer J, Dard M, et al. Function of linear and cyclic RGD-containing peptides in osteoprogenitor cells adhesion process. Biomaterials. 2002;**23**(2):585-596. DOI: 10.1016/S0142-9612(01)00145-4

[80] Crivelli B, Bari E, Perteghella S, Catenacci L, Sorrenti M, Mocchi M, et al. Silk fibroin nanoparticles for celecoxib and curcumin delivery: ROS-scavenging and anti-inflammatory activities in an in vitro model of osteoarthritis. European Journal of Pharmaceutics and Biopharmaceutics. 2019;**137**:37-45. ISSN 0939-6411. DOI: 10.1016/j.ejpb.2019.02.008

[81] Schkarpetkin D, Reise M, Wyrwa R, Völpel A, Berg A, Schweder M, et al. Development of novel electrospun dual-drug fiber mats loaded with a combination of ampicillin and metronidazole. Dental Materials. 2016;**32**:951-960

[82] Widjaja LK, Bora M, Chan PN, Lipik V, Wong TT, Venkatraman SS. Hyaluronic acid–based nanocomposite hydrogels for ocular drug delivery applications. Journal of Biomedical Materials Research. Part A. 2015;**102**:3056-3065. DOI: 10.1002/jbm.a.34976

[83] Ubani-Ukoma U, Silva BO, Okubanjo OO, Aribaba OT, Ilomuanya MO, Igbokwe NH. In vitro release from antibiotic-loaded silicone hydrogel contact lenses for the treatment of ocular bacterial infections. Nigerian Journal of Pharmaceutical Research [S.l.]. 2019;**15**(1):1-8. ISSN 2635-3555 Available from: http://nigjpharmres.com/ojs/index.php/NigJPharmRes/article/view/107

[84] Dhanda DS, Tyagi P, Mirvish SS, Kompella UB. Supercritical fluid technology based large porous celecoxib-PLGA microparticles do not induce pulmonary fibrosis and sustain drug delivery and efficacy for several weeks following a single dose. Journal of Controlled Release. 2013;**168**(3):239-250. DOI: 10.1016/j.jconrel.2013.03.027

[85] Ahmed S, Govender T, Khan I, et al. Experimental and molecular modeling approach to optimize suitable polymers for fabrication of stable fluticasone nanoparticles with enhanced dissolution and antimicrobial activity. Drug Design, Development and Therapy. 2018;**12**:255-269. DOI: 10.2147/DDDT.S148912

[86] Yavuz B, Morgan JL, Herrera C, Harrington K, Perez-Ramirez B, Li Wang PJ, et al. Sustained release silk fibroin discs: Antibody and protein delivery for HIV prevention. Journal of Controlled Release. 2019;**301**:1-12. DOI: 10.1016/j.jconrel.2019.03.001

[87] Bischoff I, Tsaryk R, Chai F, Fürst R, Kirkpatrick CJ, Unger RE. In vitro evaluation of a biomaterial-based anticancer drug delivery system as an alternative to conventional post-surgery bone cancer treatment. Materials Science & Engineering. C, Materials for Biological Applications. 2018;**93**:115-124. DOI: 10.1016/j.msec.2018.07.057

[88] Wang C, Feng N, Chang F, Wang J, Yuan B, Cheng Y, et al. Injectable cholesterol-enhanced stereocomplex polylactide tthermogel loading chondrocytes for optimized cartilage regeneration. Advanced Healthcare Materials. 2019;**8**:1900312. DOI: 10.1002/adhm.201900312

[89] Daniell H, Mangu V, Yakubov B, Park J, Habibi P, Shi Y, et al. Investigational new drug enabling angiotensin oral-delivery studies to attenuate pulmonary hypertension. Biomaterials. 2020;**233**:119750. ISSN 0142-9612. DOI: 10.1016/j.biomaterials.2019.119750

[90] Ilomuanya MO, Seriki ZA, Ubani-Ukoma UN, Oseni AO, Silva BO. Silver sulphadiazine- xanthan gum-hyaluronic acid composite hydrogel for wound healing: Formulation development and in vivo evaluation. Nigerian Journal of Pharmaceutical Research. 2020a;**16**(1):21-29. DOI: 10.4314/njpr.v16i1.3

[91] Devlin B, Nuttall J, Wilder S, Woodsong C, Rosenberg Z. Development of dapivirine vaginal ring for HIV prevention. Antiviral Research. 2013;**100**(Suppl):S3-S8. DOI: 10.1016/j.antiviral.2013.09.025

[92] Ilomuanya MO, Adebona AC, Wang W, Sowemimo AA, Eziegbo C, Silva BO, et al. Development and characterization of collagen-based electrospun scaffolds containing silver sulphadiazine and *Aspalathus linearis* extract for potential wound healing applications. SN Applied Sciences. 2020b;**2**:881. DOI: 10.1007/s42452-020-2701-8

[93] Sengupta S, Banerjee S, Sinha B, Mukherjee B. Improved skin penetration using in situ nanoparticulate diclofenac diethylamine in hydrogel systems: In vitro and in vivo studies. AAPS PharmSciTech. 2016;**17**(2):307-317. DOI: 10.1208/s12249-015-0347-4

[94] Dalsin JL, Hu BH, Lee BP, Messersmith PB. Mussel adhesive protein mimetic polymers for the preparation of nonfouling surfaces. Journal of the American Chemical Society. 2003;**125**(14):4253-4258. DOI: 10.1021/ja0284963

[95] Löpez GP, Ratner BD, Tidwell CD, Haycox CL, Rapoza RJ, Horbett TA. Glow discharge plasma deposition of tetraethylene glycol dimethyl ether for fouling-resistant biomaterial surfaces. Journal of Biomedical Materials

Research. 1992;**26**(4):415-439. DOI: 10.1002/jbm.820260402

[96] Ishihara K, Ziats NP, Tierney BP, Nakabayashi N, Anderson JM. Protein adsorption from human plasma is reduced on phospholipid polymers. Journal of Biomedical Materials Research. 1991;**25**(11):1397-1407. DOI: 10.1002/jbm.820251107

[97] Holland NB, Qiu Y, Ruegsegger M, Marchant RE. Biomimetic engineering of non-adhesive glycocalyx-like surfaces using oligosaccharide surfactant polymers. Nature. 1998;**392**(6678):799-801. DOI: 10.1038/33894

[98] Murthy N, Campbell J, Fausto N, Hoffman AS, Stayton PS. Bioinspired pH-responsive polymers for the intracellular delivery of biomolecular drugs. Bioconjugate Chemistry. 2003;**14**(2):412-419. DOI: 10.1021/bc020056d

[99] Jeong B, Kim SW, Bae YH. Thermosensitive sol-gel reversible hydrogels. Advanced Drug Delivery Reviews. 2002;**54**(1):37-51. DOI: 10.1016/S0169-409X(01)00242-3

[100] Shimoboji T, Larenas E, Fowler T, Hoffman AS, Stayton PS. Temperature-induced switching of enzyme activity with smart polymer-enzyme conjugates. Bioconjugate Chemistry. 2003;**14**(3):517-525. DOI: 10.1021/bc025615v

[101] Kim B, La Flamme K, Peppas NA. Dynamic swelling behavior of pH-sensitive anionic hydrogels used for protein delivery. Journal of Applied Polymer Science. 2003;**89**(6):1606-1613. DOI: 10.1002/app.12337

[102] Lowman AM, Morishita M, Kajita M, Nagai T, Peppas NA. Oral delivery of insulin using pH-responsive complexation gels. Journal of Pharmaceutical Sciences. 1999;**88**(9):933-937. DOI: 10.1021/js980337n

[103] Zhang S. Emerging biological materials through molecular self-assembly. Biotechnology Advances. 2002;**20**(5-6):321-339. DOI: 10.1016/S0734-9750(02)00026-5

[104] Caplan MR, Lauffenburger DA. Nature's complex copolymers: Engineering design of oligopeptide materials. Industrial and Engineering Chemistry Research. 2002;**41**(3):403-412. DOI: 10.1021/ie010149z

[105] Caplan MR, Schwartzfarb EM, Zhang S, Kamm RD, Lauffenburger DA. Control of self-assembling oligopeptide matrix formation through systematic variation of amino acid sequence. Biomaterials. 2002;**23**(1):219-227. DOI: 10.1016/S0142-9612(01)00099-0

[106] Hartgerink JD, Beniash E, Stupp SI. Peptide-amphiphile nanofibers: A versatile scaffold for the preparation of self-assembling materials. Proceedings of the National Academy of Sciences of the United States of America. 2002;**99**(8):5133-5138. DOI: 10.1073/pnas.072699999

[107] Wright ER, McMillan RA, Cooper A, Apkarian RP, Conticello VP. Thermoplastic elastomer hydrogels via self-assembly of an elastin-mimetic triblock polypeptide. Advanced Functional Materials. 2002;**12**(2):149-154. DOI: 10.1002/1616-3028

[108] Petka WA, Harden JL, McGrath KP, Wirtz D, Tirrell DA. Reversible hydrogels from self-assembling artificial proteins. Science. 1998;**281**(5375):389-392. DOI: 10.1126/science.281.5375.389

[109] Leja K, Lewandowicz G. Polymer biodegradation and biodegradable polymers – A review. Polish Journal of Environmental Studies. 2010;**19**(2):255-266

[110] Vroman I, Tighzert L. Biodegradable polymers. Materials. 2009;**2**(2):307-344. DOI: 10.3390/ma2020307

[111] Castner DG, Ratner BD. Biomedical surface science: Foundations to frontiers. Surface Science. 2002;**500**(1-3):28-60. DOI: 10.1016/S0039-6028(01)01587-4

[112] Tathe A, Ghodke M, Nikalje AP. A brief review: Biomaterials and their application. International Journal of Pharmacy and Pharmaceutical Sciences. 2010;**2**(4):19-23

[113] Middleton JC, Tipton AJ. Synthetic biodegradable polymers as orthopedic devices. Biomaterials. 2000;**21**(23):2335-2346. DOI: 10.1016/S0142-9612(00)00101-0

[114] Leong KW, Langer R. Polymeric controlled drug delivery. Advanced Drug Delivery Reviews. 1988;**1**(3):199-233. DOI: 10.1016/0169-409X(88)90019-1

[115] Allen TM. Liposomal drug delivery. Current Opinion in Colloid & Interface Science. 1995;**1**(5):645-651. DOI: 10.1016/S1359-0294(96)80103-8

[116] Chonn A, Cullis PR. Recent advances in liposomal drug-delivery systems. Current Opinion in Biotechnology. 1995;**6**(6):698-708. DOI: 10.1016/0958-1669(95)80115-4

[117] Finkelstein EI, Chao PHG, Hung CT, Bulinski JC. Electric field-induced polarization of charged cell surface proteins does not determine the direction of galvanotaxis. Cell Motility and the Cytoskeleton. 2007;**64**(11):833-846. DOI: 10.1002/cm.20227

[118] Papahadjopoulos D, Allen TM, Gabizon A, Mayhew E, Matthay K, Huang SK, et al. Sterically stabilized liposomes: Improvements in pharmacokinetics and antitumor therapeutic efficacy. Proceedings of the National Academy of Sciences of the United States of America. 1991;**88**(24):11460-11464. DOI: 10.1073/pnas.88.24.11460

[119] Liu Z, Jiao Y, Wang Y, Zhou C, Zhang Z. Polysaccharides-based nanoparticles as drug delivery systems. Advanced Drug Delivery Reviews. 2008;**60**(15):1650-1662. DOI: 10.1016/j.addr.2008.09.001

[120] Augst AD, Kong HJ, Mooney DJ. Alginate hydrogels as biomaterials. Macromolecular Bioscience. 2006;**6**(8):623-633. DOI: 10.1002/mabi.200600069

[121] Tønnesen HH, Karlsen J. Alginate in drug delivery systems. Drug Development and Industrial Pharmacy. 2002;**28**(6):621-630. DOI: 10.1081/DDC-120003853

[122] Leonard M, De Boisseson MR, Hubert P, Dalençon F, Dellacherie E. Hydrophobically modified alginate hydrogels as protein carriers with specific controlled release properties. Journal of Controlled Release. 2004;**98**(3):395-405. DOI: 10.1016/j.jconrel.2004.05.009

[123] Martins S, Sarmento B, Souto EB, Ferreira DC. Insulin-loaded alginate microspheres for oral delivery—effect of polysaccharide reinforcement on physicochemical properties and release profile. Carbohydrate Polymers. 2007;**69**(4):725-731. DOI: 10.1016/j.carbpol.2007.02.012

[124] Matricardi P, Di Meo C, Coviello T, Alhaique F. Recent advances and perspectives on coated alginate microspheres for modified drug delivery. Expert Opinion on Drug Delivery. 2008;**5**(4):417-425. DOI: 10.1517/17425247.5.4.417

[125] Morrison WR, Karkalas J. Starch. In: Dey PM, editor. Methods in Plant Biochemistry: Carbohydrates. Vol. 2. 1990. p. 323. DOI: 10.1016/b978-0-12-461012-5.50001-x

[126] Marques AP, Reis RL, Hunt JA. The biocompatibility of novel

starch-based polymers and composites: In vitro studies. Biomaterials. 2002;**23**(6):1471-1478. DOI: 10.1016/S0142-9612(01)00272-1

[127] Illum L, Fisher AN, Jabbal-Gill I, Davis SS. Bioadhesive starch microspheres and absorption enhancing agents act synergistically to enhance the nasal absorption of polypeptides. International Journal of Pharmaceutics. 2001;**222**(1):109-119. DOI: 10.1016/S0378-5173(01)00708-6

[128] Hennink WE, Franssen O, Van Dijk-Wolthuis WNE, Talsma H. Dextran hydrogels for the controlled release of proteins. Journal of Controlled Release. 1997;**48**(2-3):107-114. DOI: 10.1016/S0168-3659(97)00047-3

[129] Dhaneshwar SS, Kandpal M, Gairola N, Kadam SS. Dextran: A promising macromolecular drug carrier. Indian Journal of Pharmaceutical Sciences. 2006;**68**(6):705-714. DOI: 10.4103/0250-474X.31000

[130] Alemzadeh I. The study on microbial polymers: Pullulan and PHB. Iranian journal of chemistry and chemical engineering. 2009;**28**(1):13-21. Available from: http://www.ijcce.ac.ir/article_6910.html

[131] Lloyd LL, Kennedy JF, Methacanon P, Paterson M, Knill CJ. Carbohydrate polymers as wound management aids. Carbohydrate Polymers. 1998;**37**(3):315-322. DOI: 10.1016/S0144-8617(98)00077-0

[132] Ilomuanya MO, Elesho RF, Amenaghawon AN, Velusamy V, Akanmu AS. A development of trigger sensitive hyaluronic acid/palm oil-based organogel for in vitro release of HIV/AIDS microbicides using artificial neural networks. Future Journal of Pharmaceutical Sciences. 2020c;**6**(1). DOI: 10.1186/s43094-019-0015-8

[133] Rinaudo M. Chitin and chitosan: Properties and applications. Progress in Polymer Science (Oxford). 2006;**31**(7):603-632. DOI: 10.1016/j.progpolymsci.2006.06.001

[134] Khor E, Lim LY. Implantable applications of chitin and chitosan. Biomaterials. 2003;**24**(13):2339-2349. DOI: 10.1016/S0142-9612(03)00026-7

[135] Zdrahala RJ, Zdrahala IJ. Biomedical applications of polyurethanes: A review of past promises, present realities, and a vibrant future. Journal of Biomaterials Applications. 1999;**14**(1):67-90. DOI: 10.1177/088532829901400104

Chapter 11

SeDeM-ODT Expert System: A Solution to Challenges in Characterization of Pharmaceutical Powders and Powdered Material

Amjad Khan, Shabnam Nazir, Hamad S. Alyami and Aman Ullah

Abstract

In the field of pharmaceutical sciences, material characterization has been a focus of research as properties of the powder ingredients govern characteristics of the finished dosage form. It has been a tedious and time-consuming job to develop a correlation between the characteristics of powder material and final dosage form. Extensive experimentation is carried out at different stages of formulation development to optimize the final blend and produce a product fulfilling official requirements. Various approaches have been applied for the purpose with varying degree of applications. SeDeM-ODT expert system is a novel pre-formulation technique developed for characterization of powder material of varying nature. Experimental and quantitative determination of various parameters provides a basis for SeDeM-ODT expert system. The system predicts suitability of powder material (APIs and excipients) for tablet preparation by direct compression technology and disintegration behavior of the resultant dosage form. It provides a basis for selection of excipients, both quantitatively and qualitatively. The present study covers area of powder characterization at pre-formulation level of pharmaceutical product development. SeDeM-ODT expert system reduces lead time for pre-formulation studies and provides formulations with minimum number of excipients. SeDeM-ODT expert system has been successfully applied for material characterization (APIs and excipients) before processing and after processing.

Keywords: direct compression, powder characterization, pharmaceutical powders, pre-formulation study, SeDeM-ODT expert system

1. Introduction

Tablets are the most preferred dosage form with respect to patient acceptability, flexibility in dose adjustment, easy manufacturing, and better stability [1, 2]. Irrespective of the nature of the drug and its manufacturing technique, tablets should meet some strict requirements in terms of mechanical strength,

IntechOpen

disintegration, and drug release [3–5]. A variety of techniques are available for tablet preparation among which direct compression is mostly preferred due to simplicity, cost-effectiveness, and less number of involved steps [6, 7]. However it can be applied only to the powder blend having optimum rheological characteristics, mechanical strength, and disintegration behavior, i.e., the powder blend should flow efficiently, and the resultant tablet should have sufficient mechanical strength with acceptable disintegration behavior [8–10]. All these characteristics are interlinked, and usually improvement of one characteristic can adversely affect the other. Hit and trial is the mostly applied method for optimization of powder characteristics which is laborious and material consuming. Optimization of powder blend is carried out, mostly, in the last stage of formulation development (following pre-formulation studies). Usual reported time for formulation development is in the range of 14–20 days, which can further extend in certain cases. There was a need for a technique that can avoid the experimentation for optimization of powder characteristics and help in excipient selection, i.e., select proper quantity of an excipient with desired characteristics. SeDeM-ODT expert system is a pre-formulation tool and has solved most of the problems associated with material characterization at pre-formulation level. SeDeM-ODT expert system minimizes experimentation and facilitates the process of formulation development by helping in excipient selection (in terms of desired characteristics and required quantity).

2. SeDeM-ODT expert system

SeDeM-ODT expert system is novel pre-formulation technique applied for development of a solid dosage form (tablets) by direct compression technology [11]. The system characterizes powder substance on the basis of various parameters related to flow, compressibility, and disintegration behavior. Physical profile of powder substance is developed, suggesting its suitability for direct compression and bucco-dispersibility [12, 13]. SeDeM expert system can be segmented into the following:

SeDeM expert system: It determines suitability of the powder substance for direct compression only [11, 13]. That is, the system characterizes powder substance with respect to its rheological characteristics and compressibility.

SeDeM-ODT expert system: It is the newest version of SeDeM expert system and characterizes the powder substances with respect to rheological characteristics, compressibility, and disintegration behavior, simultaneously [14]. Three extra parameters related to characterization of disintegration behavior are included in the SeDeM-ODT expert system.

The SeDeM-ODT expert system has been introduced with the aim of designing oro-dispersible tablets (ODTs) by direct compression [14]. This system is unique as it provides an oro-dispersible tablet formulation by direct compression, i.e., it links prediction of suitability of powder for direct compression and rapid disintegration of the tablets. SeDeM-ODT expert system is used for evaluation of critical quality attributes of powder substance, having an impact on the final product. Quality by Design guidelines ICH-Q8 [15] provides the basis for SeDeM-ODT expert system.

SeDeM-ODT expert system also calculates the amount of excipients with certain characteristic required for the correction of a particular property in order to make a final blend suitable for direct compression [14]. Several parameters have been selected that must be fulfilled by the formulation (excipients) to ensure successful and robust processing by direct compression technology.

2.1 Parameters determined for characterization of powder material

On the basis of physical characteristics and functionality of the ingredients, various parameters are grouped into six factors, as follows:

2.1.1 Dimension factor

Parameters included in this factor affect the size of the tablet and its ability to pile up. Results of these parameters are also used in the mathematical calculation of other indices related to powder compressibility. Parameters included in this group are:

- Bulk density
- Tapped density

2.1.2 Compressibility factor

The factor comprised of the parameters related to compressibility of powder and includes the following:

- Inter-particle porosity
- Carr's index
- Cohesion index

2.1.3 Flow ability/powder flow factor

This factor governs flow ability of the powder during compression and includes the following:

- Hausner ratio
- Angle of repose
- Flow ability

2.1.4 Lubricity/stability factor

Lubricity during compression and stability of the compressed tablets are affected by the parameters included in this factor. These are the following:

- Loss on drying
- Hygroscopicity

This incidence factor shows the rheological properties and stability of the powder and depends upon the intrinsic moisture and hygroscopicity of the material [14]. The low value of this incidence factor shows that the product will absorb moisture from the atmosphere, worsening its rheological properties (flow and compression) and consequently altering product stability. In case of values below the acceptable limit, the following measures are taken:

- Drying of material to reduce its loss on drying.
- Product should be processed in a controlled environment at low humidity.

2.1.5 Lubricity/dosage factor

Parameters included in this factor affect the lubricity and dosage of the tablet and comprised of the following:

- Particles having size below 50 μm
- Homogeneity index

2.1.6 Disgregability

Parameters included in disgregability factor govern disintegration behavior of the final product and are specified for fast dispersible tablets. Parameters included in this factor are as follows:

- Effervescence test
- Disintegration time with disk
- Disintegration time without disk

Incidence factor	Parameter	Symbol	Unit	Equation	Limits	Applied factor
Dimension	Bulk density	Da	g/mL	Da = P/Va	0–1	10 V
	Tapped density	Dc	g/mL	Dc = P/Vc	0–1	10 V
Compressibility	Inter-particle porosity	Ie	—	Dc – Da/Dc × Da	0–1.2	10 V/1.2
	Carr's index	Ic	%	100 (Dc–Da)/Dc	0–50	V/5
	Cohesion index	Icd	N	*Experimental	0–200	V/20
Flow ability/powder flow	Hausner ratio	IH	—	Dc/Da	3–1	(30–10 V)/2
	Angle of repose	(α)	o	$\tan^{-1}$ (h/r)	0–50	10 – (V/5)
	Powder flow	t″	S	Experimental	0–20	10 – (V/2)
Lubricity/stability	Loss on drying	%HR	%	Experimental	0–10	10 – V
	Hygroscopicity	%H	%	Experimental	0–20	10 – (V/2)
Lubricity/dosage	Particles <50	%Pf	%	Experimental	0–50	10 – (V/5)
	Homogeneity index	Iϴ	—	*Fm/100 + ΔFmn*	0–2 × 10^{-2}	500 V
Disgregability	Effervescence time	DE	Min	Experimental	0–5	(5 – V) × 2
	D. time with disk	DCD	Min	Experimental	0–3	(3 – V) × 3.333
	D. time without disk	DSD	Min	Experimental	0–3	(3 – V) × 3.333

D. time with disk: Disintegration time with disk.
D. time without disk: Disintegration time without disk.
**Experimental; The parameter was determined experimentally.*

Table 1.
Basic parameters of SeDeM-ODT expert system divided into different incidence factors.

Table 1 shows the basic parameters determined according to the SeDeM-ODT expert system along with their symbols, units, and classification into different incidence factors.

3. Methodology for powder characterization by SeDeM-ODT expert system

To determine the suitability of powder/powder blend for direct compression and bucco-dispersibility, SeDeM-ODT expert system needs 15 parameters to be found out. The individual parameters of SeDeM-ODT expert system are determined according to their respective pharmacopoeial methods, reported methods, or calculation on the basis of other basic parameters.

Characterization of powder according to the SeDeM-ODT expert system [11, 14] involves the following:

- Determination/calculation of basic parameters
- Conversion of experimental values to "r" values by applying specific factors and graphical presentation of results
- Calculation of various indices on the basis of "r" values

3.1 Determination/calculation of basic parameters of SeDeM-ODT expert system

SeDeM-ODT expert system is based on 15 basic parameters [14] which are determined experimentally or calculated on the basis of other included parameters. Procedures for the determination of basic parameters are given below:

3.1.1 Bulk density

Bulk density of the powder substance is determined according to the USP using graduated cylinder method [16]. The volume of the weighed amount of powder is determined using a graduated cylinder, and the density is calculated using the following equation:

$$D = \frac{m}{V} \tag{1}$$

where D is the density of the powder (g/mL), m is the weight of the powder (g), and v is the volume of the powder (mL).

3.1.2 Tapped density

Tapped density of powdered material is determined according to the USP by tapping known volume of powder taken in a graduated cylinder and noting the volume reduction [16]. Tapping can be carried out manually or using mechanical tappers.

3.1.3 Inter-particle porosity

Values of bulk density and tapped density are used for the calculation of inter-particle porosity [17], using the following equation:

$$\mathrm{Ie} = \frac{\mathrm{Dc} - \mathrm{Da}}{\mathrm{Dc} \times \mathrm{Da}} \tag{2}$$

where Ie is the inter-particle porosity, Dc is the tapped density (g/mL), and Da is the bulk density (g/mL).

3.1.4 Carr's index

Carr's index is calculated on the basis of tapped density and bulk density of powder [16]:

$$\mathrm{C.I.} = \frac{\mathrm{Dc} - \mathrm{Da}}{\mathrm{Dc}} \times 100 \tag{3}$$

where C.I. is the Carr's index of the powder (%), Dc is the tapped density of the powder (g/mL), and Da is the bulk density of the powder (g/mL).

3.1.5 Cohesion index

Cohesion index is the crushing strength of powder compressed, preferably in an eccentric press under maximum pressure without capping and lamination [11]. The mean crushing strength is calculated for at least 10 compacts, indicating the cohesion index of the powder. The raw powder is tested for compressibility, and in case of failure, 3.5% of the following mixture is added to the mix:

- Talc 2.36%
- Aerosil® 200 0.14%
- Magnesium stearate 1.00%

3.1.6 Hausner ratio

Hausner ratio is calculated from bulk density and tapped density of powder [16] according the equation given below:

$$\mathrm{Hr} = \frac{\mathrm{Dc}}{\mathrm{Da}} \tag{4}$$

where Hr is the Hausner ratio of the powder, Dc is the tapped density of the powder (g/mL), and Da is the bulk density of the powder (g/mL).

3.1.7 Angle of repose

Angle of repose is determined by funnel method [18]. The test powder is allowed to flow from a glass funnel fitted at certain height, and angle of repose was determined using the equation:

$$\propto = \tan^{-1}\left(\frac{H}{r}\right) \tag{5}$$

where α is the angle of repose of powder (°), H is the height of the cone formed by powder (cm), and r is the radius of the base of cone formed by powder (cm).

3.1.8 Powder flow

Powder flow is determined, in accordance with the European Pharmacopeia, by measuring the time required for the powder (100 g) to flow through the orifice of a glass funnel fitted at certain height [19].

3.1.9 Loss on drying

Loss on drying is determined gravimetrically according to the USP [20], using a halogen moisture analyzer. The powder (1 g) is loaded into the pan of moisture analyzer and heated for specified time at 100°C, and the value of percent loss is noted.

3.1.10 Hygroscopicity

Hygroscopicity is measured by placing the accurately weighed amount of powder in a climatic chamber at 75 ± 5% relative humidity for 24 h at ambient temperature. The material is analyzed after 24 h for percent weight gain by reweighing [13], indicating its hygroscopicity.

3.1.11 Particle size distribution

Sieve shaker fitted with standard sieves of pore size 850, 600, 425, 300, and 250 μm is used for the determination of particle size distribution. The powder (100 g) is loaded on the top sieve and the sieve shaker is vibrated for 10 min. The percent amount of the powder retained over each mesh is calculated [21].

3.1.12 Homogeneity index

Homogeneity index is determined according to the European Pharmacopoeia [21]. The powder (100 g) is loaded to a sieve shaker fitted with sieves of 850, 500, 425, 300, 250, and 50 μm pore size, and the sieve shaker is vibrated for 10 min. The percent amount of powder retained over each sieve and that passed through a 50 μm sieve is calculated. Homogeneity index of the material is calculated using the equation mentioned below:

$$I\theta = \frac{Fm}{100 + \Delta Fmn} \tag{6}$$

where Iθ is the relative homogeneity index and Fm is the percentage of particles in the majority range.

If the percentage is higher than that calculated in the complete sieve test, it is because some of the particles become adhered to the product retained in the sieves during the grain size test, and the percentage of particles below 50 μm particles found may be lower than the true figure. The following equation (Eq. (7)) is then applied to the data obtained.

$$
\begin{aligned}
I\theta = {} & Fm/100 + (dm - dm - 1)\ Fm - 1 + (dm + 1 - dm)\ Fm + 1 \\
& + (dm - dm - dm - 2)\ Fm - 2 + (dm + 2 - dm)\ Fm + 2 + \ldots \\
& + (dm - dm - dm - n)\ Fm - n + (dm + n - dm)\ Fm + n
\end{aligned}
\tag{7}
$$

where Iθ is the relative homogeneity index and particle size homogeneity in the range of the fractions studied; Fm is the percentage of particles in the majority range; Fm − 1 is the percentage of particles in the range immediately below the majority range; Fm + 1 is the percentage of particles in the range immediately above the majority range; n is the order number of the fraction studied under a series, with respect to the major fraction; dm is the diameter of the particles in the major fraction; dm − 1 is the mean diameter of the particles in the fraction of the range immediately below the majority range; and dm + 1 is the mean diameter of the particles in the fraction of the range immediately above the majority range.

The major fraction (Fm) corresponds to the interval from 0.100 to 0.212 mm, because it falls in the middle of the other fractions of the table. This interval is calculated as the proportion in which the powder particles are found in each fraction considered in the table (as described above). Those particles located in the major fraction (Fm) in a proportion of 60% are considered to represent the minimum acceptable value of 5. The distributions of the other particles are considered to be Gaussian. The limits for the homogeneity index are set between 0 and 0.02.

3.1.13 Effervescence time

Effervescence test for powder compact is determined as per official monograph [22]. The powder is compressed into tablets under maximum pressure without any capping and lamination. One tablet is placed in a beaker containing 200 mL of purified water at ambient temperature. Time taken by the tablet to disperse completely is taken as its effervescence time. Tablet is said to be dispersed completely when there is no agglomerate of the particles. In the context of SeDeM expert system, effervescence does not mean conventional acid-base reaction rather refers to dispersion of the compact in water. Effervescence time is an indicator for oro-dispersible tablets. When tablet disaggregates in less than 5 min, it is considered suitable for oral disintegration.

3.1.14 Disintegration time with disk

The powder is compressed under maximum pressure without any capping or lamination and subjected to the determination of disintegration time using USP disintegration apparatus. Disintegration time with disk is determined for at least six tablets, using de-ionized water as a medium held at 37 ± 2°C [23], and their mean is calculated (n = 6).

3.1.15 Disintegration time without disk

Disintegration time is determined as described in the previous section without any disk [23].

All the basic parameters of SeDeM-ODT expert system, along with symbols, units, and acceptable limits, are listed in **Table 1**.

3.2 Conversion of experimental values (V) to radius values (r) and graphical presentation of results

Results of SeDeM-ODT expert system are graphically presented as SeDeM-ODT diagram built on the basis of basic parameters. Values obtained from the

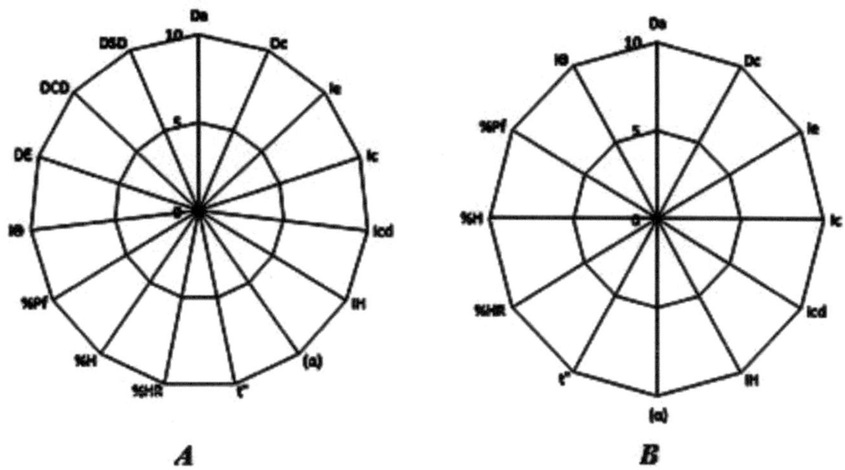

Figure 1.
Diagrammatic presentation of (A) SeDeM-ODT and (B) SeDeM expert system. Da, bulk density; %HR, loss on drying; dc, tapped density; %H, hygroscopicity; Ie, inter-particle porosity; %Pf, particle size; IC, Carr's index; Iθ, homogeneity index; ICd, cohesion index; DE, effervescence test; IH, Hausner ratio; DCD, disintegration time with disk; A, angle of repose; DSD, disintegration time without disk; t , flow ability.

experimental determination or calculations of various parameters are converted to "r" values by applying specific factors, representing radii of the diagram. The diagram is formed by connecting radius values with linear segment [13], having 0 as a minimum value, 10 as maximum value, and 5 as minimum acceptable value as shown in **Figure 1**. The resultant diagram indicates suitability of the material to be compressed by direct compression.

3.3 Calculation of various indices

Optimum mechanical strength, disintegration behavior, and rheological characteristics of powder are estimated on the basis of the following indices [11, 14] calculated using "r" values of the basic parameters.

3.3.1 Parametric index

Parametric index is the ratio of number of parameters having "r" values equal to or greater than 5 to the total number of parameters determined during the study. Parametric index was calculated using the following equation:

$$I.P = \frac{No.P \geq 5}{No.Pt} \tag{8}$$

where I.P. is the parametric index, No. P ≥ 5 is the number of parameters with "r" values equal to or more than 5, and No. Pt is the total number of parameters determined.

Acceptability limit corresponds to a score of 5.

3.3.2 Parameter profile index

Parameter profile index is the average of "r" values of all the parameters determined in the study, and its acceptable limit corresponds to a score of 5.

IPP = Average of "r" value of all parameters

3.3.3 Good compressibility and bucco-dispersibility index

Good compressibility and bucco-dispersibility index (IGCB) is the product of parameter profile index and reliability factor:

$$I.G.C.B = I.P.P \text{ x} f \tag{9}$$

where f is the reliability factor.

Inclusion of more parameters in the study will increase reliability factor. Its values are as follows:

- For infinite number of parameters, $f = 1$ (maximum value)
- For 15 parameters, $f = 0.971$
- For 12 parameters, $f = 0.952$
- For 08 parameters, $f = 0.900$

3.4 Determination of acceptable limit values for each parameter of SeDeM-ODT expert system

Certain limit values are set for each parameter included in SeDeM-ODT expert system on the basis of experimental results and values described in the *Handbook of Pharmaceutical Excipients* [24]. The rationale for establishing limit values for each parameter is given below.

Limit values of bulk density, tapped density, inter-particle porosity, and Carr's index are calculated from the extreme values of these parameters given in the *Handbook of Pharmaceutical Excipients* and official monograph.

Limit of Icd is obtained by compressing powder into tablet under maximum compression force to get tablets without capping. Maximum hardness at which tablets are compressed without any capping is taken as upper limit, while 0 is taken as lower limit. 0 shows that powder cannot be compressed into tablet.

Limits for angle of repose, IH, and powder flow were set as per official monograph. **Table 2** shows correlation of flow characteristics of powder to various rheological parameters on the basis of the USP [20].

Limits for hygroscopicity are based upon the *Handbook of Pharmaceutical Excipients* [24]. As per published literature [25–27], rheological and compression

Flow characteristics	Carr's index	Hausner ratio	Angle of repose
Excellent	≤10	1.00–1.11	25–30
Good	11–15	1.12–1.18	31–35
Fair—aid not needed	16–20	1.19–1.25	36–40
Passable—may hang up	21–25	1.26–1.34	41–45
Poor—must agitate, vibrate	26–31	1.35–1.45	46–55
Very poor	32–37	1.46–1.59	56–65
Very very poor	>38	>1.6	>66

Table 2.
Relationship between flow characteristics and various rheological parameters.

Sieve size (mm)	Fraction	Average diameter of particles of fraction	Corresponding diameter (dm ... dm ± n)	Difference of dm with major fraction
0.355–0.500	Fm + 2	427	dm + 2	271
0.212–0.355	Fm + 1	283	dm + 1	127
0.100–0.212	Fm	156	dm	0
0.050–0.100	Fm −1	75	dm −1	81
<0.050	Fm −2	25	dm −2	131

Table 3.
Particle size distribution for the determination of homogeneity index.

problems are encountered when the ratio of the fine particles exceeds 25% of the formulation.

Size distribution of the particles provided a basis for assigning limit values to homogeneity index. **Table 3** indicates the size of the sieve (in mm), average particle size in each fraction, the difference in average particle size in the fraction between 0.100 and 0.212, and others.

As the sieve range 0.100–0.212 mm falls in the middle of other factions, it corresponds to major fraction. A proportion of 60% in major fraction (Fm) is considered to be the minimum acceptable value, that is, 5. Distribution of particles into other fractions is considered to be Gaussian. Limit of homogeneity index is 0–0.02.

Initially, relative humidity was calculated based on the establishment of three intervals because the percentage relation obtained from the measurement of the humidity of the substance does not follow a linear relation with respect to the correct behavior of the dust. Humidity below 1% makes the powder too dry, and electrostatic charge is induced, which affects the rheology. Furthermore, low humidity percentages do not allow compression of the substance (moisture is necessary for compacting powders). Moreover, more than 3% moisture causes caking, in addition to favoring the adhesion to punches and dyes. Consequently, it was considered that this parameter should present optimal experimental values from 1 to 3%. Nevertheless, experience using the SeDeM diagram has demonstrated no significant variations in the results, so the previous three intervals of relative humidity can be simplified to the calculation of the parameter; thus, finally, the linear criterion of treatment of results is adopted.

4. Practical applications of SeDeM-ODT expert system

The SeDeM/SeDeM-ODT expert system is based on the experimental study and quantitative determination of the characterization parameters of powdered substances, with the aim to determine suitability for producing tablets by direct compression technology. Additionally, this expert system also provides formulations with a minimum number of excipients and reduces the lead time during formulation development [11]. Some of the reported applications of SeDeM-ODT expert system are summarized below:

4.1 Formulation development by direct compression technology

Direct compression is the most preferred technique for tablet manufacturing due to simplicity, material safety, and cost-effectiveness. Direct compression technique cannot be applied for every formulation because of some strict requirements in terms rheological characteristics and compressibility [2, 3]. Intensive experimentation is carried out to get a final powder blend suitable for direct compression. SeDeM-ODT expert system has been applied for characterization of powder to predict its suitability for direct compression. The main advantage of the expert system is to avoid extra experimentation during formulation development, reducing time and cost of formulation development [11]. Various mathematical equations are used for powder characterization, and a data base is developed which facilitates the selection of excipients having desired characteristics, at pre-formulation level.

Johny et al. applied SeDeM expert system in formulation development of orodispersible tablets of ibuprofen by direct compression [11]. They developed formulation after characterization of API (ibuprofen) and 21 disintegrants. Various parameters were determined for all the 21 disintegrants, according to the standard protocols, converted to "r" values by applying specific factors, and presented as SeDeM diagram (**Figure 2**). Deficiencies were found out for each disintegrant and were solved by proper selection of other excipients.

In another study SeDeM expert system was applied for formulation development of effervescent tablets of domperidone by direct compression [28]. During the study SeDeM profile was developed for domperidone, effervescent pair (citric acid, tartaric acid, and sodium bicarbonate), and two diluents (Tablettose-80 and

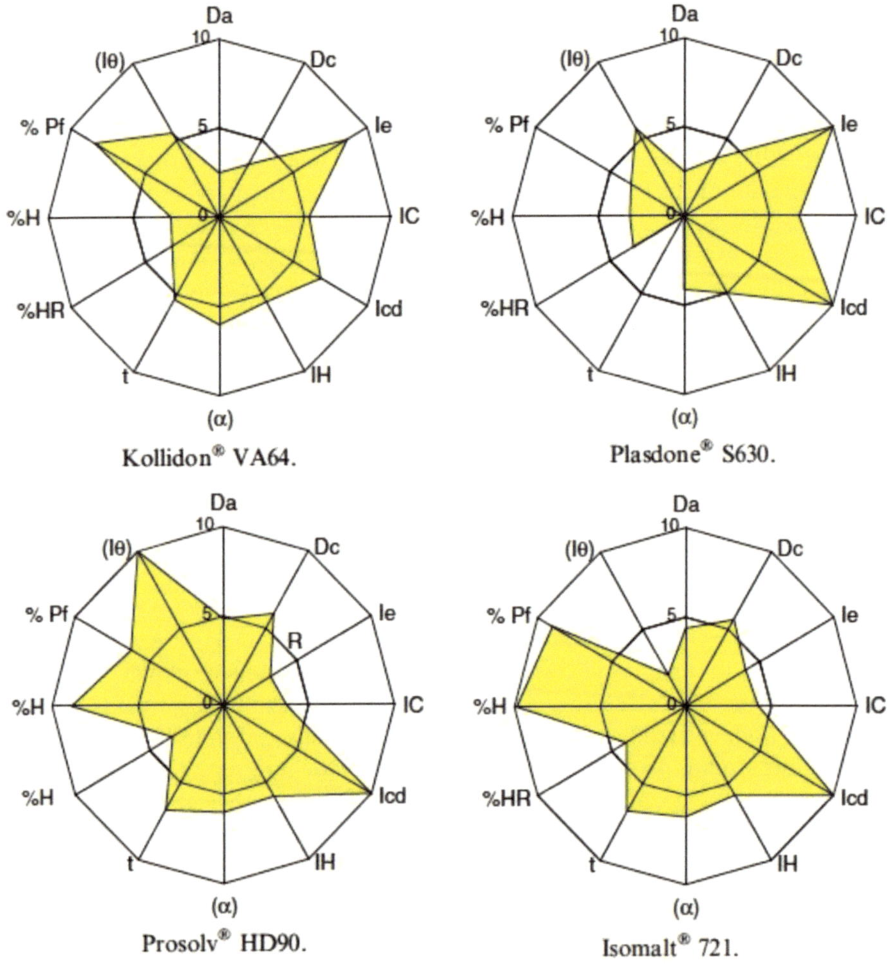

Figure 2.
SeDeM diagram of various disintegrants [11].

microcrystalline cellulose). The model drug, domperidone, was characterized, according to the established procedure, and was found deficient in dimension, compressibility, and flowability/powder flow factors. Index of good compressibility (IGC) value of domperidone was below the acceptable limit. Combination of diluents was used to get a diluent system (**Figure 3**) capable of compensating lower IGC value of domperidone. The developed formulations resulted in tablets fulfilling the official requirements without any stability issue with minimum experimental work.

In a study SeDeM expert system was applied for establishing a design space and determination of critical quality attributes during formulation development of captopril SR matrix by direct compression [29].

Cefuroxime axetil and paracetamol have poor rheological characteristics and compressibility. Inderbir and Pradeep [30] applied the SeDeM expert system for formulation of these two APIs by direct compression. Both the APIs were characterized following standard procedure, and excipients were selected on the basis of mathematical calculations [14].

4.2 Determination of the amount of excipient required for the compression of an API

Josep et al. developed a mathematical Equation [14] for the calculation of the amount of diluent required for the preparation of tablets by direct compression containing glucosamine salt (750 mg). Glucosamine is used in high dose (750 mg/tablet) and presents poor rheological characteristics and compressibility. Six direct compression diluents were characterized according to the SeDeM expert system, and mathematical equation was applied for the calculation of the amount of excipient to compensate the deficiencies. The theoretical model was validated by studying the calculated amounts experimentally.

$$CP = 100 - \left(\frac{RE - R}{RE - RP}\right) \times 100) \quad (10)$$

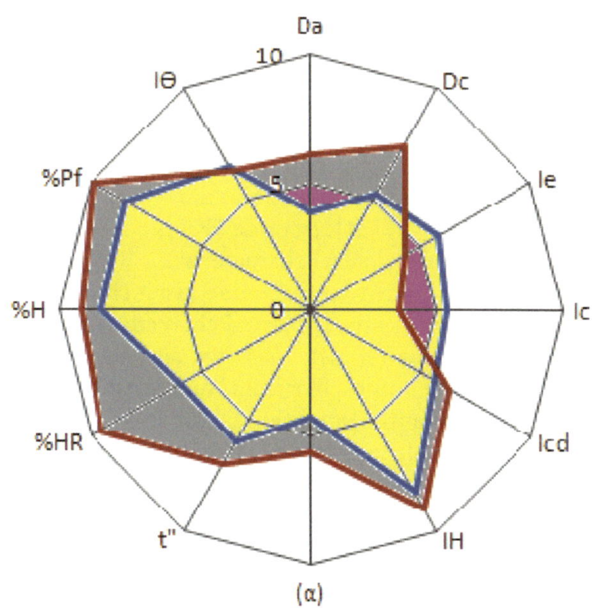

Figure 3.
SeDeM diagram of microcrystalline cellulose and Tablettose-80 [28].

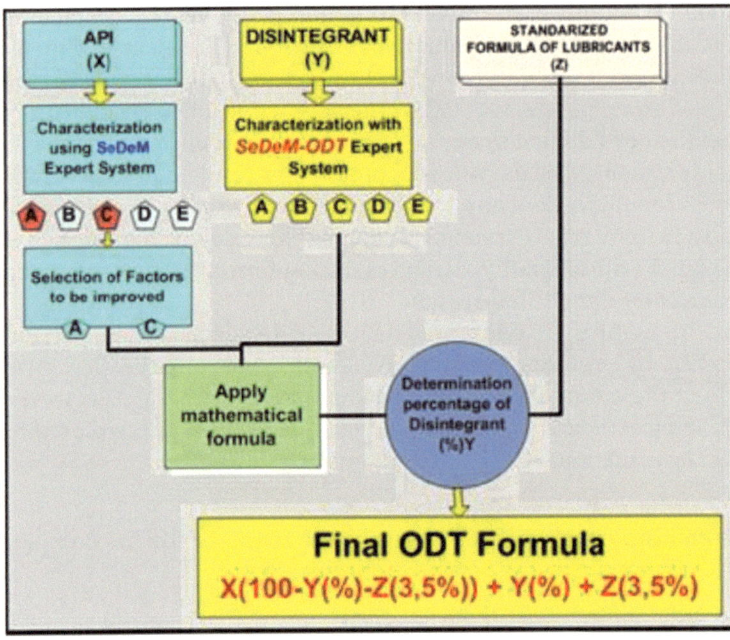

Figure 4.
Strategy proposed by SeDeM expert system to develop orally disintegrating tables [14].

where CP is the % of corrective excipient, RE is the mean-incidence radius value (compressibility) of the corrective excipient, R is the mean-incidence radius value to be obtained in the blend, RP is the mean-incidence radius value (compressibility) of the API to be corrected, and R is the 5 as 5 is the minimum value that is regarded as necessary in order to achieve good compression [14].

Figure 4 presents a strategy for the development of orally disintegrating tablets by direct compression by applying the proposed equation (Eq. (10)).

4.3 Elucidation of the effect of processing on characteristics of powder substances

SeDeM-ODT expert system has been applied for elucidation of the effect of processing parameters on characteristics of powder substance. Amjad et al. applied SeDeM-ODT expert system for predicting the effect of taste masking on the rheological characteristics, mechanical strength, and disintegration behavior of highly water-soluble drug (Itopride HCl) [31]. Itopride HCl is a bitter-tasting, highly water-soluble drug with poor rheological characteristics. Taste of Itopride HCl was masked by water-based wet granulation technique using HPMC as taste masking polymer. Itopride HCl powder was the subjected characterization as per SeDeM-ODT expert system, before and after taste masking, and results were compared (**Figure 5**) to evaluate the effect on rheological characteristics, disintegration behavior, and mechanical strength. Dimension factor and flowability/powder flow factors were below the acceptable limit. Comparison of results before and after taste masking showed that taste masking significantly improved the mechanical strength and rheological characteristics and decreased the disintegration behavior of powder. It was concluded that in order to formulate by direct compression, the formulation will require large amount of disintegrant to overcome increase in mechanical strength after taste masking.

Amjad [32] has applied SeDeM-ODT expert system for the optimization of process variables of roller compaction. He studied ribavirin powder and powder

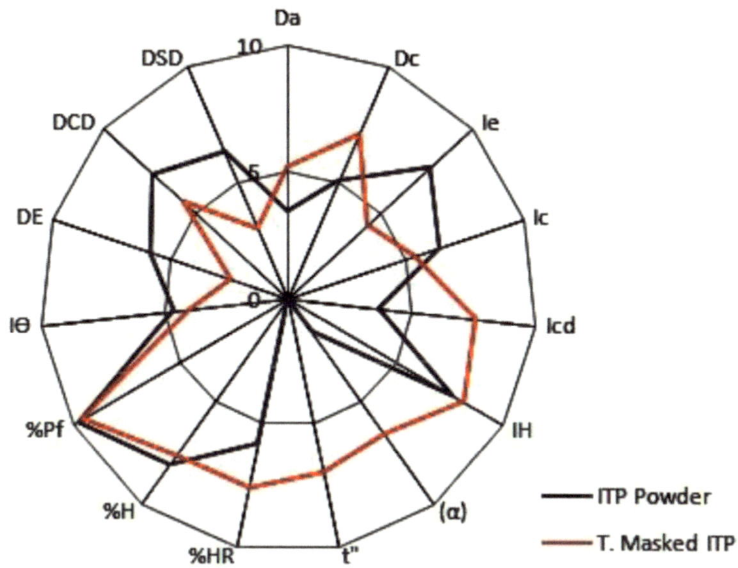

Figure 5.
SeDeM-ODT diagram for Itopride HCl before and after taste masking [33].

blend containing ribavirin and other ingredients included in granule formulation. Powder blend was compacted under varying degree of experimental conditions, and selected in the optimal conditions with better granule characteristics. He claimed that it decreased experimental work and resulted in granules suitable for compression and encapsulation.

4.4 Prediction of behavior of a new pharmaceutical ingredient (APIs and excipients)

SeDeM-ODT expert system has been applied for the determination of suitability of new powdered substances for direct compression. The powder substance may be a new API or excipients which are intended to be used in formulation of compact solid dosage forms.

Sune-Negre et al. used the SeDeM method to characterize an active product ingredient in powder form (API SX-325) and to determine whether it is suitable for direct compression [12], applying the profile to the SeDeM diagram. Twelve parameters were determined for the powdered raw material according to the standard protocols, presented as SeDeM diagram, indicating suitability of the material for direct compression. Findings of the study implied deficient rheological characteristics and poor stability. The product was declared hygroscopic on the basis of SeDeM profile and tended to capture moisture, worsening rheological characteristics and impairing its stability. Various precautionary measures were suggested for prevention of negative effects like drying of the material and tablet preparation in an environment of controlled humidity (relative humidity below 25%).

Sune-Negre et al. applied SeDeM expert system for characterization of 51 directly compressible excipients [12]. On the basis of the results, directly compressible excipients were classified into different groups with different rheological and compressibility capability, and a periodic table of directly compressible excipients was developed, as shown in **Figure 6**.

They showed that the best excipient for direct compression should have an index of good compressibility of 8.832 [12]. SeDeM expert system has been applied for the

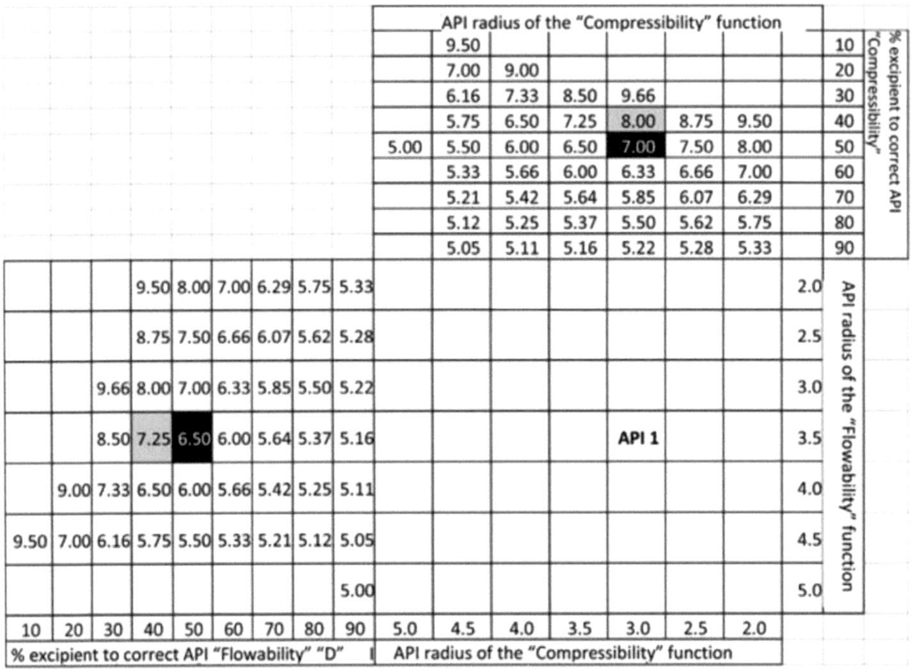

Figure 6.
Correction of "compressibility" and "flowability" of APIs with excipient [12].

determination of reproducibility of various batches of pharmaceutical ingredients (APIs and excipients). Various batches were characterized according to the SeDeM expert system, and reproducibility was estimated on the basis of consistency of the results [13].

4.5 Optimization of powder characteristics

Josep et al. [33] applied SeDeM expert system for the optimization of Hausner ratio and relative humidity. The proposed optimization did not involve any conceptual change in the parameters considered or did a significant change in the results obtained compared with the previous calculation methodology initially established, meaning that the conclusion obtained by applying this method is equivalent [33].

4.6 Quality control of batches of pharmaceutical powders (APIs and excipients) prepared by the same procedure

SeDeM expert system can be applied for the determination of reproducibility of manufacturing process of pharmaceutical powder substance (API and excipients). By establishing specifications for different parameters as per SeDeM-ODT expert system, variation among different batches of a product produced by the same manufacturing process can be determined [34]. **Figure 7** shows the SeDeM diagram of different batches of glucosamine sulfate, prepared by the same manufacturing procedure.

4.7 Differentiation of excipients with same functionality

SeDeM-ODT expert system has been applied for the differentiation of excipients having the same chemical nature and function on the basis of physical

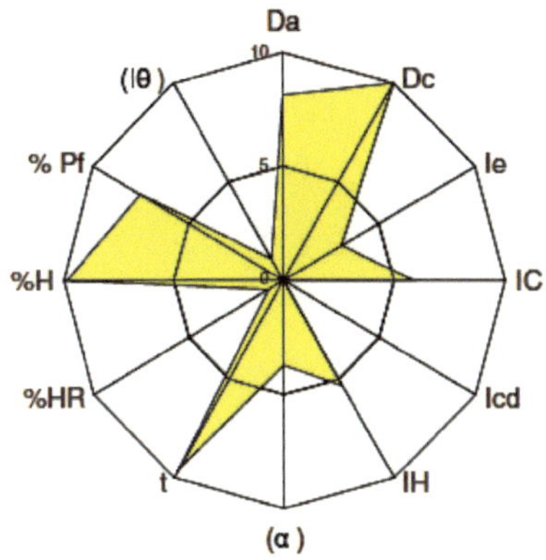

SeDeM Diagram for glucosamine salt batch 4/0008.

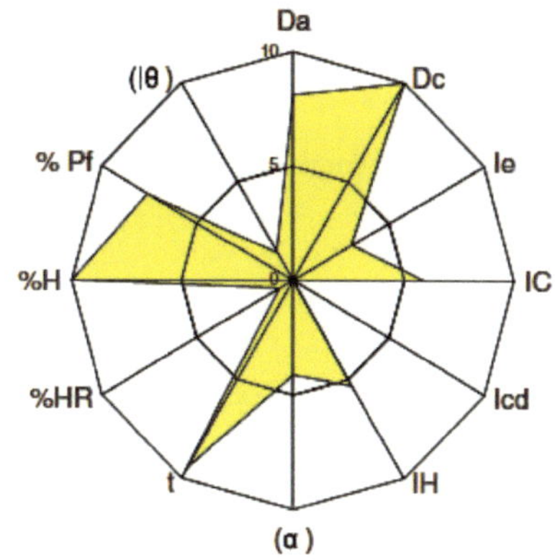

SeDeM Diagram for glucosamine salt batch 4/0011.

Figure 7.
SeDeM diagram of different batches of glucosamine sulfate [35].

characteristics [35]. For example, various disintegrants and diluents were characterized on the basis of the expert system [13], and the suitable one is selected for a particular formulation. Various parameters are determined according to the SeDeM expert system, deficiencies are defined, and an adequate substance can be selected to get a final blend suitable for direct compression. In a study [11] several lactose were characterized and differentiated on the basis of SeDeM expert system.

5. Conclusions

SeDeM-ODT expert system is a novel tool for the characterization of powder substances on the basis of their physical parameters. The system has been successfully applied for the determination of rheological characteristics, mechanical strength, and disintegration behavior of pharmaceutical powders (APIs and excipients) and determination of suitability for direct compression and bucco-dispersibility. SeDeM-ODT expert system facilitates the process of excipient selection and calculation of their relative proportion in oral solid dosage form. A data base can be developed for various excipients which will help in the selection of excipients having desired characteristics. It avoids extra experimentation for optimization of various characteristics of powder blend, reducing the cost and time span of formulation development process. This method characterizes the individual components of a formulation and applies a mathematical analysis to determine the exact amount of each ingredient in the final formulation. This innovative tool is consistent with the current requirements of regulatory health authorities such as the FDA and ICH, whereas data generated on the basis of the system can contribute to the concept of Quality by Design. SeDeM-ODT expert system has certain limitations, and misleading results are possible in certain cases. Suitability of material for direct compression is decided on the basis of index of good compressibility value, which is based on "r" values of the individual parameters. Substances having high "r" values will raise IGC value and vice versa. So suitability of a material should not be judged on the basis of IGC value. The incidence factor value should be considered, and outliers in "r" value of the individual factors, if present, should be properly addressed. Overall SeDeM-ODT expert system is an unmatched tool for material characterization at pre-formulation level and has significantly decreased time span and cost of pharmaceutical formulation development process.

Acknowledgements

We are thankful to Mr. Zahir Rahman (B. Pharm, M. Phil) Plant Manager, Ferozsons Laboratories Ltd., Nowshera, for his technical input and moral support.

Conflict of interest

The authors claim no conflict of interest.

Author details

Amjad Khan[1*], Shabnam Nazir[1], Hamad S. Alyami[2] and Aman Ullah[3]

1 Department of Pharmacy, Kohat University of Science and Technology (KUST), Kohat, Pakistan

2 Department of Pharmaceutics, Najran University, Najran, Saudi Arabia

3 Department of Pharmacy, Abasyn University, Peshawar, Pakistan

*Address all correspondence to: dr.amjad@kust.edu.pk

References

[1] Podczeck F, Al-Muti E. The tensile strength of bilayered tablets made from different grades of microcrystalline cellulose. European Journal of Pharmaceutical Sciences. 2010;**41**: 483-488

[2] Kasa P, Bajdik J, Zsigmond Z, Pintye-Hodi K. Study of the compaction behaviour and compressibility of binary mixtures of some pharmaceutical excipients during direct compression. Chemical Engineering and Technology. 2009;**48**:859-863

[3] Mazel V, Diarra H, Busignies V, Tchoreloff P. Comparison of different failure tests for pharmaceutical tablets: Applicability of the Drucker–Prager failure criterion. International Journal of Pharmaceutics. 2014;**470**:63-69

[4] Ofori-Kwakye K, Mfoafo KA, Kipo SL, Kuntworbe N, Boakye-Gyasi ME. Development and evaluation of natural gum-based extended release matrix tablets of two model drugs of different water solubilities by direct compression. SPJ. 2015;**24**(1):82-91. DOI: 10.1016/j.jsps.2015.03.005

[5] Mangwandi C, Zainal NA, Tao LJ, Glocheux Y, Albadarin AB. Investigation of influence of process variables on mechanical strength, size and homogeneity of pharmaceutical granules produced by fluidised hot melt granulation. Powder Technology. 2015; **272**:173-180

[6] Akseli I, Ladyzhynsky N, Katz J, He X. Development of predictive tools to assess capping tendency of tablet formulations. Powder Technology. 2013;**236**:139-148

[7] Sen M, Rogers A, Singh R, Chaudhury A, John J, Ierapetritou MG, et al. Flowsheet optimization of an integrated continuous purification processing pharmaceutical manufacturing operation. Chemical Engineering Science. 2013;**102**:56-66

[8] Vaidya MP, Avachat AM. Investigation of the impact of insoluble diluents on the compression and release properties of matrix based sustained release tablets. Powder Technology. 2011;**214**:375-381

[9] Desai PM, Liew CV, Heng PWS. Assessment of disintegration of rapidly disintegrating tablets by a visiometric liquid jet-mediated disintegration apparatus. International Journal of Pharmaceutics. 2013;**442**:65-73

[10] Singh R, Sahay A, Muzzio F, Ierapetritou M, Ramachandran R. A systematic framework for onsite design and implementation of a control system in a continuous tablet manufacturing process. Computers and Chemical Engineering. 2014;**66**:186-200. DOI: 10.1016/j.compchemeng.2014.02.029

[11] Sune-Negre JM, Perez-Lozano P, Minarro M, Roig M, Fuster R, Hernandez C, et al. Application of the SeDeM diagram and a new mathematical equation in the design of direct compression tablet formulation. European Journal of Pharmaceutics and Biopharmaceutics. 2008;**69**: 1029-1039

[12] Suñé-Negre JM, Roig M, Fuster R, Hernández C, Ruhí R, García-Montoya E, et al. New classification of directly compressible (DC) excipients in function of the SeDeM diagram expert system. International Journal of Pharmaceutics. 2014;**470**:15-27

[13] Aguilar-Díaz JE, García-Montoya E, Pérez-Lozano P, Suñe-Negre JM, Miñarro M, Ticó JR. The use of the SeDeM diagram expert system to determine the suitability of diluents–disintegrants for direct compression and their use in formulation of ODT.

European Journal of Pharmaceutics and Biopharmaceutics. 2009;**73**:414-423

[14] Aguilar-Díaz JE, García-Montoya E, Suñe-Negre JM, Pérez-Lozano P, Miñarro M, Ticó JR. Predicting orally disintegrating tablets formulations of ibuprofen tablets: An application of the new SeDeM-ODT expert system. European Journal of Pharmaceutics and Biopharmaceutics. 2012;**80**:638-648

[15] ICH Topic Q8, Note for Guidance on Pharmaceutical Development, EMEA/CHMP/167068/2004; 2004

[16] The United States Pharmacopeia (USP-38/NF-33), General Chapter: <616> Bulk density and tapped density of powder page No. 420. Rockville, MD: United States Pharmacopeial Convention, Inc.; 2015

[17] Font Q, Medicamenta P. Guía teórico práctica para farmacéuticos y medicos. 6th ed. Vol. 1. Barcelona: Labor Ed.; 1962. pp. 340-341

[18] The United States Pharmacopeia (USP-38/NF-33), General Chapter: <1174> Powder Flow, page No. 1326. Rockville, MD: United States Pharmacopeial Convention, Inc.; 2015

[19] European Pharmacopoeia. 5th ed. Vol. 5.0. Strasbourg (France): Council of Europe; 2005. p. 2.9.16

[20] The United States Pharmacopeia (USP-38/NF-33), General Chapter: <731> Loss on drying, page no. 513. Rockville, MD: United States Pharmacopeial Convention, Inc.; 2015

[21] European Pharmacopoeia. 5th ed. Vol. 5.1. Strasbourg (France): Council of Europe; 2005. p. 2.9.38

[22] European Pharmacopoeia. 7th ed. Vol. 7.2. Strasbourg (France): General Manograph <0478> Council of Europe; 2011

[23] The United States Pharmacopeia (USP-38/NF-33), General Chapter: < 701> Disintegration, page no. 483. Rockville, MD: United States Pharmacopeial Convention, Inc.; 2015

[24] Rowe RC, Sheshkey PJ, Quinn ME. Handbook of Pharmaceutical Excipients. 6th ed. London: Pharmaceutical press; 2009

[25] Alves-Silva I, Sá-Barreto LCL, Lima EM. M.S.S. Cunha-Filho pre formulation studies of itraconazole associated with benznidazole and pharmaceutical excipients. Thermochimica Acta. 2014;**575**:29-33

[26] Pilcer G, Amighi K. Formulation strategy and use of excipients in pulmonary drug delivery. International Journal of Pharmaceutics. 2010;**392**:1-19

[27] Jannin V, Rodier JD, Musakhanian J. Polyoxylglycerides and glycerides: Effects of manufacturing parameters on API stability, excipient functionality and processing. International Journal of Pharmaceutics. 2014;**466**:109-121

[28] Khan A, Iqbal Z, Rehman Z, Nasir F, Khan A, Ismail M, et al. Application of SeDeM expert system in formulation development of effervescent tablets by direct compression. SPJ. 2014;**22**: 433-444

[29] Millán SD, Suné-Negre JM, Pérez-Lozano P, Sarrate R, Fàbregas A, Carrillo C, et al. The use of the SeDeM diagram expert system for the formulation of captopril SR matrix tablets by direct compression. International Journal of Pharmaceutics. 2014;**461**:38-45

[30] Singh I, Kumar P. Pre formulation studies for direct compression suitability of cefuroxime and Paracetamol: A graphical presentation using SeDeM diagram. Acta Poloniae Pharmaceutica. 2012;**69**:87-93

[31] Khan A, Iqbal Z, Ibrahim M, Nasir F, Ullah Z. Prediction of the effect of taste masking on disintegration behavior, mechanical strength and rheological characteristics of highly water soluble drug (Itopride HCl); an application of SeDeM-ODT experts system. Powder Technology. 2015;**284**: 411-417. DOI: 10.1016/j.powtec.2015.06.062

[32] Khan A. Optimization of the process variables of roller compaction, on the basis of granules characteristics (flow, mechanical strength, and disintegration behavior): An application of SeDeM-ODT expert system. Drug Development and Industrial Pharmacy. 2019;**45**(9): 1537-1546. DOI: 10.1080/03639045.2019.1634094

[33] Suñé-Negre JM, Pérez-Lozano P, Roig M, Fuster R, Hernández C, Ruhí R, et al. Optimization of parameters of the SeDeM diagram expert system: Hausner index (IH) and relative humidity (%RH). European Journal of Pharmaceutics and Biopharmaceutics. 2011;**79**:464-472

[34] Perez P, Sune-Negre JM, Minarro M, Roig M, Fuster R, Garcıa-Montoya E, et al. A new expert systems (SeDeM diagram) for control batch powder formulation and pre formulation drug products. European Journal of Pharmaceutics and Biopharmaceutics. 2006;**64**:351-359

[35] Aguilar-Diaz JE, Garcıa-Montoya E, Perez-Lozano P, Sune-Negre JM, Minarro M, Tico JR. SeDeM expert system a new innovator tool to develop pharmaceutical forms. Drug Development and Industrial Pharmacy. 2014;**40**(2):222-236. DOI: 10.3109/03639045.2012.756007

Section 3

Advanced Functional Materials for Concrete and Ceramics

Chapter 12

Self-Healing Concrete and Cementitious Materials

Tanvir Qureshi and Abir Al-Tabbaa

Abstract

Concrete is one of the most used materials in the world with robust applications and increasing demand. Despite considerable advancement in concrete and cementitious materials over last centuries, infrastructure built in the present world with these materials, such as dams, roads, bridges, tunnels and buildings requires intensive repair and maintenance throughout its design life. Self-healing concrete and cementitious materials, which have the ability to recover after initial damage, have the potential to address these challenges. Self-healing technology in concrete and cementitious materials can mitigate the unnecessary repair and maintenance of built infrastructure as well as overall CO_2 emission due to cement production. This chapter provides the state-of-the-art of self-healing concrete and cementitious materials, mainly focusing on autogenic or intrinsic self-healing using fibre, shrinkable polymers, minerals and supplementary cementitious materials, and autonomic self-healing using non-traditional concrete materials such as microscale to macroscale capsule as well as vascular systems with polymeric, mineral and bacterial agents.

Keywords: concrete, autogenic self-healing, autonomic self-healing, healing process, mineral, polymer, microstructure

1. Introduction

Concrete is the most used and efficient construction material in the world. It is durable, can resist high compressive stress, is cheaper than most of the construction materials and can be moulded in a wide variety of shapes. Despite that concrete cracks due to its weakness in tension, shrinkage, fatigue loading, and under the action of environmental conditions. These microcracks can reduce concrete's toughness, increase permeability, which can ultimately lead to the reduction of concrete's structural integrity, durability and life span. Self-healing concrete in that context offers an actual solution.

Any process whereby concrete recovers its performance after initial damage is termed self-healing in concrete [1]. A typical self-healing in cementitious materials is presented in **Figure 1**. The concept of concrete self-healing has evolved from that found in biological life forms, that is, plants and animals that naturally exhibit self-healing performance when any damage appears.

According to Schlangen and Joseph (Cited in [2]), the strength of concrete gradually decreases when the first repair is required. Also, commonly, a second repair is required in concrete after 10–15 years. However, the initial repair period

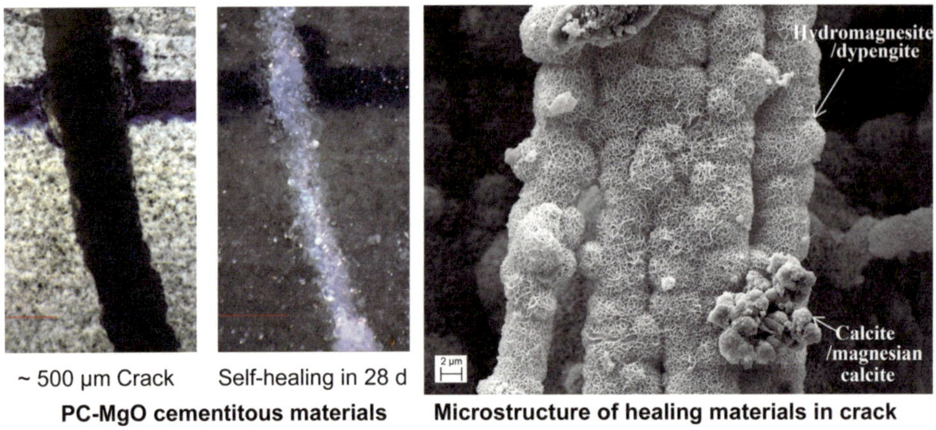

Figure 1.
Example of self-healing concrete and cementitious systems (Adopted from [3]).

can be extended considerably with the application of self-healing technology in concrete. Self-healing leads to a longer material lifetime, and it involves no repair and maintenance costs.

This chapter presents the state-of-the-art of self-healing in concrete and cement-based materials. It discusses advancements in this field and limitations. The next section (Section 2) presents the concept of self-healing in concrete and measurement techniques. Then the chapter describes major developments in different self-healing concrete field.

2. Self-healing concrete systems and measurement techniques

The self-healing system in concrete is principally divided into two types, autogenic and autonomic [1]. Autogenic self-healing in concrete is an intrinsic material-healing property wherein the self-healing process initiates from the generic materials present. For example, cementitious materials exhibit a self-repairing ability due to the rehydration property of unhydrated cement remaining on the crack surface. In contrast, a self-healing process that involves the incorporation of material components that are not traditionally used in the concrete is termed autonomic self-healing [1].

Figure 2 presents the developed autogenic and autonomic self-healing systems. One of the principal causes of autogenic self-healing is the hydration of unhydrated cement remaining in the matrix. Then again, the volume of healing products formed in this process is limited. Hence, the autogenic self-healing is effective within the crack width up to 50–150 μm [4]. Autogenic self-healing performance is higher in early age due to high content of unhydrated cement, and parameters such as compressive stress [5] to restrict crack and wet-dry cycles [6] can increase the healing performance. Autogenic healing performance can also be enhanced using fibres to restrict crack opening and the use of superplasticizer in engineered cementitious composite (ECC) to reduce w/c ratio [6]. Cardiff University research group introduced polyethylene terephthalate (PET) tendons [7], a shrinkable polymer activated with a heating system inside the concrete structural element to compress and close the crack enhancing the autogenous healing process. Considerable enhancement in healing performance is also possible to achieve using optimum supplementary cementitious materials (SCMs) and smart expansive minerals [3, 8–22]. Autonomic self-healing in concrete, in contrast to the autogenous healing

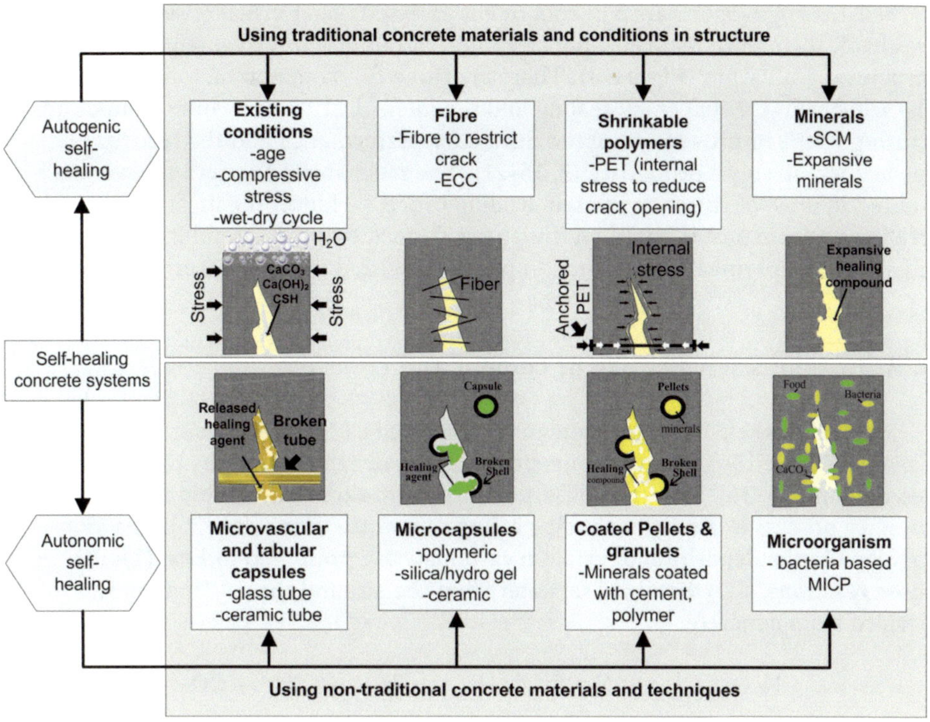

Figure 2.
Self-healing concrete systems.

process, requires the release of the healing agent from reserved encapsulation or a continuous vascular network. Common encapsulating shell materials are glass [23, 24] and polymers [1, 25, 26]. Healing agents in autonomic self-healing are epoxy resins, cyanoacrylates (super glues), alkali-silica solutions [23, 24, 27, 28], methyl methacrylate [24, 28], expansive minerals [16, 29], hydrogel [30] and bacteria-based microorganisms [31–33].

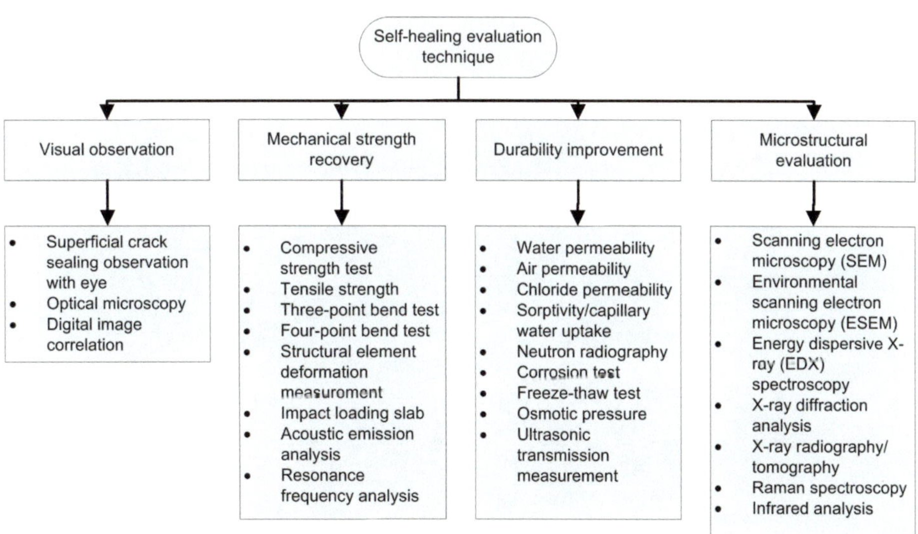

Figure 3.
Self-healing performance in concrete measurement techniques.

Self-healing performance in concrete is assessed using visual observation, mechanical strength recovery, permeability, durability improvement and micro-structural evaluation (**Figure 3**). There are three fundamental factors in evaluating the self-healing: visual crack sealing and the identification of healing compounds causing it, the improvement of the durability performance and the recovery of mechanical strength properties [3, 15–21]. The mechanical strength recovery is limited in most of the concrete self-healing process. Hence, the most reliable self-healing performance is based on the physical crack closure, durability improvement, that is, permeability reduction parameters, and microstructural evaluations.

3. Autogenous self-healing of cement and concrete

Autogenous self-healing in cement was spotted early in the twentieth century by Lauer and Slate [34], and the concept was gradually established by different researchers [35, 36]. The crystallisation of calcium carbonate within the crack is the primary process in autogenous self-healing of matured concrete [35]. Reactions involved in the deposition of calcium carbonate are presented in Eqs. (1)–(3). In those reactions, CO_2 dissolved in water from the air, and the calcium ion Ca^{2+} is derived from concrete.

$$H_2O + CO_2 \leftrightarrow H_2CO_3 \leftrightarrow H^+ + HCO_3^- \leftrightarrow 2H^+ + CO_3^{2-} \quad (1)$$

$$Ca^{2+} + CO_3^{2-} \leftrightarrow CaCO_3 \left(pH_{WATER} > 8\right) \quad (2)$$

$$Ca^{2+} + HCO_3^- \leftrightarrow CaCO_3 + H^+ \left(7.5 < pH_{water} < 8\right) \quad (3)$$

Reasons for autogenous self-healing proposed by different researchers [36] are: (i) Further reaction of the unhydrated cement, (ii) expansion of the concrete in the crack flanks, (iii) crystallisation of calcium carbonate, (iv) closing of the cracks by fine particles existing in the water and (v) closing of the cracks by spilling off loose concrete particles resulting from the cracking. This five action model is schematically presented in **Figure 4**.

The understanding and improvement of autogenous self-healing have developed in four major directions (**Figure 2**). These are: (i) manipulation of existing

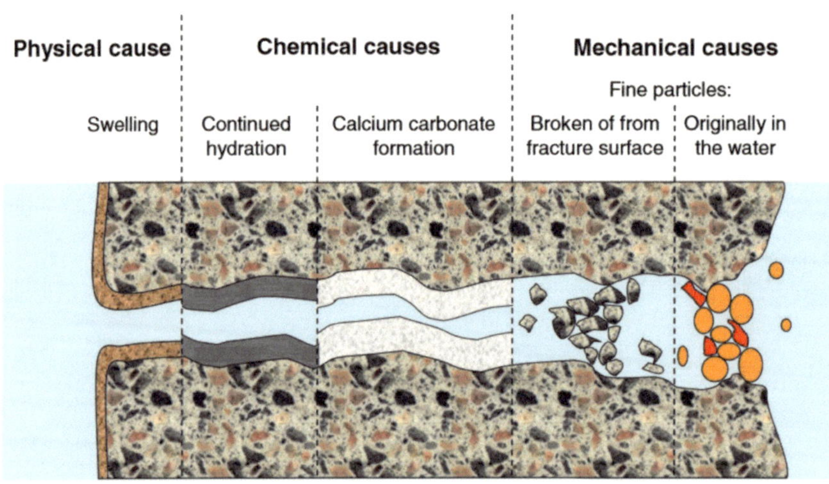

Figure 4.
A model of five steps taking place within three processes, physical, chemical and mechanical (Reproduced from [1]).

conditions, such as age, compressive stress and curing condition (e.g. wet-dry cycle); (ii) fibres to restrict cracks (e.g. ECC); (iii) shrinkable polymers to initiate internal stress after cracking to shrink the cracks and (iv) cement-compatible mineral additives.

3.1 Existing condition influence in autogenous self-healing

Autogenous self-healing of concrete is significantly influenced by its age, internal stress and curing conditions. Early age concrete naturally heals rapidly due to autogenous healing. Concrete prisms with cracks up to 50 μm were autogenously healed under 0.1, 1 and 2 Mpa compressive stresses [5] (**Figure 5a**). The crack face comes into contact by the impelled compressive stress. Hence, the concrete specimens cured under any amount of compressive stress healed much better than specimens cured under no compression stress (**Figure 5b**). Only a specific amount of compression is required to keep the crack faces in contact. Samples that are submerged in water during curing recovered their strength. In contrast, specimens stored in 95% RH for 3 months did not heal at all. This is due to insufficient hydration in the high humid condition, which is not enough to trigger the healing process.

3.2 Fibre action in autogenous self-healing

Fibres can restrict the propagation of crack width, and smaller crack width is favourable for enhanced autogenous healing in concrete. Fibre is a common feature in Fibre-Reinforced Composite Concrete (FRCC) and ECC. Randomly distributed fibres can bridge over cracks, which can decrease the crack width and block the migration of aggressive agents (e.g. chloride ions and CO_2) [6, 37]. These properties improve the autogenous self-healing capacity of concrete and composites. A series of wetting and drying cycles on ECC was carried out by [6] to mimic self-healing performance in outdoor environments. Through self-healing, crack-damaged ECC recovered 76–100% of its initial resonant frequency value and attained a distinct rebound in stiffness. The tensile strain capacity after self-healing recovered close to 100% that of virgin specimens without any preloading. This was found even for the specimens deliberately pre-damaged with microcracks by loading up to 3% tensile strain. It takes about four to five wet-dry cycles to attain the full benefit of self-healing. The use of high cement content, low water-to-cement ratio also increases the autogenous self-healing capacity of ECC. However, FRCC, ECC and

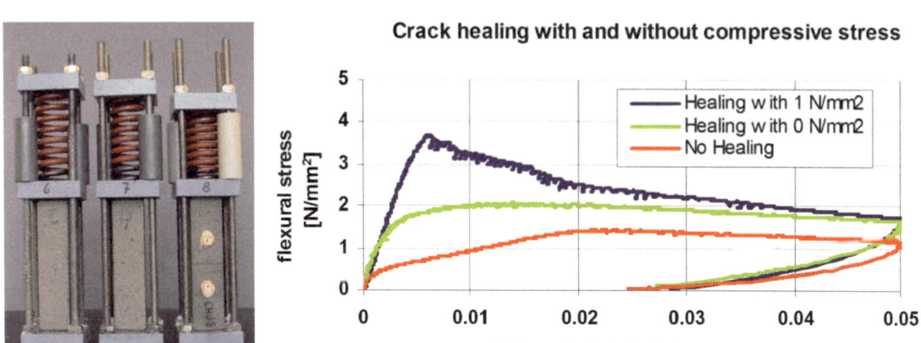

(a) (b)

Figure 5.
(a) Application of compression and (b) stress-displacement curve of specimens after healing with and without applied compressive stress. (Both figures reproduced from [5]).

HFRCC are costly and maintaining homogeneity of fibres in the matrix for consistent self-healing is challenging.

3.3 Shrinkable polymers action in autogenous self-healing

The shrinkable polymers such as PET can shrink when activated by heating in a specific condition. This shrinkage stress can be used for pre-stressing the concrete thus bringing crack-tip closure for efficient healing. Cardiff University self-healing research team is working with the original crack-closure system for cementitious materials using shrinkable polymer tendons [7]. The system involves the incorporation of unbonded pre-oriented polymer tendons in cementitious beams (**Figure 6**). Crack closure is achieved by thermally activating the shrinkage mechanism of the restrained polymer tendons (PTs) after the cement-based material has undergone initial curing. Upon activation, the polymer tendon completely closes the preformed macrocracks and imparts significant stress across the crack faces. This enhances the autogenous self-healing process in concrete.

3.4 Mineral admixture in autogenous self-healing

Supplementary cementitious materials (SCMs) and expansive minerals compatible with cement can improve the self-healing capacity of concrete. Depending on minerals, it can serve either or both functionalities, that is, to *remain considerably un-hydrated* after the initial mixing stage, and to *produce compatible expansive hydrated compounds* that can heal cracks [19]. Both these functionalities contribute to the autogenous healing process. A summary of mineral additives use for self-healing is illustrated in **Table 1**. SCMs such as fly ash, silica fumes and blast-furnace slag, and expansive minerals such as MgO, calcium sulphoaluminate (CSA), lime, bentonite clay and crystalline additive (CA), have been mostly used for improving the concrete autogenous self-healing performance.

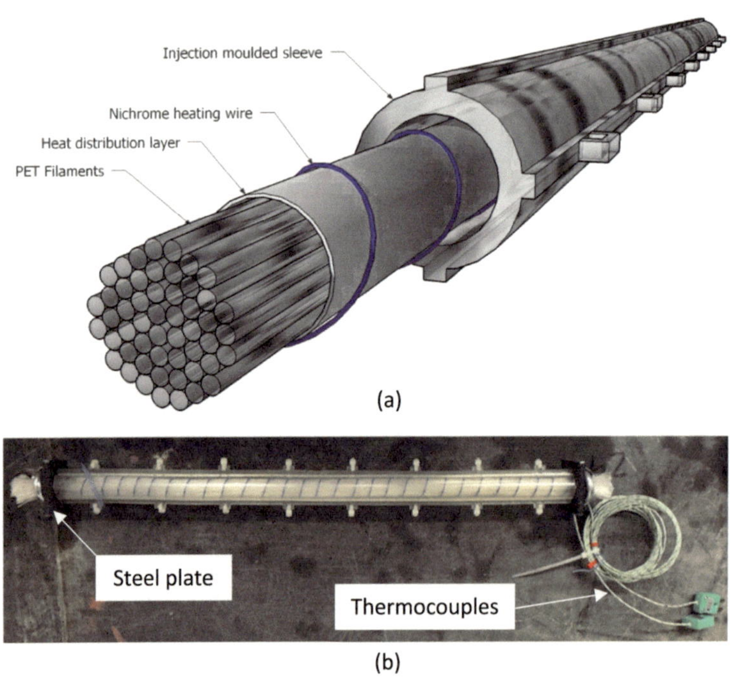

Figure 6.
(a) Schematic of shape memory PET polymer tendon, and (b) photo of the setup (Both reproduced from [7]).

Minerals	Composition	Damage type	Curing condition	Performance (healed crack width in time etc.)	Source
CSA, [a]H, [b]A, [c]L, Mont.	Up to 10% (concrete)	3 PB, mechanical	Water	160–220 μm in 33d Calcite, CASH	[8]
CSA	4.44 and 15.24% of cement (concrete)	Tension force	Still/ continuous flow water	Reduced flow in 100 μm cracks, continuous flow is efficient	[38]
CSA, CA, [a]H, [b]A, [c]L, Mont.	PC with 10% CSA and 1.5% CA	Sp. tensile test	Water	100–400 μm in 56 d Calcite	[9]
Silica, [d]CEA, bentonite, CA	8% individual combination up to 14%	Compression, sp. tensile	Water, wet-dry, air, freeze–thaw	220 μm in 2 weeks [d]CEA (individually efficient) silica, bent., CA (combination is efficient)	[11]
FA, SF, CA	OPC, OPC + 30%FA, OPC + 10%SF, OPC + 1%CA	Splitting tensile test	Water	50 μm in 12d larger cracks heal efficiently with SF	[39]
FA	15–20% with PC (paste)	Shrinkage microcracks	Water	Meso-macro pores at 91, 182 and 364 d	[40]
FA	5–15% wt. of sand (concrete)	Freeze–thaw	Water	Improve [e]DME over 90% in 28d	[41]
BFS	OPC + 50% BFS	Mechanical	Water	Product formation is three times faster for CEM I	[42]
FA, slag	30–40% of cement (mortar)	Shrinkage	Water	Improvement in compressive strength	[43]
[c]L, slag, FA	30, 50% FA; 50, 75, 85% slag (paste/ mortar)	3 PB, mechanical	Water	200 μm in 42d	[12]
Slag	66% of cement (paste)	Sliced, mechanical	$Ca(OH)_2$ solution	60% of 10 μm in 240 h C-S-H, ettringite, hydrogenate etc.	[44]
Bentonite	Nanoclay in mortar as internal water reservoir	Mechanical	Water	Enhanced hydration for self-healing	[45]
Bentonite, slag, [c]L	2% PVA by vol. Length = 8 mm, dia = 40 μm	4 PB	Water, wet-dry cycle, air	Nanoclay improves the reloading deflection capacity	[46]
Quicklime, FA	(3%) on fly ash-PC cement pastes	Mechanical	Water	Increased SiO_2 solubility extra $Ca(OH)_2$	[14]
Expanded clay LWAs	Na-MFP and PC coated (mortar)	Mechanical	Water	Absorption decrease sodium, phosphorous and fluoride, CH	[47]
CSA	PVA coated, up to 10% by wt. of cement (mortar, 1:3)	3 PB	Water	<100 μm in 11d, 100–200 μm in 14d, >200 μm in 16d	[48]
CA: cement + sand + microsilica	1–2% of cement	4 PB	Water, open air	60% cracks sealed under open air condition	[49]

Minerals	Composition	Damage type	Curing condition	Performance (healed crack width in time etc.)	Source
MgO	4–12% of cement	Drying shrinkage, 3 PB	Water	<500 μm in 28d durability improved	[3]

[a]H = hauyne.

[b]A = anhydrite.

[c]L = lime/limestone powder.

[d]CEA = chemical expansive agent.

3/4 PB = Three/four-point bending, OPC = ordinary Portland cement, CASH = calcium aluminosilicate hydrate, CSA = calcium sulphoaluminate, CA = crystalline additive, FA = fly ash, SF = silica fume, [e]DME = dynamic modulus of elasticity, LWAs = lightweight aggregates, Na-MFP = sodium mono fluorophosphate (Na_2FPO_3, Na-MFP)

Table 1.
Advancement in autogenous self-healing of cementitious materials with mineral additives (Adopted from [17]).

3.4.1 SCMs to enhance autogenous self-healing

Fly ash (FA) and silica fume (SF) and blast furnace slag (BFS) are mostly used as SCMs in the OPC system to improve concrete self-healing performance [12, 13, 37–43].

The substitution of FA 15–20% in OPC paste system has increased the volume of C-S-H gel and reduced meso-macropores, increasing the autogenous self-healing performance [40]. Watanabe et al. [41] replaced about 5–15% wt. of sand with FA in concrete and found a better dynamic modulus of elasticity recovery at 5% replacement and improving trend at 15% under the non-destructive ultrasonic test method. While freezing and thawing decreased dynamic modulus to 80% of the initial state, curing in water recovered it to over 93–98% after 28 days.

FA and SF, and a crystalline additive (CA) mineral were used for improving the self-healing performance of concrete [39]. CA was composed of 35.58% CaO, 16.81% SiO_2, 15.22% Na_2O, 1.98% Fe_2O_3, 1.93% Al_2O_3 and 1.29% MgO. Four different mixes (OPC, OPC + 30%FA, OPC + 10%SF, and OPC + 1%CA) were compared. Larger cracks (0.05–0.30 mm) healed better with SF additives. Microcracks in the range of 0–0.05 mm in CA additive mixes completely healed within 12 days.

The blast furnace slag (BFS) was used individually and in combination with FA and other minerals for improving self-healing properties. Fibre-reinforced cement composition with a local waste BFS and limestone powder (LP) in a mix proportion of 1:1.2:2 (C:BFS:LP), 0.5 w/b-ratio and 0.018% total mass of superplasticizer demonstrated improved self-healing performance [13]. The specimens cured under water recovered 65–105% deflection capacity compared to virgin specimens, while specimens cured in the air recovered only 40–60%. Small 25-μm cracks were healed efficiently, while larger cracks such as 60 μm were not healed completely. A higher proportions of BSF (50%) substitution in OPC decreases the formation of the healing material at an early age, which alters after 22 days [42]. However, optimum self-healing ability for the mixing content of slag and FA were 30 and 40%, respectively [43].

A considerable proportion (up to 70% of total weight) of slag and two classes fly ash (FA) were used as SCMs in ECC for improving autogenous self-healing performance [50]. Microscopic observation showed that slag-ECC healed up to 100-μm width crack. On the other hand, both F- and C-Class FA containing ECC sealed up to 50- and 30-μm width cracks, respectively. A microstructural investigation on the self-healed materials revealed that it was mostly composed of calcite and C-S-H gels and that composition varied with the supplementary minerals used (**Figure** 7).

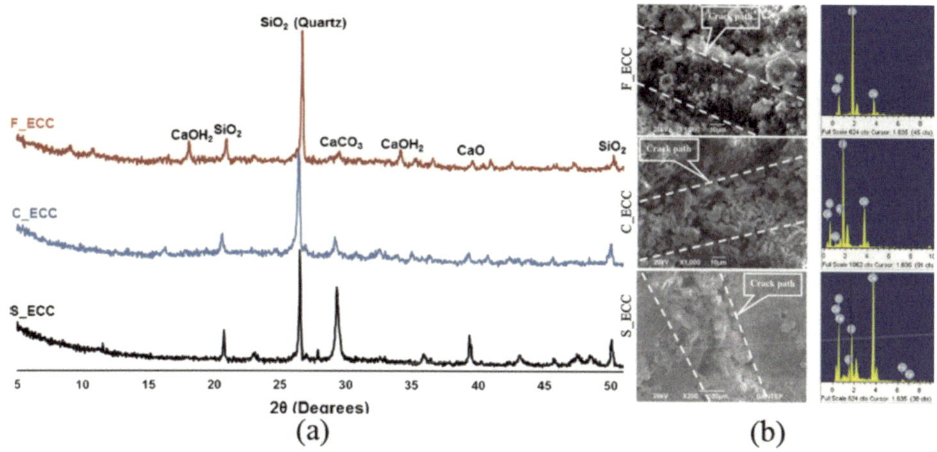

Figure 7.
Self-healing materials, (a) XRD and (b) SEM image with EDX element detection (Both reproduced from [50]).

A higher amount of healing products of slag-ECC formed due to the higher pH value of pore solution and CaO content.

3.4.2 Expansive minerals to improve autogenous self-healing

Several types of expansive minerals can enhance autogenous self-healing performance of concrete. Calcium sulphoaluminate (CSA) is one of the popular expansive minerals used for improving healing capacity in concrete [8, 9]. A self-healing agent (SHA) composed of silicon oxide (71.3%) and sodium aluminium silicate hydroxide [$Na_{0.6}Al_{4.70}Si_{7.32}O_{20}(OH)_4$] (15.4%) along with various types of carbonates such as $NaHCO_3$, Na_2CO_3 and Li_2CO_3 (etc.), and minerals such as bentonite clay (montmorillonite), feldspar and quartz was also used as an expansive self-healing agent [8]. Cracks of about 150 μm were healed within 33 days in the concrete with SHA, forming alumina silicate and modified gehlenite phases (CASH: calcium aluminosilicate hydrate). The reported healing mechanism was a swelling effect initiated by montmorillonite, and then expansion and re-crystallisation triggered by aluminosilicate with calcium ion. Ferrara et al. [51] used an active silica-based crystalline admixture (CA) as an expansive agent in cement and sand to improve the self-healing potential of raw concrete structures. Crack sealing of over 70–80% was required for reasonable mechanical performance to be recovered, such as stiffness (larger than 20%). The healing compounds formed by the crystalline admixture are similar to cement hydration products such as ettringite and calcium silicate hydrates.

Magnesium oxide (MgO), bentonite clay and quicklime were used in different proportions to enhance the autogenous self-healing capacity of concrete and cementitious materials [3, 16–21]. Substitution of PC with up to 12.5–15% by a mix of the three expansive mineral agents, MgO 5–7.5%, bentonite clay 2.5–5%, and quicklime 2.5–5%, results in optimum enhancement of the autogenous self-healing in the cement mix [17, 18]. A typical crack healing image is presented in **Figure 8** that shows how efficiently the expansive mineral containing PC mix sealed 170-μm crack in 28 days. The flexural strength recovery and crack sealing efficiency of early age (1 day) cracked specimen was enhanced up to 48 and 39%, respectively, in an expansive mineral containing cement mix, compared to the 100% PC cement mix. The permeability (gas permeability coefficient) decreased by about 70% in the

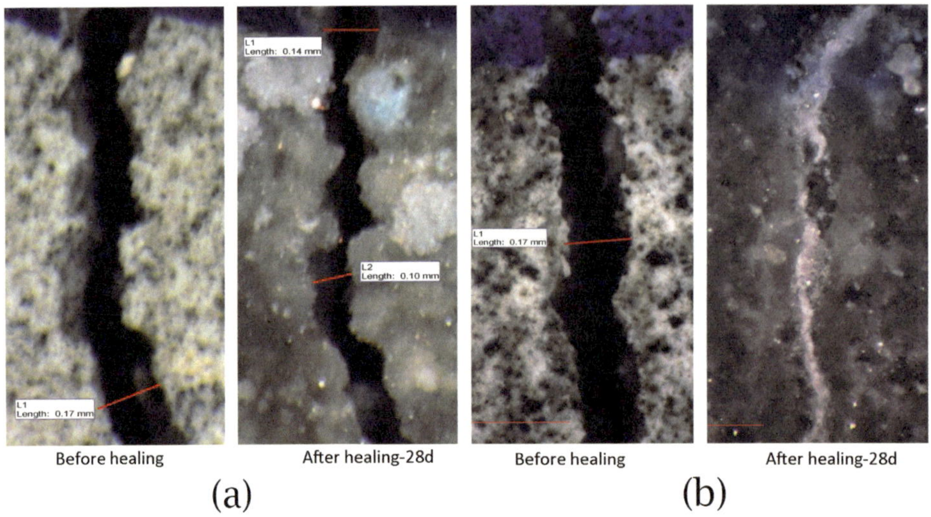

Figure 8.
The typical crack sealing pattern in 28 days: (a) 100% PC cement mix and (b) cement with expansive minerals (Reproduced from [17]).

expansive mineral containing mix compared to the 100% PC cement mix. Besides common healing compounds, calcite, portlandite, ettringite and C-S-H, MgO formed brucite, other magnesium hydro-carbonate products. Although, the healing capacity of cementitious materials decreases with the increase in the age of cement paste mix at crack formation, expansive minerals improved the autogenous self-healing capacity of PC mixes at all ages compared to the 100% PC paste [18].

Expansive minerals combination, that is, MgO, bentonite clay and quicklime can improve the autogenous self-healing capacity of drying shrinkage cracks in the cementitious materials. The maximum healable drying shrinkage cracks width in

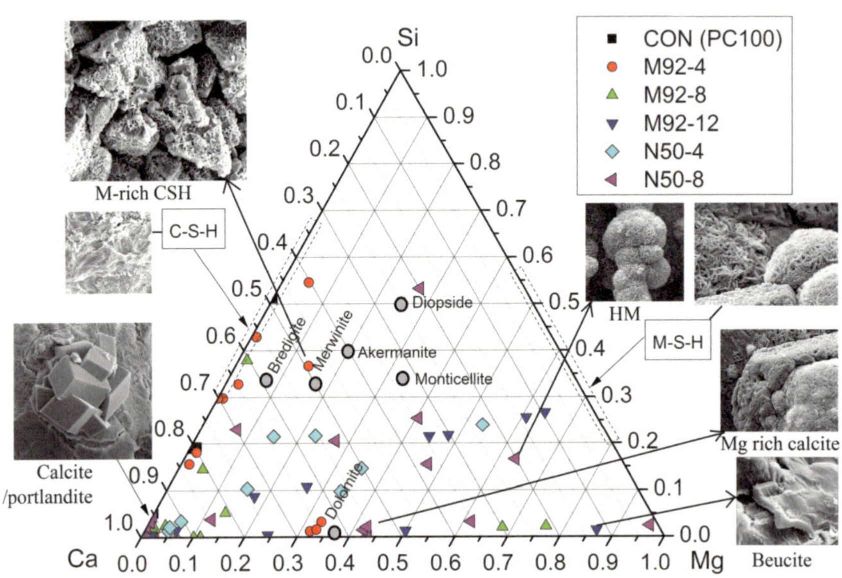

Figure 9.
Ternary diagrams of healing compounds EDX computed atomic mass percentage formed in PC-MgO cement mixes (Reproduced from [2]).

100% PC and PC-expansive minerals mixes were up to 160 and 400–500 μm, respectively, after 28 days healing in water [3, 19]. Contained expansive minerals, such as reactive MgO can enhance healing compounds within the crack (**Figure 9**) to effectively heal the crack.

Expansive minerals can also improve the self-healing capacity of ECCs [46, 52]. Bentonite (Na-Montmorillonite) as a nanoclay was mixed with slag and limestone powder and used in ECC to improve its self-healing performance [46]. An ECC-MgO system resulted in higher flexural strength recovery of pre-cracked prismatic specimens cured under accelerated autoclaved conditions compared to their pre-cracked ECC without MgO [52]. The combined effect of fibre to restrict crack and the expansive minerals to heal the crack is promising.

4. Autonomic self-healing system in concrete

In the autonomic self-healing system, different kinds of active healing agents are encapsulated into the concrete or composites. Popular encapsulation systems are microvascular glass tube network [23, 24] and microcapsules [1, 25, 26]. **Table 2** presents an overall conception of encapsulation materials and technical developments for the autonomic self-healing process. Typically a mobile liquid healing agent is always required. Less viscosity of healing agents is expected so that it can enrich a longer crack path in the damage zone, including microcracks [54]. Healing agents also should possess the ability to make a strong bond between the crack faces.

4.1 Autonomic microvascular and tabular capsules for self-healing

Capillary glass tubes are a popular choice for the microvascular network or tabular system to carry the healing agent into the concrete matrix [23, 24, 27, 28]. Diameters of the glass tubes typically range from 0.8 mm [23] to 4 mm [55]. A cyanoacrylate (<5 cP viscosity) enclosed in capillary tubes (0.8 mm inner diameter and 100 mm length), with 50 μl capacity and sealed the end with silicon considerably recovered flexural stiffness in beams [23]. Mihashi et al. [28] used embedded glass pipes with two types of healing agent, alkali-silica based and two-part epoxy resin. Considerable strength recovery performance was noted with both types of the healing agent within the crack range between 300 and 500 μm. Nevertheless, efficient mixing of two-component resin inside the crack was a challenging issue.

Cardiff University researchers have investigated the type of healing agent, delivery technique, mortar mix design and the quantity of steel reinforcement used [27]. They used three popular healing agents, (i) epoxy resins following [28], (ii) cyanoacrylates following [23] and (iii) alkali-silica solutions following [28]. During the first and second loading cycles under a three-point bend test, both primary and secondary healing occurs. Low-viscosity (typically 5 cP) single-agent cyanoacrylate adhesive resulted in optimum self-healing due to its efficient infiltration into microcracks. However, healing agents carried into the cracks are limited due to the capillary action [27]. This limitation can be eliminated with the use of an open-ended system.

The most recent advancement of a vascular network system in concrete was used in a filed trail of a road improvement scheme by Materials for Life (M4L) project [56]. The vascular network systems with shape memory polymer tendons (PET) were combined in large-scale structural elements (**Figure 10**). The self-healing performances were promising in this field trial.

		Shell material	Core material	Øi (µm)	Øo (µm)	Wall thickness (µm)	Length (mm)	Mixed in
Capsule for self-healing	Spherical	Expanded clay	Na_2FPO_3	x	4000	x	x	√
		Expanded clay	Bacteria	x	1000–4000	x	x	√
		Expanded clay	$CaC_6H_{10}O_6$	x	1000–4000	x	x	√
		Diatomaceous earth	Bacteria	x	_	x	x	√
		Gelatin	Acrylic resin	—	125–297	—	x	—
		Gelatin	Epoxy	—	50	—	x	√
		Gelatin	Tung oil	—	50	—	x	√
		Gelatin	$Ca(OH)_2$	—	50	—	x	√
		Wax	Retarder agent	—	120	—	x	√
		Paraffin	Water	—	900	—	x	—
		Cement + paraffin	SAP	—	—	—	x	—
		UF	Epoxy	—	120	4	x	√
		UFF	Epoxy	—	20–70	—	x	—
		PU	Na_2SiO_3	—	40–800	—	x	√
		Silica gel	MMA/ TEB	—	4.15	—	x	√
		Silica	Epoxy	—	—	—	x	√
		Silica Gelatin + acacia gum	Na_2SiO_3 Mineral oil+ Na_2SiO_3	– –	5000 300–700	– 5–20	x x	/ √
	Cylindrical	Glass	CA	800	1000	100	100	/
		Glass	CA	800, 1500, 3000	—	—	75, 75, 100	/
		Glass	epoxy	3000–4000	5000–7000	—	250	/ –

		Shell material	Core material	Øi (μm)	Øo (μm)	Wall thickness (μm)	Length (mm)	Mixed in
		Glass	CA	3200	4000	400	200	/
		Glass	CA	—	100	–	63.5	√
		Glass	CA, epoxy, polyacrylate, PU, bacteria	2000–3000	2200–3350	100	20–80	/
		Ceramics	PU	2500–3500	3000–4000	250	15–50	/
		Perspex	Epoxy	—	—	—	—	/
		Plant fibre	—	—	40–188	—	—	—
		PP with wax concentric glass capsule	MMA MgO, bentonite, lime	– 6150	– 11,400	– 450	– –	/ √
	Pellets	Cement PVA PVA	Na_2FPO_3, Na-MFP MgO CSA	– – –	~4000 600–4000 500	– 10–50 12–73	x x 500	√ √ √
Vascular network for self-healing	Tubular	Glass	Alkali silica, epoxy	800	2000	600	x	/
		Glass	CA	3000	4000	500	x	/
		Glass	Epoxy	4800	6000	600	x	/
		Glass	CA	3200	4000	400	x	/
		Glass	Foam, epoxy, silicon, CA	1500	—	—	x	/
		Spiral twisted wire with EVA	Epoxy	2000	3400	700	x	/
		Porous concrete	Epoxy	—	25,000–35,000	—	x	/

Table 2.
Autonomic self-healing: Encapsulation materials and techniques used ('–' means 'not reported', 'x' means 'not applicable', '√' means 'yes' and '/' means 'no'). (upgraded from [53]).

(a) (b)

Figure 10.
(a) Vascular network in concrete slab panel and (b) vascular network combination with PET in field trial (Reproduced from [56]).

4.2 Autonomic microcapsule self-healing system

Microcapsules are developed to avoid challenging issues in tubes-based capsulation systems incorporation in bulk concrete production. In this healing technique, microcapsules preserving reactive healing agents are ruptured by the forces imposed on capsules' shell due to the cracks propagation in the matrix. The released healing agent then reacts with the cementitious matrix crack surface to form healing compounds that bridge the gap and eventually heal the cracks.

The compatibility of microcapsules with bulk concrete depends on a wide variety of factors. Major influencing factors are the size and volume fraction of microcapsules used, the capsules' mechanical properties and interlock properties between the capsules and the surrounding materials [57]. The shape of the embedded capsule is another major factor that should be considered for compatibility issues. Spherically shaped capsules provide a more controlled and enhanced release of the healing agent upon breakage. It also reduces the stress concentrations around the void left from the empty capsule. However, a tubular capsule can cover a larger internal area of influence on the concrete for the same volume of a healing agent (higher surface area to volume ratio).

Yang et al. have investigated methyl methacrylate (MMA) as a monomer and triethylborane (TEB) as the healing agent and the catalyst [25]. In the investigation, about 50.2 and 66.8% reduction in permeability has been achieved within 3 and 30 days, respectively. Microscopic imaging confirms that some ruptured microcapsules existed and filled the cracks of the sample after 80% ultimate compressive strength at 28 days.

About 2% crystalline sodium silicate in polyurethane-encapsulated microcapsules with a diameter ranging from 40 to 800 μm increased 24% mechanical load recovery compared to 12% in the control samples [58]. However, the compressive strength of the composite reduced by 12% compared to that of the control mix. In the concrete containing microcapsules, sodium silicate reacts with calcium hydroxide of cement and produces a calcium-silica-hydrate (C-S-H) gel that heals the cracks partially. The C-S-H further reacts with dissolved CO_2 in water and sodium oxide, which produced calcium carbonate. This is similar to the main hydration phase of cement, which causes strengthening.

Sodium silicate encapsulated in double-walled polyurethane/urea-formaldehyde (PU/UF) was reported in [59]. The addition of 2.5 and 5% microcapsules resulted in about 24 and 35% healing efficiency based on the crack depth measurements.

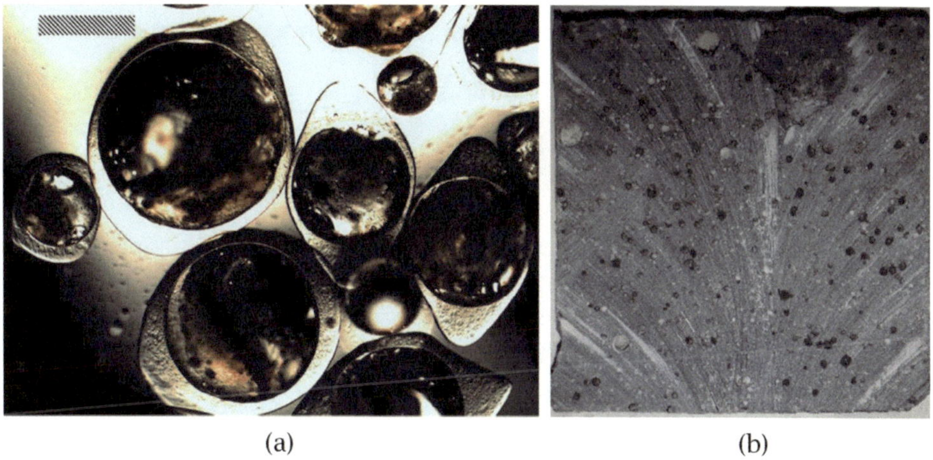

Figure 11.
(a) Microscopic image of microcapsules (scale bars correspond to 500 μm) and (b) ruptured microcapsules appearing as 'wet' spots on the digital image of the split face (Both reproduced from [60]).

Further advancement with sodium silicate encapsulated in gelatin and gum arabic shell materials (**Figure 11**) was found in recent studies [57, 60]. These microcapsules survive mixing with cement and rupture successfully upon crack formation and release sodium silicate solution. Although increasing microcapsules volume fractions in a ~24% reduced the mechanical properties, the crack sealing was just under 100%. Besides, the crack depth and sorptivity coefficient were decreased by 70 and 54%, respectively. These microcapsules were also successfully implemented in the filed trail of a road improvement scheme by M4L project [61].

The colloidal silica solution capsules up to 16 vol% in PC grout increased the sealing efficiency from ~20% for the only PC to ~85% in 28 days [62]. However, monodisperse photo-polymerised acrylate shell with hydrophilic mineral core microfluidic droplets are further advancement in the self-healing microcapsule field [63].

4.3 Coated minerals (pellets and granules) for self-healing

Although the direct addition of potential minerals to the concrete mix improves autogenous self-healing performance, protecting those minerals in initial mixing may further enhance the healing process. With this in mind, pellets of potential healing mineral agents have been used for improved concrete self-healing. Sisomphon et al. [47] used expanded lightweight clay aggregates (LWAs) impregnated with a solution of sodium mono fluorophosphate (Na_2FPO_3, Na-MFP) and coated by cement paste layers. The entire mechanism is schematically presented in **Figure 12a**. Pellets with expansive minerals such as a reactive MgO were spray-coated (10–50 μm) with polyvinyl alcohol (PVA) to produce PVA-coated MgO pellets for self-healing concrete applications (**Figure 12b**). A PVA-coated granulated CSA (calcium sulpho aluminate)-based expansive mineral was used for improving the self-healing performance of cementitious materials [48]. Replacement of CSA pellets was up to 10% by wt. of cement and mortar was prepared with 1:3 cement-to-sand ratio and w/c = 0.5. Cracks in the range of 0.1–0.2 mm were healed completely within 14 days whereas larger crack >0.2 healed within 16 days.

Granules of expansive self-healing agent coated with an extra layer of cement compounds were investigated by [64]. The self-healing concept is schematically

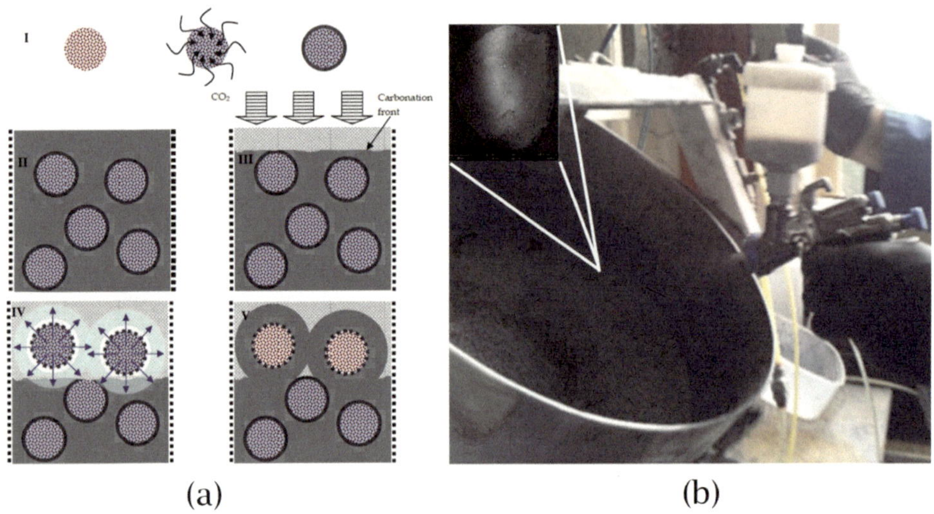

Figure 12.
(a) Impregnation of LWAs to prepare pellet and self-healing concept: I-V (Reproduced from [49]), and (b) spraying PVA coating solution on the MgO pellets in the disc pelletizer and a microscopic image of a pellet covered by PVA (Adopted from [29]).

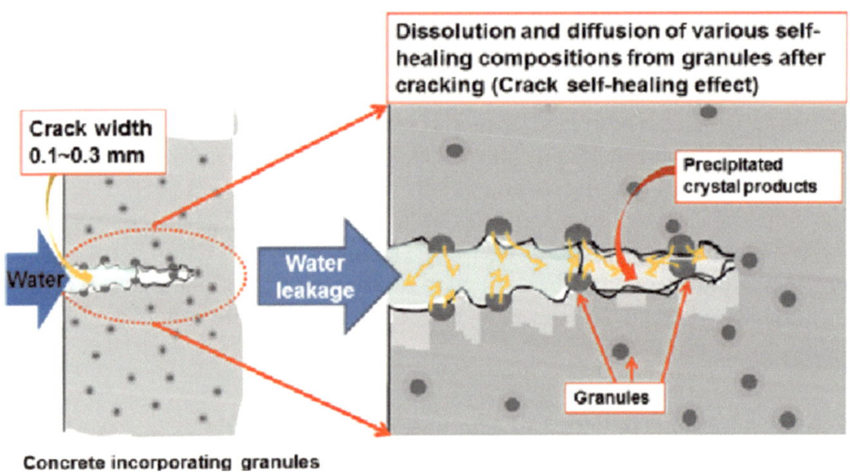

Figure 13.
Concept of self-healing concrete with granules containing expansive mineral agents (Reproduced from [64]).

presented in **Figure 13**. The fundamental concept is that the surface of the coating may hydrate during initial production and mixing while the core healing mineral agent remains unhydrated; this may then dissolute and diffuse into the crack surface after crack propagation and form new products for self-healing.

4.4 Bacteria-based self-healing in concrete (bioconcrete)

Alkali-resistant endospore-forming bacteria that precipitate calcite through biological metabolism are used for self-healing in concrete. Examples of these bacteria are *B. cohnii*, *B. pseudofirmus* and *B. sphaericus*. The process involved in calcite production is termed as microbiologically induced calcite precipitation (MICP) [32]. There are two conventional MICP processes: firstly, the urease system, which is initiated by the hydrolysis of urea by the bacteria, secreted enzyme urease (urea aminohydrolase) as a catalyst [33] and secondly, calcium lactate-based MICP [65].

In the urea-based MICP process, hydrolysis of urea with urease results in ammonia and carbonate ions, which increase the pH value into the bacteria cell. Researchers have experimented with urea as a mineral precursor for bio-cementation using bacteria [33, 66]. In the presence of $CaCl_2$ as a source of Ca^{2+}, high pH content bacteria cause $CaCO_3$ crystal precipitation from the solution. Typically, bacteria shell made with various ions are negatively charged to attract positive cautions Ca^{2+} ions surrounding the cell wall, which reacts with CO_3^{2-} and precipitate $CaCO_3$ around the cell [66].

Calcium lactate ($CaC_6H_{10}O_6$) is a crystalline salt, typically produced from the reaction of lactic acid with calcium carbonate or hydroxide. This was used as an alternative of urea-$CaCl_2$, as a precursor for bacterial metabolism in concrete to avoid ammonia production in hydrolysis reactions. According to [65], metabolic absorption and breakdown of calcium lactate with bacteria lead to the precipitation of $CaCO_3$.

Bacteria cannot survive long if they are mixed directly with fresh cement. The survivability of bacterial spores was optimized in [65], through the technique of packing bacterial spores and organic mineral precursor compounds in porous expanded clay particles before mixing in the concrete matrix. The pellets (2–4 mm) were principally made with the three components of a solid mixture, and they were used as a replacement of some of the similar size coarse aggregate. A high concentration of calcite precipitation has been found in concrete specimens with bacteria incorporated expended clay particles, which efficiently acted in crack-plugging and reduced permeability (**Figure 14**). About to micron sized (0.15 mm width), cracks were sealed. However, the main drawback in the bacterial pellet process is the

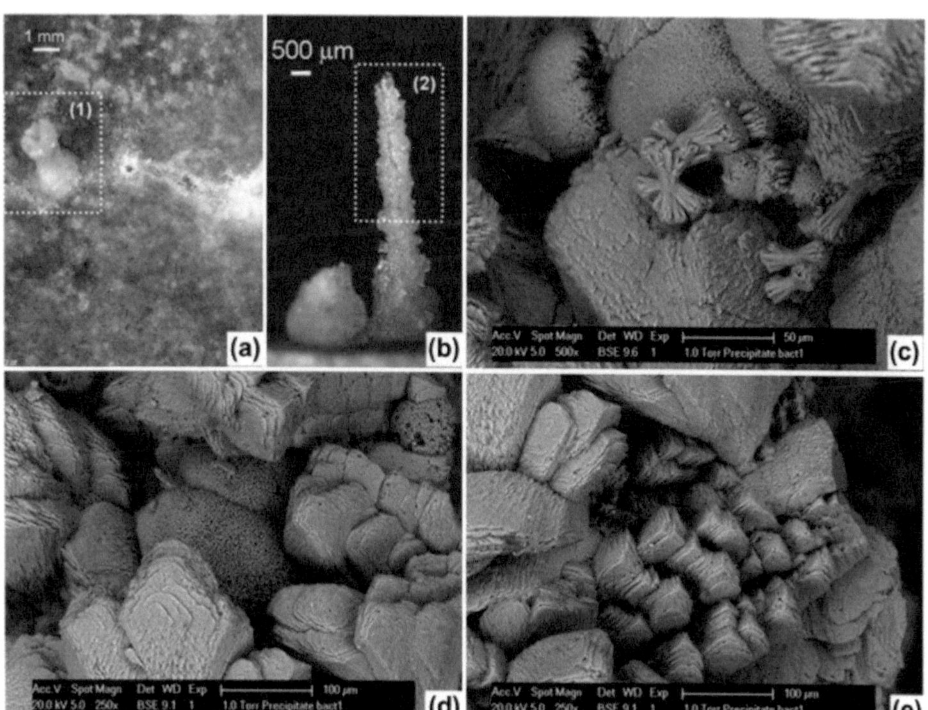

Figure 14.
Microscopic images of bacteria based self-healing concrete, (a) Stereomicroscopic image of crack sealing, (b) Stereomicroscopic close-up image of massive columnar precipitate (c–e) ESEM images of top part of massive columnar precipitate indicated in image by dotted square (Reproduced from [65]).

negative impact on the mechanical performance of concrete. About 50% of the total aggregate volume requires replacing with bacterial pellets for satisfactory self-healing performance, which negatively impacts the mechanical strength of concrete.

An encapsulation of bacterial spores inside microcapsules is a recent advancement in this field [26]. These microcapsules were reported flexible in humid/water conditions and becoming brittle in the dry environment. With their bacterial encapsulation systems, about 970-μm width cracks were healed successfully, which was four times greater than for non-bacterial mixes. Nevertheless, bacterial activity reduces dramatically with the increase in the pH (>12) value in concrete.

5. Conclusions

Concrete being one of the most-used construction development materials, early damage and failure within a structure's design lifetime is a threat to infrastructure industries. A self-healing concrete has great potential to mitigate this challenge. Self-healing in concrete can be broadly classified into two categories: autogenic and autonomic healing [1].

The autogenous self-healing capacity of concrete could be enhanced through restricting crack growth, wet-dry cycle, using SCM's such as GGBS, fly ash, and silica fume, and using expansive minerals such as MgO, bentonite clay, quicklime, CSA and crystalizing mineral agents. However, the effectiveness of autogenous self-healing is considerably dependant on the remaining unhydrated cement or mineral in the concrete. This is hitherto restricted to smaller healable crack widths, more extended healing periods and the strength recovery.

Autonomic healing in concrete, in contrast to autogenous healing, requires the release of the self-healing triggering agent from reserved encapsulation or a continuous supply network. This is to further improve the self-healing efficiency of concrete compared to the autogenous healing process. Popular autonomic self-healing systems are microencapsulation, microvascular and pellets with different autonomic healing agents such as epoxies, cyanoacrylates, methyl methacrylate, alkali-silica solutions, minerals and microorganisms.

The self-healing concrete technology can be adopted in developing smart and resilient infrastructure development. Different self-healing concrete technology can be utilized depending on different applications. The greatest challenges of all self-healing technology in the concrete industry remain the difficulties in widespread uptake, the additional costs involved and the validation of long-term durability performances. Field trials such as those initiated by the University of Cambridge, Cardiff University and the University of Bath through Materials for Life (M4L) and Resilient Materials for Life (RM4L) research projects are significantly crucial for self-healing concrete validation in large scale.

Acknowledgements

The authors are grateful for collaboration and support from the Engineering and Physical Sciences Research Council (EPSRC) research projects 'Materials for Life (M4L)' and 'Resilient Materials for Life (RM4L)'.

Conflict of interest

There is no conflict of interest.

DOI: http://dx.doi.org/10.5772/intechopen.92349

Author details

Tanvir Qureshi[1]* and Abir Al-Tabbaa[2]

1 University of Toronto, Toronto, Canada

2 University of Cambridge, Cambridge, UK

*Address all correspondence to: tanvir.qureshi@utoronto.ca

References

[1] de Rooij M, van Tittelboom K, de Belie N, Schlangen E, editors. Self-Healing Phenomena in Cement-Based Materials: State-of-the-Art Report of RILEM Technical Committee 221-SHC: Self-Healing Phenomena in Cement-Based Materials. Vol. 11. Springer Science & Business Media; 2013. ISBN: 978-94-007-6624-2

[2] Erik S, Joseph C. Self-healing processes in concrete. In: Self-healing materials: Fundamentals, Design Strategies, and Applications. Weinheim: Wiley-vch; 2009. pp. 141-198

[3] Qureshi TS, Al-Tabbaa A. Self-healing of drying shrinkage cracks in cement-based materials incorporating reactive MgO. Smart Materials and Structures. 2016;**25**:084004

[4] Li V, Yang E. Self-healing in concrete materials, self heal. mater. An Altern. Approach to 20 centuries. Materials Science. 2007;**100**:161

[5] ter Heide N, Schlangen E. Self-healing of early age cracks in concrete. In: First International Conference on Self Healing Materials. Noordwijk, The Netherlands; 2007. pp. 1-12

[6] Yang Y, Lepech MD, Yang E-H, Li VC. Autogenous healing of engineered cementitious composites under wet–dry cycles. Cement and Concrete Research. 2009;**39**:382-390

[7] Teall O, Pilegis M, Davies R, Sweeney J, Jefferson T, Lark R, et al. A shape memory polymer concrete crack closure system activated by electrical current. Smart Materials and Structures. 2018;**27** (7):075016

[8] Ahn T-H, Kishi T. Crack self-healing behavior of cementitious composites incorporating various mineral admixtures. Journal of Advanced Concrete Technology. 2010;**8**:171-186

[9] Sisomphon K, Copuroglu O, Koenders EAB. Self-healing of surface cracks in mortars with expansive additive and crystalline additive. Cement and Concrete Composites. 2012; **34**:566-574

[10] Huang H, Ye G, Shui Z. Feasibility of self-healing in cementitious materials—By using capsules or a vascular system? Construction and Building Materials. 2014;**63**:108-118

[11] Jiang Z, Li W, Yuan Z. Influence of mineral additives and environmental conditions on the self-healing capabilities of cementitious materials. Cement and Concrete Composites. 2015;**57**:116-127

[12] Van Tittelboom K, Gruyaert E, Rahier H, de Belie N. Influence of mix composition on the extent of autogenous crack healing by continued hydration or calcium carbonate formation. Construction and Building Materials. 2012;**37**:349-359

[13] Qian S, Zhou J, de Rooij MR, Schlangen E, Ye G, van Breugel K. Self-healing behavior of strain hardening cementitious composites incorporating local waste materials. Cement and Concrete Composites. 2009;**31**:613-621

[14] Antiohos SK, Papageorgiou A, Papadakis VG, Tsimas S. Influence of quicklime addition on the mechanical properties and hydration degree of blended cements containing different fly ashes. Construction and Building Materials. 2008;**22**:1191-1200

[15] Kanellopoulos A, Qureshi T, Al-Tabbaa A. Encapsulated mineral precursors for self-healing cement based composites. In: 5th International Conference on Self-Healing Materials; 2015

[16] Qureshi TS, Kanellopoulos A, Al-Tabbaa A. Encapsulation of

expansive powder minerals within a concentric glass capsule system for self-healing concrete. Construction and Building Materials. 2016;**121**:629-643

[17] Qureshi T, Kanellopoulos A, Al-Tabbaa A. Autogenous self-healing of cement with expansive minerals-I: Impact in early age crack healing. Construction and Building Materials. 2018;**192**:768-784

[18] Qureshi T, Kanellopoulos A, Al-Tabbaa A. Autogenous self-healing of cement with expansive minerals-II: Impact of age and the role of optimised expansive minerals in healing performance. Construction and Building Materials. 2019;**194**:266-275

[19] Qureshi TS. The Role of Expansive Minerals in the Autogenous and Autonomic Self-Healing of Cement Based Materials. Cambridge, UK: University of Cambridge; 2016

[20] Qureshi T, Al-Tabbaa A. The effect of magnesia on the self-healing performance of Portland cement with increased curing time. In: 1st International Conference on Ageing of Materials & Structures; Delft, The Netherlands; 2014. pp. 635-642

[21] Qureshi T, Al-Tabbaa A. Influence of expansive minerals on the self-healing of cement paste and mortar systems. In: Fifth International Conference on Self-Healing Materials; 2015

[22] Kanellopoulos A, Qureshi TS, Al-Tabbaa A. Glass encapsulated minerals for self-healing in cement based composites. Construction and Building Materials. 2015;**98**:780-791

[23] Li VC, Lim YM, Chan Y-W. Feasibility study of a passive smart self-healing cementitious composite. Composites. Part B, Engineering. 1998; **29**:819-827

[24] Dry CM. Design of self-growing, self-sensing, and self-repairing materials for engineering applications. In: Smart Materials. International Society for Optics and Photonics. 2001; **4234**:23-29

[25] Yang Z, Hollar J, He X, Shi X. A self-healing cementitious composite using oil core/silica gel shell microcapsules. Cement and Concrete Composites. 2011; **33**:506-512

[26] Wang JY, Soens H, Verstraete W, de Belie N. Self-healing concrete by use of microencapsulated bacterial spores. Cement and Concrete Research. 2014; **56**:139-152

[27] Joseph C, Jefferson AD, Isaacs B, Lark RJ, Gardner DR. Experimental investigation of adhesive-based self-healing of cementitious materials. Magazine of Concrete Research. 2010; **62**:831-843

[28] Mihashi H, Kaneko Y, Nishiwaki T, Otsuka K. Fundamental study on development of intelligent concrete characterized by self-healing capability for strength. Trans. Japan Concrete Institute. 2001;**22**:441-450

[29] Alghamri R, Kanellopoulos A, Litina C, Al-Tabbaa A. Preparation and polymeric encapsulation of powder mineral pellets for self-healing cement based materials. Construction and Building Materials. 2018;**186**:247-262

[30] de Belie N, Gruyaert E, Al-Tabbaa A, Antonaci P, Baera C, Bajare D, et al. A review of self-healing concrete for damage management of structures. Advanced Materials Interfaces. 2018;**5**: 1800074

[31] Jonkers HM, Thijssen A, Muyzer G, Copuroglu O, Schlangen E. Application of bacteria as self-healing agent for the development of sustainable concrete. Ecological Engineering. 2010;**36**:230-235

[32] Reeksting BJ, Hoffmann TD, Tan L, Paine K, Gebhard S. In-depth profiling of calcite precipitation by environmental bacteria reveals fundamental mechanistic differences with relevance to application. bioRxiv. 2019:850883

[33] Siddique R, Chahal NK. Effect of ureolytic bacteria on concrete properties. Construction and Building Materials. 2011;**25**:3791-3801

[34] Lauer K, Slate F. Autogenous healing of cement paste. ACI Journal Proceedings. 1956;**52**:1083-1098

[35] Edvardsen C. Water permeability and autogeneous healing of cracks in concrete. ACI Materials Journal. 1999; **96**:448-454

[36] Reinhardt H-W, Jooss M. Permeability and self-healing of cracked concrete as a function of temperature and crack width. Cement and Concrete Research. 2003;**33**:981-985

[37] Li VC. On engineered cementitious composites (ECC). Journal of Advanced Concrete Technology. 2003;**1**(3): 215-230

[38] Hosoda A, Komatsu S, Ahn T, Kishi T, Ikeno S, Kobayashi K. Self healing properties with various crack widths under continuous water leakage. Concrete Repair, Rehabilitation and Retrofitting II. 2009:221-227

[39] Jaroenratanapirom D, Sahamitmongkol R. Effects of different mineral additives and cracking ages on self-healing performance of mortar. In: Proceedings of the 6th Annual Concrete Conference; Petchaburi, Thailand; 2010

[40] Termkhajornkit P, Nawa T, Yamashiro Y, Saito T. Self-healing ability of fly ash-cement systems. Cement and Concrete Composites. 2009;**31**:195-203

[41] Watanabe T, Fujiwara Y, Hashimoto C, Ishimaru K. Evaluation of self healing effect in fly-ash concrete by ultrasonic test method. International Journal of Modern Physics B. 2011;**25**: 4307-4310

[42] Olivier K. Experimental studies of self-healing cementitious materials incorporating mineral admixtures. Proceedings of the Fourth International Conference on Self-Healing Materials. 2013:21-24

[43] Zhou ZH, Li ZQ, Xu DY, Yu JH. Influence of slag and fly ash on the self-healing ability of concrete. Advances in Materials Research. 2011;**306-307**: 1020-1023

[44] Huang H, Ye G, Damidot D. Effect of blast furnace slag on self-healing of microcracks in cementitious materials. Cement and Concrete Research. 2014; **60**:68-82

[45] Valcke S, de Rooij M, Nijland TG, Fisher H, Mendoza S. Carriers of Self-Healing Agents in Concrete and Their Effect on the Microstructure. TNO report No. TNO-034-DTM-2009-04262 for IOP project of healing built-in building materials. 2009

[46] Qian SZ, Zhou J, Schlangen E. Influence of curing condition and precracking time on the self-healing behavior of engineered cementitious composites. Cement and Concrete Composites. 2010;**32**:686-693

[47] Sisomphon K, Copuroglu O, Fraaij A. Application of encapsulated lightweight aggregate impregnated with sodium monofluorophosphate as a self-healing agent in blast furnace slag mortar. Heron. 2011;**56**:17-36

[48] Lee Y-S, Ryou J-S. Self healing behavior for crack closing of expansive agent via granulation/film coating method. Construction and Building Materials. 2014;**71**:188-193

[49] Ferrara L, Krelani V, Moretti F. On the use of crystalline admixtures in cement based construction materials: From porosity reducers to promoters of self healing. Smart Materials and Structures. 2016;**25**:084002

[50] Sahmaran M, Yildirim G, Erdem TK. Self-healing capability of cementitious composites incorporating different supplementary cementitious materials. Cement and Concrete Composites. 2013;**35**:89-101

[51] Ferrara L, Krelani V, Carsana M. A "fracture testing" based approach to assess crack healing of concrete with and without crystalline admixtures. Construction and Building Materials. 2014;**68**:535-551

[52] Sherir MAA, Hossain KMA, Lachemi M. The influence of MgO-type expansive agent incorporated in self-healing system of engineered cementitious composites. Construction and Building Materials. 2017;**149**: 164-185

[53] van Tittelboom K, de Belie N. Self-healing in cementitious materials—A review. Materials. 2013;**6**:2182-2217

[54] van Tittelboom K, de Belie N, van Loo D, Jacobs P. Self-healing efficiency of cementitious materials containing tubular capsules filled with healing agent. Cement and Concrete Composites. 2011;**33**(4):497-505

[55] Thao T, Johnson T, Tong QS, Dai PS. Implementation of self-healing in concrete—Proof of concept. IES Journal Part A: Civil & Structural Engineering. 2009;**2**:116-125

[56] Davies R, Pilegis M, Kanellopoulos A, Sharma T, Teall O, Gardner D, et al. Multi-scale cementitious self-healing systems and their application in concrete structures. In: 9th International Concrete Conference. Dundee; 2016

[57] Kanellopoulos A, Giannaros P, Al-Tabbaa A. The effect of varying volume fraction of microcapsules on fresh, mechanical and self-healing properties of mortars. Construction and Building Materials. 2016;**122**:577-593

[58] Pelletier MM, Brown R, Shukla A, Bose A. Self-healing Concrete with a Microencapsulated Healing Agent. Kingston: University of Rhode Island; 2011

[59] Mostavi E, Asadi S, Hassan MM, Alansari M. Evaluation of self-healing mechanisms in concrete with double-walled sodium silicate microcapsules. Journal of Materials in Civil Engineering. 2015;**27**:04015035

[60] Kanellopoulos A, Giannaros P, Palmer D, Kerr A, Al-Tabbaa A. Polymeric microcapsules with switchable mechanical properties for self-healing concrete: Synthesis, characterisation and proof of concept. Smart Materials and Structures. 2017;**26**: 045025

[61] Al-Tabbaa A, Litina C, Giannaros P, Kanellopoulos A, Souza L. First UK field application and performance of microcapsule-based self-healing concrete. Construction and Building Materials. 2019;**208**:669-685

[62] Litina C. Development and Performance of Self-Healing Blended Cement Grouts with Microencapsulated Mineral Agents. Cambridge, UK: University of Cambridge; 2015

[63] Souza LR, Kanellopoulos A, Al-Tabbaa A. Synthesis and characterization of acrylate microcapsules using microfluidics for self-healing in cementitious materials. In: Fifth International Conference on Self-Healing Materials. Durham, USA; 2015. pp. 1-4

[64] Kishi T. Development of crack self-healing concrete by cost beneficial

semi-capsulation technique. In: Third International Conference on Sustainable Construction Materials and Technologies. Kyoto, Japan; 2013. pp. 1-9

[65] Wiktor V, Jonkers HM. Quantification of crack-healing in novel bacteria-based self-healing concrete. Cement and Concrete Composites. 1 August 2011;**33**(7):763-770

[66] van Tittelboom K, de Belie N, de Muynck W, Verstraete W. Use of bacteria to repair cracks in concrete. Cement and Concrete Research. 2010; **40**:157-166

Chapter 13

Exploits, Advances and Challenges in Characterizing Self-Healing Materials

Camillus Sunday Obayi and Paul Sunday Nnamchi

Abstract

Characterization is an indispensable tool for understanding the structure–property-processing relationship in all material classes. However, challenges in self-healing materials characterization arise from the preparation routes, material types, damage mechanism and applications. Here, the discourse surveys the exploits, advances and challenges encountered within various characterization methods that have been exploited to reveal the damage-restoring processes in some material classes, namely metals, polymers, ceramics, concretes and coatings. Since there is no unified characterization procedure for the different classes of materials displaying self-repairing capabilities, the outcome of this discourse contributes to the advancement of knowledge about understanding self-healing testing procedures. An overview of methods, challenges and prospects toward self-healing property standardization at different length scales has been discussed.

Keywords: material characterization, properties, processing, challenges, test standardization, self-healing materials

1. Introduction

Characterization is a vital tool in elucidating the characteristics of material systems and ascertaining their suitability for various applications. Characterization is also an indispensable technique for quantifying self-repairing capabilities of various material classes. A self-healing material is a material designed to mimic nature, self-repair and restore its partial or original characteristics, thereby prolonging its service life like a biological system after encountering some form of damage induced by its service environment [1, 2].

Developing a man-made material capable of sensing and self-repairing in response to changes in the operating environment is a challenging task. This is because it takes a great exploit to design and process a synthetic material to mimic very closely the nature-designed mechanism of self-repair obtained in complex biological architectures. Numerous design approaches and preparatory routes have been attempted by research community to create self-healing abilities in different material classes for many applications. The material classes exploited so far include polymers, metals, ceramics, cements, coatings and composites [3].

Another huge task is quantifying the material's capability to self-repair and its suitability for a specific application via characterization techniques.

Several techniques have been employed to evaluate and quantify self-healing capacities and effectiveness in each of above material classes. Nevertheless, challenges arise in the characterization of self-healing materials as evaluation methods are not only materials and application specific, but also depend on the mechanism of the self-healing process. More so, self-healing processes take place on a very small length scale requiring sophisticated experimental procedures and equipment to unravel the mechanism of self-healing. Furthermore, it is important to establish a uniform testing procedure and standard for different material groups for better understanding of the concept [4]. This chapter surveys the main techniques applied to reveal the damage-restoring mechanisms in some material classes, but begins with a brief survey of preparatory routes and mechanisms of self-repair in the different materials classes.

2. Self-healing materials

Self-healing is the capability of a material to recover from any kind of damage automatically without any external intervention as obtainable in biological systems or with external stimulation such as heat, light, electrical stimulus and solvent. The materials that exhibit self-healing without any external intervention or stimuli are said to be autonomic while self-healing that involves human or external influence to induce healing is said to be nonautonomic in nature [5, 6]. One of the major problems encountered in the use of materials in diverse fields is how to ensure their durability and minimize structural failures [7]. A self-healing material is therefore an artificial material designed with built-in ability to detect failure and respond automatically to restore partial or full properties or function of the structure after encountering in-service damage [3, 7].

This in-service damage, which is usually in form of micro-scratches, surface and internal cracks, voids or other defects [8, 9], is majorly responsible for failure in materials systems. Over time, these micro-cracks accumulate and grow until catastrophic failure of the entire product or system occurs. Since this source of failure normally initiates at the nanoscale level and progresses subsequently to the micro- and macroscale levels until failure occurs, an ideal self-healing material would without any external influence prevent initiation of failure at these small length scales or repair already nucleated damage, thereby restoring the original material properties in a shortest time [3, 7]. Since the greater of the in-service damage encountered in material systems is usually in form of micro-cracks, voids or other defects, the objective of designing self-healing materials is to impart them with the capabilities to prevent the initiation of micro-cracks and voids or fill and seal them automatically.

For so many years now, the strategies of fabricating synthetic materials with the capability to self-heal like a biological system or as envisaged above have been exploited greatly. This huge interest is anchored on benefits of self-healing in materials. These benefits include enhancing materials' service lifetime, reduction in replacement costs and improvement in product safety [7]. Great advances have been witnessed in creation of self-healing materials since the birth of the concept. The concept has been exploited in almost all materials classes including polymers, metals, ceramics, cements, coatings and composites [3]. The design strategies and processing routes involved in creating self-healing capabilities in these material classes are different, just like their self-healing reactions to damage encountered during their lifetimes. The next subsection takes a look at the creation of self-healing abilities in these systems, the prevalent mode of failures and mechanisms of self-repairing.

3. Preparatory routes and mechanisms of self-healing

3.1 Self-healing polymers and polymer matrix composites

The self-healing concept has been most successful in the development of self-healing polymer-based systems [10, 11]. This emanates from the fast diffusion rate and high plasticity due to open-molecular structures in polymers, which facilitate diffusion of healing agents to fill and seal voids or micro-cracks [10] even at room temperature. Unlike metallic and ceramic systems, polymeric systems are light weight, chemically stable and can be easily processed [4]. These properties are exploited in developing efficient self-repairing polymers and polymer-based fiber-reinforced composites, which have applications in transportation, electronics, defense, biomedicine and construction industries [7, 12, 13].

Based on the strategies exploited to achieve self-repair, polymers are generally grouped into extrinsic and intrinsic self-healing systems [14]. Intrinsic self-repair is achieved by synthesis of smart polymers containing functional groups with the inherent ability to reversibly polymerize or cross-link their bonds in the presence of a stimulus like light or heat [15] and by so doing act as healing agents. The processes for obtaining extrinsic self-healing include (a) embedding microcapsules containing curable healing agents into polymer networks; and (b) incorporation of healing agents into polymer networks via microvascular channels [5, 16].

The microcapsule in (a) could be in form of capsule containing healing agent and catalyst or twin microcapsules each containing a monomer/resin and its hardener [5, 16, 17] while that in (b) can be in form of fibrous composite architecture impregnated with a microvessel filled with reactive healants [18]. Unlike extrinsic routes where healing agents are consumed during the curing process and are not replenished, intrinsic approaches have the advantage of multiple healing of damage in the same area owing to reversible polymerization [19]. Self-healing has been exploited and accomplished in thermoplastic, thermosetting and elastomeric systems.

The mode of damage often encountered in polymers and structural composites is in form of matrix micro-cracking, fiber breakage or delamination and fiber-matrix debonding [7, 20, 21]. Self-healing mechanism or recovery or recuperation takes place when a damage/crack is encountered and is healed by intrinsic polymerization or polymerization of healing agent as crack ruptures the capsules as in (a) or by favorable reaction kinetics and post-polymerization as in (b).

3.2 Self-healing metals

When compared to other material systems, it is much difficult to achieve self-healing in metals [22–24]. This is as a result of their high melting temperatures and strong atomic bonds, which limit diffusion of healing agents/solute atoms to sites of damage at low temperatures [22, 23]. There is also further restriction due to the relative small size and volume of the solute atoms. As a result, rate of mass transport to fill damage sites is intrinsically low at the usual low operating temperatures [23, 25].

The major factor limiting the useful life of metals is the occurrence of internal damage such as voids and cracks during processing or service. These defects usually initiate as nano- or micro-cracks in the bulk or on the surface, grow and propagate and eventually lead to failure. The self-healing process in metals in response to crack initiation follows the sequence of diffusion or release or transport of healing agents or atoms into the void or crack to fill and seal it, thereby restoring partially or fully the mechanical properties such as fatigue strength, stiffness and fracture toughness.

The approaches that have been proposed and attempted in developing self-healing in metallic systems according to [22] include: (a) precipitation-induced self-healing approach at low and high temperatures [26]; (b) dispersion of nano-shaped memory alloy (SMA) in off-eutectic metal matrix [27–29]; (c) SMA-clamp and melt [30]; (d) solder tubes/capsules; (e) coating agent [31] and (f) electro-healing [32, 33]. Blazej Grabowski and C. Cem Tasan [22] classified these concepts into two based on the healing length scale as (i) healing of nanoscale voids (which includes approaches a and b) and (ii) healing of macroscale cracks (which includes approaches c, d, e and f) [22]. One (I) and two (II) above were earlier classified into damage prevention and damage management by Van der Zwag et al. [34], respectively. This implies that healing at nanoscale targets prevention of macroscale damage while healing at macroscale focuses on management of macroscale damage to prevent total failure. Manuel [30] further classified approaches a and b as solid-state healing; approaches c, d and e as liquid-state healing and approach f as electrolyte-assisted healing. The self-healing concepts in metals and metal matrices are summarized in **Figure 1** and details of the features of these concepts are available in Ref. [22].

3.3 Self-healing ceramics

Ceramics are very important engineering material and are widely applied in electrical, magnetic, chemical, nuclear and biomedical fields [35]. However, ceramics have major shortcomings of being inherently porous and brittle in nature. As a result, ceramics have low strengths and fracture toughness as the components are prone to catastrophic failure by crack damage even at subcritical loading [36, 37]. Thus, self-healing concept in ceramics targets induced healing of structural defects (cracks and pores) in order to prolong lifetime [38].

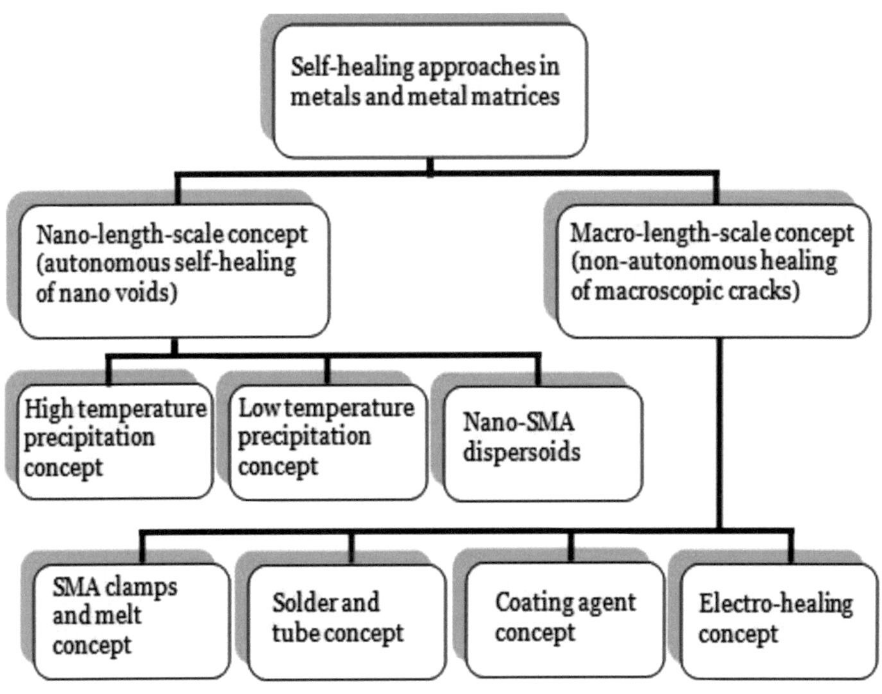

Figure 1.
Self-healing concepts in metals and metal matrices adapted from Ref. [22].

Self-healing in brittle ceramics has also witnessed extensive studies like polymers, but healing in ceramics is difficult to achieve at temperatures below 1000°C [35]. This is because self-repairing in ceramics takes place readily via solid-state diffusion, which requires high activation energy. This thermally activated solid-state reaction has the disadvantage of inhibiting long-range transport of material required to heal macro-cracks. More so, the healing of nano-cracks is inhibited by crack surface relaxation phenomena triggered by the ionic and covalent bonding character in ceramics [39]. Although, processing of self-healing ceramic materials is regarded as a high-temperature healing process, processing at lower temperatures is being pursued.

Some healing routes have been adopted to repair crack damages in ceramics at elevated temperatures. The important ones include crack closure enabled by diffusion-controlled sintering; crack opening rebonding promoted by viscous flow of glass phase; filling the crack opening space with products of oxidation reaction as obtainable in silicon carbide composites [35, 40] and healing of multicomponent and multiphase ceramic materials through local particle rearrangement-induced eutectic melt or phase transition [41]. Damage mechanisms and crack healing have been widely studied in various ceramic materials including single crystalline, polycrystalline and amorphous glasses.

3.4 Self-healing concrete materials

Concrete is the most popular cement-based material and most widely used constructional material [42, 43]. Concrete is regarded as a composite material made up of water (H_2O), cement, fine and coarse aggregates. It has many good qualities such as availability and affordability of its constituent materials, versatility, durability and low maintenance [42–45]. Concrete exhibits superior compressive strength, but low tensile strength. In order to improve its tensile strength, it is reinforced with steel bars [43]. However, the major limitation of concrete is its high susceptibility to cracking [43, 44]. The causes of cracks at both micro- and macro levels include preparation processes, temperature differences, shrinkage, fatigue loads and settlement of structures [42, 44]. The cracks serve as channels for water, dissolved particles in fluids and unwanted acidic gasses to penetrate the concrete [42, 45], attacking the concrete and corroding the steel reinforcement [44, 45]. Thus, the susceptibility of concrete to crack leading to structural failures is a major concern and has remained unsolved in industry [43].

However, concrete has been known to exhibit natural or autonomous self-healing to a certain extent under a long-term hydration [45]. It has been proved that some initial cracks in concrete can be suddenly closed when un-hydrated cement reacts with carbon dioxide dissolved in water, producing calcium carbonate, $CaCO_3$ [46, 47]. Thus, self-healing in cementitious materials can be obtained naturally or artificially [42]. The blocking of cracks through natural routes occurs owing to the following: expansion of hydrated cementitious matrix; precipitation of $CaCO_3$; presence of impurities in H_2O and further hydration of unreacted cement [42].

The artificial approach toward the development of self-healing cementitious materials targets enhancing the natural abilities of cement-based materials by engineering artificial healing abilities [42]. The artificial route focuses on filling of cracks by use of microorganism, polymers and addition of supplementary cementing materials to the concrete mix or steel fibers [45]. Microorganisms are biological agents and they are added in cement directly or in encapsulated forms to promote the precipitation of sealing compound such as $CaCO_3$ in a crack opening [48, 49]. Crack closure can also be achieved by addition of extra cement or other additives like fly ash to initial mix design to promote continuing hydration or stimulate a reaction process releasing self-sealing products [50, 51]. Self-healing in cementitious materials has also been attempted by incorporating polymers containing healing agents and shape memory materials into cement matrices [52, 53].

3.5 Self-healing coatings

Coatings can be defined as any thin layer of covering applied to the surface of a material. The basic objective of traditional coating is to separate material surfaces, especially metal surfaces from environmental corrosive attack. Most metallic materials have the intrinsic weakness of being corrodible in aqueous service environments. Corrosion normally starts at the surface and is adjudged one of the major causes of material failures. Coating acts as a barrier, limiting the diffusion of oxidation species such as oxygen and moisture to the metal surface [54]. For effective protection, the coating must maintain its adherence, structural integrity and not break down in the presence of operating factors such as mechanical stresses, abrasions, changes in pH, surface tension and temperature [54, 55].

However, over time, these operating factors lead to formation of scratches, surface and internal micro-cracks or even delamination, requiring human intervention to prevent or stop the interaction of the coated material surface and the unfriendly environment [55, 56]. The development of self-healing/smart coatings is driven by the need for damaged protective coatings to automatically sense or respond to damages and repair without human intervention when in service. Besides corrosion sensing, smart coatings have been applied to achieve self-cleaning and antifouling functions [54]. Intrinsic and extrinsic strategies have been adopted to impart self-healing capabilities in coatings using different materials of both organic and inorganic origins [54, 56], such as polymeric compounds [57], metals [58], ceramics [59] and composites [60].

Intrinsic self-healing coating can be obtained by using organic materials that undergo reversible chemistry [56] or self-reactions [61]. Extrinsic self-repair in organic coatings can be achieved by embedding self-healing agents or corrosion inhibitors in the structure of a polymer coating. The two popular methods of doing this are encapsulating healing agent in microcapsules (microencapsulation) or storing healing agents in capillary tubes (vascular networks) [55, 62]. Healing takes place when microcapsules or capillary tubes containing the healing/anticorrosion agents are ruptured by damage and release their contents, which flow into and heal the damaged areas [55, 62, 63]. The process of healing can come in form of blocking the active sites on the exposed metal surface after encountering damage. Besides storing healing agents in open polymer structures, nano-sized containers based on inorganic systems such silica, ceramic and TiO_2 have been reported to have high storing and release abilities [64–67].

In order to overcome the limitation of low storage capacity in microencapsulation-based polymeric and inorganic self-healing systems, nano-sized core-shell and microfiber containers are being exploited [68]. The major limitation of polymeric and inorganic containers is low storage capacity, depletion and noncontinuous replenishment of the healing agent contained within it on rupturing of a microcapsule. Currently, these concepts are being extended to layer-by-layer deposition, multi-shell-core microcapsules that can contain artificial or green anticorrosive agents that enable two-in-one action of self-healing and anticorrosion [69].

4. Characterization of self-healing materials

It has been obvious from past reviews on different materials systems that the major source of failures in materials is the presence of faults/damage such as voids, cracks or other defects introduced in the materials during processing or developed during utilization. Damage initiation usually starts with one or more cracks at microscopic level [70], which gradually propagate and grow in size and eventually lead to unexpected failure of the material [70, 71]. Hence, damage can be regarded as an accumulation of micro-damages that ultimately lead to material rupture if it is not repaired in due time [71]. Great efforts are spent in designing, processing and characterization to prevent the development of the damage or manage it using different self-healing strategies enabling longer service lives, greater safety and reduced maintenance costs.

For the above, self-repairing strategies target preventing, curing, by closing or filling or sealing of these cracks or voids on the surface or in the bulk of the materials by exploiting different self-healing approaches, whereas characterization focuses on proving that self-healing has taken place by determining the extent of recovery or restoration of initial properties after encountering damage(s). Much earlier than now, the characterization of self-healing capabilities in material systems concentrated mainly on the well-known macroscale evaluation, neglecting micro- and nanoscale events, responsible for initiation of material failure [71].

On the other hand, macroscale evaluation is not always sufficient, because it cannot provide comprehensive information about self-healing at all length scales or levels as it is focused mainly on the restoration of observable properties after the damage has occurred. However, it has the advantage of easier standardization when compared to microscale methods [72] as it is observable and the procedures most times are well known. To obtain a quantitative assessment and better understanding of the materials' self-healing abilities, there is need to complement macroscopic investigation with microscopic and nanoscale measurements [4, 6, 42] as self-healing processes take place on various length scales that might require sophisticated experimental procedures and equipment to reveal the mechanism of self-healing operation.

The quantification formula for estimating the capability of a given material to self-repair is the self-healing efficiency η, defined as the ability of a given material to recover a particular property relative to the virgin or undamaged material [73]. This formula, which was initially applied to polymeric materials [74], is now commonly adopted for comparing healing efficiencies in many material classes subjected to macroscopic quasi-static and dynamic tests. The expression for healing efficiency is shown in Eq. (1).

$$\eta = \frac{f_{healed}}{f_{virgin}} \text{X } 100\% \tag{1}$$

where f is a certain property of a particular material such as tensile strength, fracture toughness, tear strength, fatigue strength, flexural strength, creep rupture strength etc.

However, unlike other testing methods, monitoring or testing of self-healing materials in most cases entails inducing controlled damage such as a crack in the material and allowing it to heal using a particular healing treatment. This is followed by testing both the healed and virgin materials to failure. The extent of recovery of properties of the healed material is compared to original or virgin material properties using Eq. (1). The sequence of characterization is shown in **Figure 2a**.

Numerous quantification methods have been used for the assessment of self-healing capabilities for the different material systems. The characterization method adopted to quantify self-healing effectiveness should take into consideration the

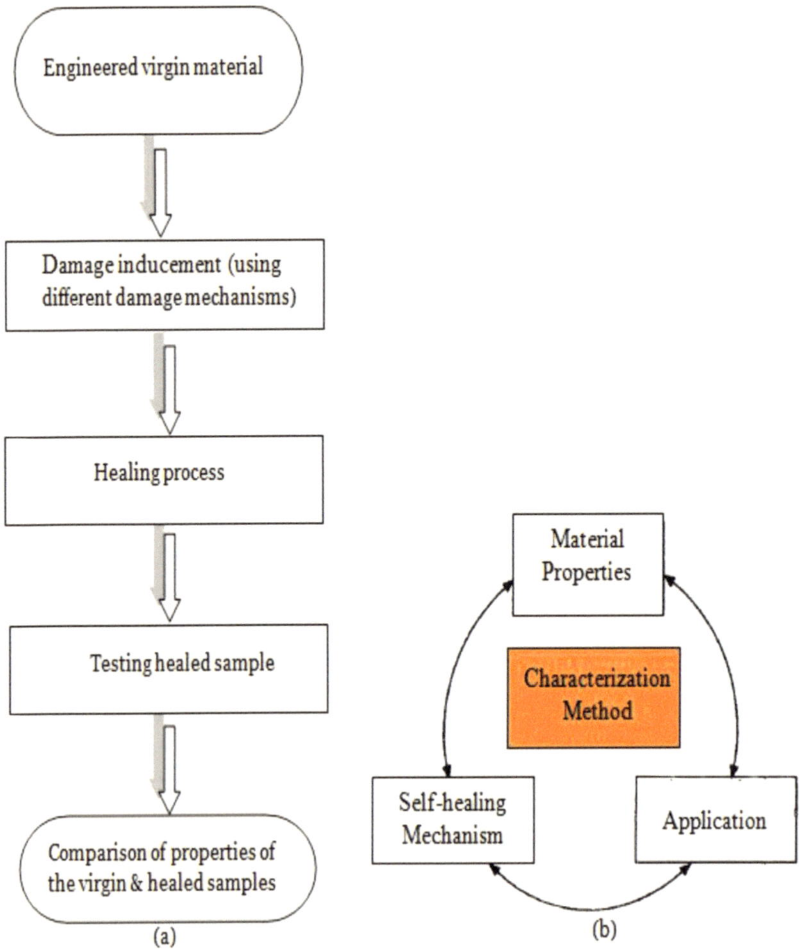

Figure 2.
(a) Steps in characterization of self-healing materials, and (b) interrelationship among material properties, self-healing mechanism, characterization method and application adapted from Ref. [6].

material properties developed, repairing mechanism and its intended application as shown in **Figure 2b** adapted from Ref. [6]. This subchapter considers quantification of healing efficiencies at different length scales in the above material classes.

4.1 General self-healing characterization techniques

4.1.1 Characterization of self-healing polymers and polymer matrix composites

Over the last few years, several testing methods have been used to assess self-repairing in polymers (thermosets, thermoplastics and elastomers) and polymer-based fiber-reinforced composites before and after repairing at macro-, micro- and nano/molecular levels. Even computational and/or predictive approaches have been attempted for deeper understanding of self-healing processes in polymer systems.

Macroscale healing evaluation leverages on fracture mechanics test procedures. Evaluation of healing requires inflicting some form of controlled crack/damage that resembles the mode of damage during utilization on the virgin polymer [73, 75, 76] by application of mechanical loads. This is accompanied by applying similar mechanical load of damage to the healed polymer and evaluation of recovery of the polymer from the fracture.

Depending on the mode of fracture, fracture evaluation loads could be in form of impact, fatigue, quasi-static fracture, tensile, compressive and flexural loads for mode I (opening) or mode III (tearing) fracture processes [71]. Also based on the type of polymeric system and the type of crack it develops under stress, the test samples have special geometries such as tapered double cantilever beam (TDCB); compact tension (CT) test specimens and width-tapered double-cantilever beam (WTDCB); double-cleavage drilled compression (DCDC); double-cantilever beam (DCB) and others [71]. These geometries and their suitability for evaluating the mechanical healing efficiencies in different polymers and polymer matrix composites are described in detail in Ref. [71].

The extent of recovery of various material properties (healing efficiency) is estimated using various forms of Eq. (1) after subjecting the damaged and undamaged samples to any or combination of the above mechanical loads as shown in Eqs. (2)–(6). The recovery of fracture toughness, flexural, tensile, impact and tear strengths has been quantified in different polymer healing systems with healing efficiencies ranging from 23 to 100% [7].

For instance, crack healing efficiency (η) for mode I type of healing can be estimated using Eq. (2) [7, 77].

$$\eta = \frac{K_{IC}^{healed}}{K_{IC}^{virgin}} X100\% \tag{2}$$

where K_{IC}^{healed} is the fracture toughness of a healed fracture specimen and K_{IC}^{virgin} is the fracture toughness of the virgin specimen.

An alternative expression for healing efficiency based on fracture energy [74, 77] is

$$\eta = \frac{G_{IC}^{healed}}{G_{IC}^{healed}} X100\% \tag{3}$$

where G_{IC}^{healed} and G_{IC}^{virgin} are the critical energy release rate from testing the healed fracture and virgin specimens, respectively.

Crack healing efficiency has also been defined in terms of fatigue life-extension [77, 78] as

$$\eta_d = \frac{N_{healed - N_{control}}}{N_{control}} \tag{4}$$

where N_{healed} and $N_{control}$ are the total number of cycles to failure for a self-healing sample and *for* similar sample without healing, respectively.

For elastomeric self-healing materials, the recovery of tear strength is used to define healing efficiency as

$$\eta = \frac{T_{healed}}{T_{virgin}} X100\% \tag{5}$$

where T_{healed} is the tear strength of the healed material and T_{virgin} is tear strength of virgin material.

Healing efficiency has also been estimated based on change in stiffness and recovery in a damaged and healed polymer [79] as

$$\eta = \frac{E_{healed}}{E_{virgin}} X100\% \qquad (6)$$

Besides the macroscale methods used for mechanical performance evaluation of self-healing materials, evaluation at smaller length scale (micro- and nanoscale levels) is necessary to reveal the underlying healing mechanisms in polymers and deepen understanding of self-healing [6]. Characterization techniques at this scale enable the monitoring of the whole process of self-healing from stage of inflicting damage to identifying interactions and confirmation of healing functionality at molecular/nanolevels [16, 80]. The techniques include imaging, spectrometric, scattering, rheological and thermal techniques [6].

4.2 Characterization of self-healing metals

Self-healing efficiency quantification of metals based on the bulk material properties can also be carried out by subjecting pre-cracked samples to various mechanical tests and self-healing efficiency determined using Eq. (1). For instance, the self-healing efficiency of metallic system based on precipitation-induced approach can be evaluated by subjecting the age-hardened alloy (virgin alloy) and its pre-cracked counterpart but filled low-melting alloy to tensile loading to fracture. Alaneme and Omosule [81] used this method to determine the self-healing efficiency of underaged Al-Mg-Si alloys and 60Sn-40Pb alloy-reinforced aluminum metal–metal composites. The self-healing efficiency, η, was estimated using relation based on the tensile strength criterion given by Eq. (7):

$$\eta_{tensile} = \frac{\sigma_{healed}}{\sigma_{virgin}} X100\% \qquad (7)$$

where σ_{virgin} is the tensile strength of the virgin specimen and σ_{healed} is the tensile strength of the healed specimen.

Other experimental techniques used to characterize solid materials subjected cyclic or creep loads can also be adapted to evaluate bulk metallic material systems.

Mechanical evaluation of materials using micro-indentation techniques is a widely accepted tool to reveal information on the surfaces of bulk hard materials [4] and it can be readily applied to study healing at micron scale. Self-mending at this length scale can be studied by inducing mechanical damage through micromachining accompanied by imaging of the repairing process. The imaging can be done using low and high imaging equipment such as optical microscope, scanning electron microscope (SEM), energy dispersive X-ray spectroscope (EDS) and environmental scanning electron microscope (ESEM) or X-ray micro-tomography instrument to provide details concerning crack propagation arrest [31] and evolution of self-healing reactions and to reveal evolved micro-structures and morphology [4]. The results obtained at micro- and nanolevels are used to buttress results at macrolevels.

Most research conducted on self-healing metals focused on either solid-state diffusion healing of micro-cracks, or shape memory alloy (SMA)-reinforced "off-eutectic" matrices. It is worthwhile to conduct tests at nanolevels to elucidate the bonding at the interface between the diffusing species and the metal matrices. Laha et al. [82] applied nanoscale investigation using Auger spectroscopy to show that boron (B) atom acts as a solute healing agent in 347-austenitic stainless steel by diffusing to the nano voids and precipitating at the void surfaces. He et al. [83] equally used positron annihilation spectroscopy to confirm that the addition of B and nitrogen (N) maximally accelerates the precipitation of copper (Cu) in iron-copper (Fe-Cu) alloy closing nano-voids in the system.

4.3 Characterization of self-healing ceramics

The current fracture mechanic tests used to assess self-healing capabilities in hard materials are also applicable to advanced ceramics, but the difficulty in creating controlled cracks in ceramic materials limits this application. Unlike other hard materials that exhibit some level of plasticity, ceramic materials are very brittle in nature and are prone to unwanted fast fracture under mechanical and thermal stresses [71]. Another limitation is that it is difficult achieve crack closure in ceramics at low temperatures. The most common procedure for creating controlled micro-cracks and quantifying healing efficiency is indentation method [84]. The healing efficiency is evaluated in terms of crack closure by comparing the control samples with the healed specimens.

For instance Nam and Hwang [40] investigated crack healing behavior of ZrO_2/SiC composite ceramics with TiO_2 additive. Cracks of about 100 μm were made on the sample surfaces using Vickers indenter. Since self-healing in ceramics is a high-temperature process, the indented samples were heat-treated to stimulate healing, but the test was conducted at room temperature. This was followed by observing the pre- and post-healed indents with X-ray diffractometer (XRD). The strength of crack closure was determined using three-point bending test. Li et al. [85] also used flexural test to evaluate the multiple healing of titanium aluminum carbide (Ti_2AlC) ceramic damaged by indentation in terms of crack propagation.

Besides using bending tests, tensile and the biaxial ball-on-three balls (B3B) tests have been used to study healing efficiencies in ceramics at room temperatures. Gao and Suo [86] assessed adhesion healing efficiency in a ceramic coating by performing tensile tests and correlating the healing time and residual stress while Harrer et al. [87] studied the healing of surface defects induced by different machining conditions on silicon nitride ceramic using biaxial ball-on-three balls.

All these tests are usually performed at room temperature while the healing process takes place at elevated temperatures >1000°C with the attendant oxidative atmosphere and thermal stresses. At high temperatures, the internal structure and mechanical performance of the ceramics could change due to the accompanying local melting and phase transformations [88]. Since mechanical performance under the above atmosphere will be different from that at low temperatures, there is the need to develop a more suitable method for the quantification of self-healing efficiency in ceramics [71] that takes into consideration the real service conditions. Attempts in this direction have been made by Ando et al. [89, 90], who determined *in situ* crack healing ability by conducting mechanical tests on some ceramics at elevated temperatures [89, 90].

4.4 Characterization of self-healing concrete

The major cause of mechanical failure in concrete is cracks. A crack not only lowers strength, but exposes the reinforcing steel components to corrosion. Self-healing targets crack closure or prevention of crack propagation to retain strength and reduce water permeability in order to maintain durability. Self-healing efficiency of concrete in hardened form has been determined by conducting tests at macro-, micro-, and nanoscale levels. Majority of the researchers evaluated the self-healing efficiency at macrostructural level, some at microscale level and very few authors at nanostructure level [42].

Just like other hard materials, quantification of healing efficiency on one hand relies on fracture mechanics tests, which involve creation of controlled cracks on concrete. The mechanical characterization procedure follows the sequence of controlled crack initiation in the matrix using the standard compression and bending tests, healing processing and retesting of the healed concrete using the same pre-cracking procedure [91]. The initiation of cracks without causing failure at certain level of stress and detection of crack development are very important in testing of concrete. To control crack initiation, some authors have applied a notch in the middle point of test sample [92] while some used three-point bending or four-point bending technique found to be more effective in crack initiation without causing failure [93–95].

Besides conducting compressive and flexural tests [45, 96, 97], the performance of concrete has also been assessed by other mechanical tests such as split tensile and toughness tests [98, 99] and stiffness tests [100]. Detection of crack initiation and its degree have been carried out using nondestructive complementary tests such as acoustic emission analysis [101], linear variable differential transformer [102] and ultrasonic pulse velocity [103, 104].

On the other hand, efficiency is also evaluated by conducting permeability tests on the pre-cracked concrete. Permeability test aims at determining how effective self-healing concrete can shield steel bars from corrosion [105]. Permeability tests are performed in simulated environments containing fluids such as chlorine or water under certain temperature [94, 105].

Microscale tests are employed to identify and characterize the deposited materials within cracks in the concrete after self-healing and are used to complement and reinforce reliability of macroscale tests [42]. These deposited materials are the calcium carbonate precipitation by different bacterial strains, hydration product as well as polymerized products. Several of these tests are conducted using the following sensitive equipment: scanning electron microscope (SEM), field emission scanning electron microscope (FESEM), and X-ray diffractometer (XRD). SEM is used to identify the morphology of the deposited materials within the cracks [106]. Self-healing performance is also assessed using Raman spectroscopy [16]. Furthermore, nanostructure test has been used to evaluate self-healing efficiency of concrete [107]. Tests conducted at nanoscale help in the determination of bonding strength at the interface between the deposited materials in the cracks and the concrete.

4.5 Characterization of self-healing coatings

The basic function of a traditional coating is to shield material surfaces, especially metal surfaces from fast corrosion in the environment. Smart coatings provide a spontaneous protection to metal surfaces upon chemical or mechanical damage [108]. This is achieved by release of inhibiting species in the coating architectures, which inhibits electrochemical interaction between the metal substrates and the environment [109]. The characterization techniques highlighted here are those traditionally used to study corrosion, but adapted for studying self-healing coatings on metals.

The testing techniques used to ascertain the self-healing properties of coatings have been generally grouped into two: electrochemical [55, 109, 110] and non-electrochemical techniques or physicochemical characterization as shown in **Figure 3** [55, 109]. Electrochemical techniques enable the quantification of self-healing efficiency by providing important information about kinetics of protection; formation of protective films and isolation of redox species [109]. The electrochemical techniques are further divided into conventional (global) electrochemical methods and localized electrochemical techniques [55, 109].

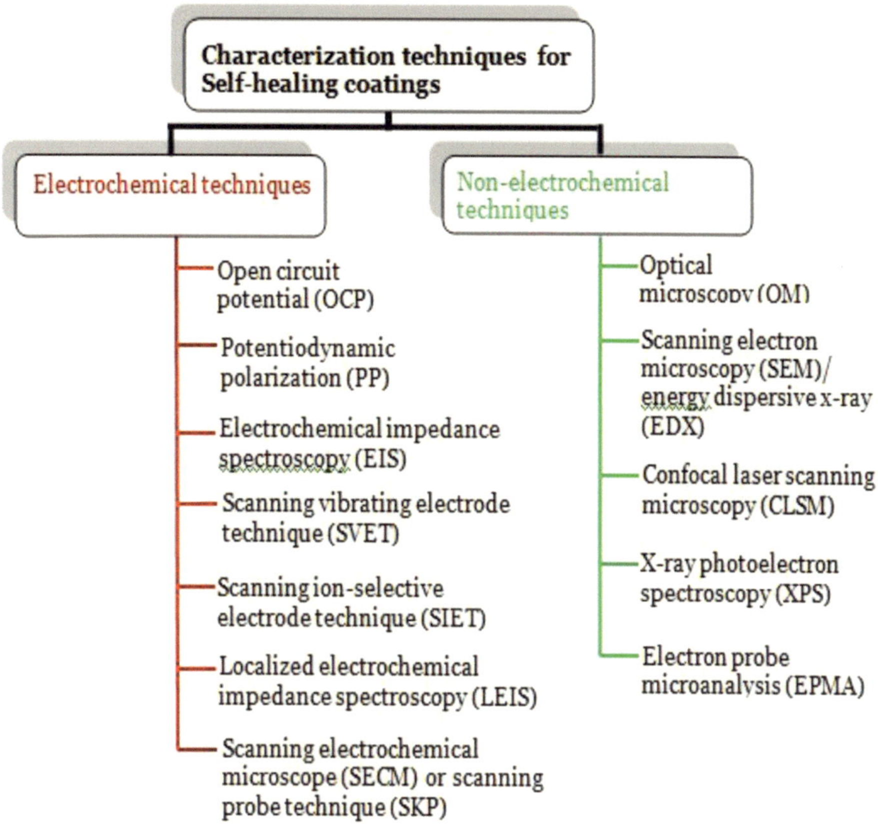

Figure 3.
Techniques for evaluating self-healing coatings adapted from [55].

The global electrochemical methods provide kinetic and deterministic information on self-healing processes, but they do not supply information about local reactions taking place at the site of damage. They include potentiodynamic polarization (PP), open circuit potential (OCP), electrochemical impedance spectroscopy (EIS) and they are the popular methods used for the study of corrosion and self-healing research [109]. For instance, PP and EIS give quantitative results about self-healing process enabling the determination of corrosion rate and protection efficiency [55]. Aramaki [111] employed the PP method to estimate the protection efficiency η_p of a protective polymer coating on an iron electrode surface using Eq. (8):

$$\eta_p = \frac{i^{\circ}_{corr} - i_{corr}}{i^{\circ}_{corr}} x100 \quad (8)$$

where i_{corr} is the corrosion current density with coating and i°_{corr} the is corrosion current density without coating.

Information about evolution of damage and healing of damage is provided by the local electrochemical methods, including micro-capillary cell, scanning vibrating electrode technique (SVET), scanning ion-selective electrode technique (SIET), scanning electrochemical microscope (SECM) or scanning probe technique (SKP) and localized electrochemical impedance spectroscopy (LEIS).

Each of these methods has advantages and limitations and should be used in combinations for detailed study of self-healing processes in coatings. Besides this, these electrochemical techniques should be complemented by conducting non-electrochemical or physicochemical analysis. Physicochemical analysis is majorly aimed at studying mechanism of self-healing coating protection—elemental analysis, interaction mechanisms between the metal substrate and the coating, morphology or phase transformation of the coating before and after healing. Surface analysis can be carried out with OM, SEM and CLSM, while energy dispersive X-ray (EDX), EPMA and XPS are useful for elemental analysis. The detailed description of the above test methods can be found in Refs. [55, 109, 112].

4.6 Challenges in the characterization of self-healing materials

Developing an artificial material to self-heal like a biological system comes as a huge task to the scientist or engineer. But more challenging is proving the material's self-healing capability and suitability for a particular application via characterization. Different approaches are used to achieve self-healing in different material classes based on their inherent properties. A self-healing material is an advanced material, a new product with new properties different from its original properties. It is important that the characteristics of the new product should be determined qualitatively or quantitatively. It is also important to monitor the changes during processing that led to the new properties. As a result, specific characterization methods are required to quantify self-healing efficiencies in each of the material classes. Most times, the equipment might not be common and sometimes unavailable unless improvised.

Unlike other characterization methods, evaluation of self-healing materials in most cases entails inducing a minimal expected mode of damage or failure during utilization on the material and using different methods to determine the extent of recovery/restoration of properties compared to original or virgin materials properties and to understand the mechanism of recovery. The methods of inducing the damage are different for the various material classes and even in the same material group. Inducing the appropriate damage simulating the real-life scenario is challenging. Also tasking is the selection of an appropriate testing procedure from an extensive range of known materials testing procedures that is adaptable and suitable for a particular self-healing concept. The field of self-healing materials is relatively new but a richly rewarding venture. The understanding of self-healing mechanisms in a variety of self-repairing material classes is still evolving. So also are the characterization methods needed to elucidate the dynamics of self-healing process. The challenges of harmonization of these methods in various research groups are yet to be resolved.

More so, the mode of damage is different and unique to the damaged material and its intended applications. Even within the same material class, there are various self-healing approaches and evaluation strategies. This makes the adopted routine of assessing the performance of the modified material and comparing its properties to the unmodified, virgin material complex. This makes it equally arduous to establish a common testing procedure for similar or for different materials classes.

5. Prospects for standardization of characterization techniques

Most of the current characterization methods used to quantify the healing performance focused on the macroscopic evaluation of recovery from macroscopically applied damage. However, early stage of damage and recovery occur at sub-macro level. Therefore, macroscopic evaluation cannot by itself be sufficient enough for self-healing quantification [4]. As the materials' failure normally starts at nano- and microscale levels, a sufficient and necessary quantification approach should take into consideration the damaging and healing events at these small length scales [71], more so, when they can easily be prevented or healed faster at the sublevels.

However, inflicting of macroscopic damage on hard self-healing materials, though far from the actual utilization conditions, can be easily replicated and holds the promise of easier standardization and comparison across similar materials [71]. Since damage initiation inherently starts at the microscopic/nanoscale level, inflicting of damage protocol can be done at microscopic level using nano- or/and micro-indentation techniques. Mechanical damage can be induced using micromachining, accompanied by imaging of the mending process using high-resolution imaging equipment such as environmental scanning electron microscope (ESEM) [4]. Currently, micro-indentation testing protocol is widely accepted and standardized tool for mechanical testing of materials [113]. *In situ* measurement techniques can also be very helpful in such environment, where dynamic and on process conditions can be captured or easily replicated. At the heart of this standardization prospect is the operator's skill.

Microscopic evaluation has limitations. One challenge is the requirement of very small volume of samples, which makes sampling difficult due to the nonhomogeneous composition of some materials [4]. Another is that it is not very suitable for testing soft materials. However, most of the self-healing material classes are hard materials. Some polymers, metals, ceramics, concrete and some coatings, especially inorganic coatings are all hard materials. Testing using micro-indentation technique complemented with macroscopic methods could be a useful step toward standardization of quantification in hard self-healing materials.

In practical situations, a single applied load affects more than one specific material's property and self-healing efficiency is regarded as the ability of a given material to recover a specific property relative to the virgin specimen. In this situation, efficiency calculation should take the combined properties of the material affected by the load into consideration and should be reported as an average or overall efficiency. The use of overall efficiency as a method of assessing performance of distinct materials using different test methods opens a path toward standardization irrespective of testing method and material class [71].

6. Future trends

The adopted routine of assessing the performance of self-healing materials has been to characterize the modified self-healing material and compare its properties to the unmodified, virgin material. An extensive literature survey indicated that it was difficult to find researchers who evaluated self-healing performance at macro-, micro- and nanostructural levels simultaneously. This is probably because the field is relatively new. Today, it is becoming obvious that an appropriate performance assessment method should take into account the damaging and healing at macroscopic as well as microscale/nanoscale levels. In order to achieve this, a combined suitable and reproducible evaluation procedure must be exploited for a better understanding of damage mechanism and healing process.

It is equally hard to find researchers who investigated several properties at the same time in one material. In real-life situations, the damage initiation and eventual failure can be caused by combined factors/loads-tensile, compressive, cyclic, bending, creep, thermal loads and others. The self-healing efficiency is defined as the ability of a given material to recover a specific property relative to the virgin or undamaged specimen [75]. For instance, a single applied load affects more than one specific material's property and efficiency calculation should take into account other properties of the materials affected by the load [71]. Therefore, the most appropriate evaluation approach should be the one that takes into account more than one property and reports combined efficiencies or overall efficiency, equal to prime average of the efficiency obtained for each material property [71].

A case could also be made to suggest or define effective predictive approaches or methods that would lead to faster evaluation and design of self-healing quantifications in materials at various length scales.

7. Concluding remarks

The development of self-healing materials comes as a huge challenge to the scientist or engineer—the challenge of synthesizing smart materials that can self-repair, elucidating the healing mechanism and proving their self-healing capabilities through characterization. There are many material classes and different approaches have been attempted to achieve self-healing capabilities in these materials. For close to two decades, several self-healing materials have been developed and many methods have been employed to assess self-healing behavior and determine healing efficiency of these materials. As the materials are different, so are the evaluation techniques utilized to characterize the healing behaviors. Besides taking into consideration the type of material, the healing-enabling preparatory route and repairing mechanism, the most suitable test method should fit the intended application. For completeness, effective characterization should be the one encompassing all length scales—macro-, micro- and nanoscale levels. Thus, an ideal quantification approach needs to take into account the macroscopic as well as microscale aspects of damaging and healing.

Among the several characterization methods utilized to investigate self-healing behavior and determine healing efficiency in metals, polymers and polymer composites, ceramics and concrete at macroscale evaluation focusing on restoration of mechano-physical properties is popular. Typically, most characterization methods in metals are carried out at the macroscale level, but healing takes place at nanoscale level. This can be a fundamental limitation in the characterization process for metals. However, evaluation techniques at micro- and nanoscale levels have been employed to link and correlate mechanical healing with underlying molecular processes in particular polymeric materials. However, testing of polymers and other material systems do come with their own challenges, including getting reliable information from testing of modified materials only with the available small-scale samples and at laboratory conditions. Long-time instability of polymers is also a problem as it has been demonstrated that the healing efficiency of extrinsic self-healing systems decreases over time.

In ceramics and concrete, initiation of controlled damage is somewhat tasking due to their inherent brittleness and low diffusion rate of healing agents. This is also likened to metals, whose self-healing mostly occurs faster at high temperature, but damage initiates at low temperature. Therefore, simulating the real condition of damage and healing simultaneously is a herculean task. On the other hand, coating is used in various conditions, but the nanoscale evaluation of indentation is carried out at controlled environments, which are different from its real application condition.

Conflict of interest

The authors declare no conflicts of interest with respect to the authorship and publication of this chapter.

Author details

Camillus Sunday Obayi* and Paul Sunday Nnamchi
Department of Metallurgical and Materials Engineering, University of Nigeria, Nsukka, Nigeria

*Address all correspondence to: camillus.obayi@unn.edu.ng

References

[1] der ZS V. Self-Healing Materials: An Alternative Approach to 20th Centuries of Materials Science. Berlin: Springer; 2007. pp. 161-193. DOI: 10.1515/ci.2008.30.6.20

[2] Hayes SA, Jones FR, Marhiya K, Zhang W. Self-healing composite materials. In: Proceedings of the 15th International Conference on Composite Materials. Durban; 27 June-1 July 2005. pp. 24-26

[3] Sitnikov NN, Khabibullina IA, Mashchenko VI, Rizakhanov RN. Prospects of application of self-healing materials and technologies based on them inorganic materials. Applied Research. 2018;**9**(5):785-793. DOI: 10.1134/S207511331805026X

[4] Lucas SS, Nellesen A, Von TM. Assessment of self-healing capabilities: A route towards standardization. In: Proceedings of the 4th International Conference on Self-Healing Materials (ICSHM2013); 17-19 June 2013. Ghent, Belgium: ICSHM; 2013

[5] Madara SR, Srath Raj NS, Selvan CP. Review of research and developments in self healing composite materials. IOP Conference Series: Materials Science and Engineering. 2018;**346**:012011. DOI: 10.1088/1757-899X/346/1/012011

[6] Bode S, Enke M, Hernandez M, Bose RK, Grande AM, Schubert US, et al. Characterization of self-healing polymers: From macroscopic healing tests to the molecular mechanism. Advances in Polymer Science. 2015;**273**:113-142. DOI: 10.1007/12_2015_341

[7] Aïssa B, Therriault D, Haddad E, Jamroz W. Self-healing materials systems: Overview of major approaches and recent developed technologies. Advances in Materials Science and Engineering. 2012:854203. DOI: 10.1155/2012/854203

[8] Pang JWC, Bond IP. Bleeding composites—Damage detection and self-repair using a biomimetic approach. Composites Part A: Applied Science and Manufacturing. 2005;**36**:183-188. DOI: 10.1016/j.compositesa.2004.06.016

[9] Jones S, Rule JD, Moore JS, Sottos NR, White SR. Life extension of self-healing polymers with rapidly growing fatigue cracks. Journal of the Royal Society, Interface. 2007;**4**:395-403. DOI: 10.1098/rsif.2006.0199

[10] Trask RS, Williams H, Bond IP. Self-healing polymer composites: Mimicking nature to enhance performance. Bioinspiration & Biomimetics. 2007;**2**:1-9. DOI: 10.1088/1748-3182/2/1/P01

[11] Sanada K, Yasuda I, Shindo Y. Transverse tensile strength of unidirectional fibre-reinforced polymers and self-healing of interfacial deboning. Plastics, Rubber and Composites. 2006;**35**:67-72. DOI: 10.1179/174328906X79914

[12] Wu DY, Meure S, Solomon D. Self-healing polymeric materials: A review of recent developments. Progress in Polymer Science. 2008;**33**:479-522. DOI: 10.1016/j.progpolymsci.2008.02.001

[13] Liu Y-L, Chuo T-W. Self-healing polymers based on thermally reversible Diels–Alder chemistry. Polymer Chemistry. 2013;**4**:2194-2205

[14] Kuhl N, Bode S, Hager MD, Schubert US. Self-healing polymers based on reversible covalent bonds. In: Hager MD, Zwaag VS, Schubert US, editors. Self-Healing Materials. Advanced Polymer Science. Switzerland: Springer International Publishing; 2016. pp. 1-58. DOI: 10.1007/12_2015_336

[15] Blaiszik BJ, Kramer SLB, Olugebefola SC, Moore JS, Sottos NR, White SR. Self-healing polymers and composites. Annual Review of Materials Research. 2010;**40**:179-211. DOI: 10.1146/annurev-matsci-070909-104532

[16] Bekas DG, Tsirka K, Baltzis D, Paipetis AS. Self-healing materials: A review of advances in materials, evaluation, characterization and monitoring techniques. Composites Part B Engineering. 2016;**87**:92-119. DOI: 10.1016/j.compositesb.2015.09.057

[17] White SR, Sottos NR, Geubelle PH, Moore JS, Kessler MR, Sriram SR, et al. Autonomic healing of polymer composites. Natural Product Letters. 2001;**409**:794-797. DOI: 10.1038/35057232

[18] Patrick JF, Hart KR, Krull BP, Diesendruck CE, Moore JS, White SR, et al. Continuous self-healing life cycle in vascularized structural composites. Advanced Materials. 2014;**26**(25):4302e8. DOI: 10.1002/adma.201400248

[19] Garcia SJ. Effect of polymer architecture on the intrinsic self-healing character of polymers. European Polymer Journal. 2014;**53**:118-125. DOI: 10.1016/j.eurpolymj.2014.01.026

[20] Mulville DR, Wolock I. Failure of polymer composites. In: Andrews EH, editor. Developments in Polymer Fracture. London: Applied Science Publishers; 1979. pp. 263-316

[21] Greenhalgh ES. Failure Analysis and Fractography of Polymer Composites. Cambridge: Woodhead Publishing Ltd; 2009. pp. 1-608

[22] Grabowski B, Tasan CC. Self-healing metals. In: Hager M, van der Zwaag S, Schubert U, editors. Self-Healing Materials. Advances in Polymer Science. Vol. 273. Cham: Springer; 2016. pp. 387-407. DOI: 10.1007/12_2015_337

[23] van Dijk N, van der Zwaag S. Self-healing phenomena in metals, special issue: Self-healing materials. Advanced Materials Interfaces. 2018;**5**:7. DOI: 10.1002/admi.201800226

[24] Ferguson JB, Schultz BF, Rohatgi PK. Self-healing metals and metal matrix composites. Journal of Metals. 2014;**66**(6):866-871. DOI: 10.1007/s11837-014-0912-4

[25] Zwaag SV. Routes and mechanisms towards self healing behaviour in engineering materials. Bulletin of the Polish Academy of Sciences: Technical Sciences. 2010;**58**(2):226-237. DOI: 10.2478/v10175-010-0022-6

[26] Alaneme KK, Bodunrin MO. Self-healing using metallic material systems-a review. Applied Materials Today. 2017;**6**:9-15. DOI: 10.1016/j.apmt.2016.11.002

[27] Rohatgi PK. Al-shape memory alloy self-healing metal matrix composite. Materials Science and Engineering A. 2014;**619**:73-76. DOI: 10.1016/j.msea.2014.09.050

[28] Baxevanis T, Parrinello AF, Lagoudas DC. On the fracture toughness enhancement due to stress-induced phase transformation in shape memory alloys. International Journal of Plasticity. 2013;**50**:158-169. DOI: 10.1016/j.ijplas.2013.04.007

[29] Grabowski B, Tasan C. Towards self-healing metals by employing optimally dispersed Ti-Ni shape memory nano-particles, project within DFG SPP 1568. In: Design and Generic Principles of Self-healing Materials. Bonn: DFG; 2014

[30] Manuel MV. Principles of self-healing in metals and alloys: An introduction. In: Ghosh G, editor. Self-Healing Materials: Fundamentals, Design Strategies, and Applications. Wiley-VCH; 2009. pp. 251-263

[31] Leser PE. Mitigation of Crack Damage in Metallic Materials, Report by National Aeronautics and Space Administration. NASA/TM–2014-218272. 2014. Available from: https://ntrs.nasa.gov/search.jsp?R=20140006911 2020-05-20T08:19:01+00:00Z

[32] Zheng XG, Shi YN, Lu K. Electro-healing cracks in nickel. Materials Science and Engineering A. 2013;**561**:52-59. DOI: 10.1016/j.msea.2012.10.080

[33] Zhou Y, Guo J, Gao M, He G. Crack healing in a steel by using electropulsing technique. Materials Letters. 2004;**58**(11):1732-1736. DOI: 10.1016/j.matlet.2003.10.049

[34] Zwaag VS. An introduction to material design principles: Damage prevention versus damage management. In: Zwaag VS, editor. Self Healing Materials. Vol. 100. Dordrecht: Springer Series in Materials Science, Springer; 2007. pp. 1-18. DOI: 10.1007/978-1-4020-6250-6_1

[35] Greil P. Generic principles of crack-healing ceramics. Journal of Advanced Ceramics. 2012;**1**(4):249-267. DOI: 10.1007/s40145-012-0020-2

[36] Ayatollahi MR, Aliha MRM. Fracture analysis of some ceramics under mixed mode loading. Journal of the American Ceramic Society. 2011;**94**(2):561-569. DOI: 10.1111/j.1551-2916.2010.04076.x

[37] Suresh S, Shih CF, Morrone A, O'Dowd NP. Mixed-mode fracture toughness of ceramic 1007 materials. Journal of the American Ceramic Society. 1990;**73**(5):1257-1267. DOI: 10.1111/j.1151-2916.1990.tb05189.x

[38] Houjou K, Ando K, Takahashi K. Crack-healing behaviour of ZrO_2/SiC composite ceramics. International Journal of Structural Integrity. 2010;**1**(1):73-84. DOI: 10.1108/17579861011023810

[39] Jarvis EA, Carter EA. A nanoscale mechanism of fatigue in ionic solids. Nano Letters. 2006;**6**(3):505-509. DOI: 10.1021/nl0525655

[40] Nam KW, Hwang JR. The crack healing behavior of ZrO_2/SiC composite ceramics with TiO_2 additive. Journal of Mechanical Science and Technology. 2012;**26**:2093-2096. DOI: 10.1007/s12206-012-0521-5

[41] Jun L, Zheng ZX, Ding HF, Jin ZH. Preliminary study of the crack healing and strength recovery of Al_2O_3-matrix composites. Fatigue and Fracture of Engineering Materials and Structures. 2004;**27**:89-97. DOI: 10.1111/j.1460-2695.2004.00725.x

[42] Muhammad NZ, Shafaghat A, Keyvanfar A, Abd. Majid MZ, Ghoshal SK, Yasouj SEM, et al. Tests and methods of evaluating the self-healing efficiency of concrete: A review. Construction and Building Materials. 2016;**112**:1123-1132. DOI: 10.1016/j.conbuildmat.2016.03.017

[43] Pour-Ghaz M, Weiss J. Detecting the time and location of cracks using electrically conductive surfaces. Cement and Concrete Composites. 2011;**33**:116-123. DOI: 10.1016/j.cemconcomp.2010.09.015

[44] Durga CSS, Ruben N. Assessment of various self healing materials to enhance the durability of concrete structures. Annales de Chimie–Science des Matériaux. 2019;**43**(2):75-79. DOI: 10.18280/acsm.430202

[45] Lucas SS, Moxham C, Tziviloglou E, Jonkers H. Study of self-healing properties in concrete with bacteria encapsulated in expanded clay. Science and Technology of Advanced Materials. 2018;**30**:93-98. DOI: 10.1016/j.stmat.2018.11.006

[46] Reinhardt H, Joos M. Permeability and self-healing of cracked concrete as a function of temperature and crack width. Cement and Concrete Research. 2003;**33**(7):981-985. DOI: 10.1016/S0008-8846(02)01099-2

[47] Hearn N. Self-sealing, autogenous healing and continued hydration: What is the difference? Materials and Structures. 1998;**31**:563-567. DOI: 10.1007/BF02481539

[48] Muynck WD, Debrouwer D, De Belie N, Verstraete W. Bacterial carbonate precipitation improves the durability of cementitious materials. Cement and Concrete Research. 2008;**38**:1005-1014. DOI: 10.1016/j.cemconres.2008.03.005

[49] Stuckrath C, Serpell R, Valenzuela LM, Lopez M. Quantification of chemical and biological calcium carbonate precipitation: Performance of self-healing in reinforced mortar containing chemical admixtures. Cement and Concrete Composites. 2014;**50**:10-15. DOI: 10.1016/j.cemconcomp.2014.02.005

[50] Termkhajornkit P, Nawa T, Yamashiro Y, Saito T. Self-healing ability of fly ash-cement systems. Cement and Concrete Composites. 2009;**31**:195-203. DOI: 10.1016/j.cemconcomp.2008.12.009

[51] Rahmani H, Bazrgar H. Effect of coarse cement particles on the self-healing of dense concretes. Magazine of Concrete Research. 2015;**67**:476-486. DOI: 10.1680/macr.14.00158

[52] Perez G, Gaitero JJ, Erkizia E, Jimenez I, Guerrero A. Characterization of cement pastes with innovative self-healing system based in epoxy-amine adhesive. Cement and Concrete Composites. 2015;**60**:55-64. DOI: 10.1016/j.cemconcomp.2015.03.010

[53] Wang X, Zhao J, Chen M, Ma L, Zhao X, Dang ZM, et al. Improved self-healing of polyethylene/carbon black nanocomposites by their shape memory effect. The Journal of Physical Chemistry. B. 2013;**117**:1467-1474. DOI: 10.1021/jp3098796

[54] Vijayan P, Al-Maadeed M. Self-repairing composites for corrosion protection: A review on recent strategies and evaluation methods. Materials. 2019;**12**(2754):1-17. DOI: 10.3390/ma12172754

[55] Fayyad EM, Almaadeed MA, Jones A, Abdullah M. Evaluation techniques for the corrosion resistance of self-healing coatings. International Journal of Electrochemical Science. 2014;**9**:4989-5011

[56] Shi X, Fernando BMD, Croll S. Concurrent physical aging and degradation of cross-linked coating systems in accelerated weathering. Journal of Coating Technology and Research. 2008;**5**:299-309. DOI: 10.1007/s11998-008-9081-0

[57] DaConceicao TF, Scharnagl N, Dietzel W, Hoeche D, Kainer KU. Study on the interface of PVDF coatings and HF-treated AZ31 magnesium alloy: Determination of interfacial interactions and reactions with self-healing properties. Corrosion Science. 2011;**53**:712-719. DOI: 10.1016/j.corsci.2010.11.001

[58] Koleva DA, Boshkov N, Bachvarov V, Zhann H, de Wit JHW, van Breugel K. Application of PEO_{113}–b-PS_{218} nano-aggregates for improved protective characteristics of composite zinc coatings in chloride-containing environment. Surface and Coating Technology. 2010;**204**(23):3760-3772. DOI: 10.1016/j.surfcoat.2010.04.043

[59] Gao J, Suo J. Modeling the barrier properties of polymer-layered silicate nanocomposites. Surface and Coating Technology. 2001;**34**(26):9189-9192. DOI: 10.1021/ma010780b

[60] Moustafa EM, Dietz A, Hochsattel T. Manufacturing of nickel/nanocontainer composite coatings. Surface and Coating Technology. 2013;**216**:93-99. DOI: 10.1016/j.surfcoat.2012.11.030

[61] Cordier P, Tournilhac F, Soulie-Ziakovic C, Leibler L. Self-healing and thermoreversible rubber from supramolecular assembly. Nature. 2008;**451**:977-980. DOI: 10.1038/nature06669

[62] Zhu DY, Rong MZ, Zhang MQ. Self-healing polymeric materials based on microencapsulated healing agents: From design to preparation. Progress in Polymer Science. 2015;**49-50**:175-220. DOI: 10.1016/j.progpolymsci.2015.07.002

[63] Toohey KS, Sottos NR, Lewis JA, Moore JS, White SR. Self-healing materials with microvascular networks. Nature Materials. 2007;**6**:581-585. DOI: 10.1038/nmat1934

[64] Maia F, Tedim J, Lisenkov AD, Salak AN, Zheludkevich ML, Ferreira MGS. Silica nanocontainers for active corrosion protection. Nanoscale. 2012;**4**:1287-1298. DOI: 10.1039/C2NR11536K

[65] Kartsonakis IA, Balaskas AC, Koumoulos EP, Charitidis CA, Kordas GC. Incorporation of ceramic nanocontainers into epoxy coatings for the corrosion protection of hot dip galvanized steel. Corrosion Science. 2012;**57**:30-41. DOI: 10.1016/j.corsci.2011.12.037

[66] Arunchandran C, Ramya S, George RP, Mudali UK. Self-healing corrosion resistive coatings based on inhibitor loaded TiO2 nanocontainers. Journal of the Electrochemical Society. 2012;**159**:C552-C559. DOI: 10.1149/2.020212jes

[67] Lamaka SV, Zheludkevich ML, Yasakau KA, Serra R, Poznyak SK, Ferreira MGS. Nanoporous titania interlayer as reservoir of corrosion inhibitors for coatings with self-healing ability. Progress in Organic Coating. 2007;**58**(2-3):127-135. DOI: 10.1016/j.porgcoat.2006.08.029

[68] Zheng Z, Schenderlein M, Huang X, Brownbill NJ, Blanc F, Shchukin D. Influence of functionalization of nanocontainers on self-healing anticorrosive coatings. ACS Applied Materials & Interfaces. 2015;**7**:22756-22766. DOI: 10.1021/acsami.5b08028

[69] Jadhav RS, Mane V, Bagle AV, Hundiwale DG, Mahulikar PP, Waghoo G. Synthesis of multicore phenol formaldehyde microcapsules and their application in polyurethane paint formulation for self-healing anticorrosive coating. International Journal of Industrial Chemistry. 2013;**4**:31. DOI: 10.1186/2228-5547-4-31

[70] Darabi MK, Abu Al-Rub RK, Little DN. A continuum damage mechanics framework for modeling microdamage healing. International Journal of Solids and Structures. 2012;**49**(3-4):492-513. DOI: 10.1016/j.ijsolstr.2011.10.017

[71] Lucas SS, Tapavicza MV, Schmid AM, Bertling J, Nellesen A. Study of quantification methods in self-healing ceramics, polymers and concrete: A route towards standardization. Journal of Intelligent Material Systems and Structures. 2016:1-22. DOI: 10.1177/1045389X16641205

[72] Zhu HH, Zhou S, Yan ZG, JU JW, Chen Q. A two-dimensional micromechanical damage-healing model on micro-crack induced damage for microcapsule-enabled self-healing cementitious composites under tensile loading. International Journal of Damage Mechanics. 2015;**24**(1):95-115. DOI: 10.1177%2F1056789514522503

[73] Mauldin TC, Kessler MR. Self-healing polymers and composites.

International Materials Review. 2010;**55**(6):317-346. DOI: 10.1179/095066010X12646898728408

[74] Wool RP, O'Connor KM. A theory of crack healing in polymers. Journal of Applied Physics. 1981;**52**(10):5953-5963. DOI: 10.1063/1.328526

[75] Williams G, Trask RS, Bond IP. A self-healing carbon fibre reinforced polymer for aerospace applications. Composites Part A: Applied Science and Manufacturing. 2007;**38**(6):1525-1532. DOI: 10.1016/j.compositesa.2007.01.013

[76] Blaiszik BJ, Sottos NR, White SR. Nanocapsules for self-healing materials. Composites Science and Technology. 2008;**68**:978-986. DOI: 10.1016/j.compscitech.2007.07.021

[77] Kessler MR. Self-healing: A new paradigm in materials design. Proceedings of the Institution of Mechanical Engineers, Part G. 2007;**221**:479-495. DOI: 10.1243%2F09544100JAERO172

[78] Brown EN, Sottos NR, White SR. Fracture testing of a self-healing polymer composite. Experimental Mechanics. 2002;**42**(4):372-379. DOI: 10.1007/BF02412141

[79] Zako M, Takano N. Intelligent material systems using epoxy particles to repair microcracks and delamination damage in GFRP. Journal of Intelligent Material Systems and Structures. 1999;**10**:836-841. DOI: 10.1106%2FYEIH-QUDH-FC7W-4QFM

[80] Yin T, Rong MZ, Zhang MQ, Yang GC. Self-healing epoxy composites - Preparation and effect of the healant consisting of microencapsulated epoxy and latent curing agent. Composites Science and Technology. 2007;**67**:201-212. DOI: 10.1016/j.compscitech.2006.07.028

[81] Alaneme KK, Omosule OI. Experimental studies of self healing behaviour of under-aged Al-Mg-Si alloys and 60Sn-40Pb alloy reinforced aluminium metal-metal composites. Journal of Minerals and Materials Characterization and Engineering. 2015;**3**:1-8. DOI: journal/PaperInformation.aspx?PaperID=52698&#abstract

[82] Laha K, Kyono J, Kishimoto S, Shinya N. Beneficial effect of B segregation on creep cavitation in a type 347 austenitic stainless steel. Scripta Materialia. 2005;**52**(7):675-678. DOI: 10.1016/j.scriptamat.2004.11.016

[83] He SM, van Dijk NH, Schut H, Peekstok ER, Zwaag SV. Thermally activated precipitation at deformation-induced defects in Fe-Cu and Fe-Cu-B-N alloys studied by positron annihilation spectroscopy. Physical Review B. 2010;**81**:094103. DOI: 10.1103/PhysRevB.81.094103

[84] Le BE. Contact response of ceramics. Comptes Rendus Mécanique. 2011;**339**(7-8):466-472. DOI: 10.1016/j.crme.2011.05.005

[85] Li S, Song G, Kwakernaak K, Der Zwaag S, Slot GW. Multiple crack healing of a Ti2AlC ceramic. Journal of the European Ceramic Society. 2012;**32**(8):1813-1820. DOI: 10.1016/j.jeurceramsoc.2012.01.017

[86] Gao J, Suo J. Effects of heating temperature and duration on the microstructure and properties of the self healing coatings. Surface and Coatings Technology. 2011;**206**(6):1342-1350. DOI: 10.1016/j.surfcoat.2011.08.059

[87] Harrer W, Danzer R, Morrell R. Influence of surface defects on the biaxial strength of a silicon nitride ceramic – Increase of strength by crack healing. Journal of the European Ceramic Society. 2012;**32**(1):27-35. DOI: 10.1016/j.jeurceramsoc.2011.07.019

[88] Yang HJ, Pei YT, Rao JC, De Hosson TM, Li SB, Song GM. High temperature healing of Ti2AlC: On the origin of inhomogeneous oxide scale. Scripta Materialia. 2011;**65**(2):135-138. DOI: 10.1016/j.scriptamat.2011.03.031

[89] Ando K, Shirai Y, Nakatani M, Kobayash Y, Sato S. (Crack-healing + proof test): A new methodology to guarantee the structural integrity of a ceramics component. Journal of the European Ceramic Society. 2002;**22**(1):121-128. DOI: 10.1016/S0955-2219(01)00236-9

[90] Ando K, Houjyou K, Chu MC, Takeshita ST, Takahashi K, Sakammoto S, et al. Crack-healing behavior of Si_3N_4/SiC ceramics under stress and fatigue strength at the temperature of healing (1000°C). Journal of the European Ceramic Society. 2002;**22**(8):1339-1346

[91] Van Tittelboom K, de Belie N, van Loo D, Jacobs P. Self-healing efficiency of cementitious materials containing tubular capsules filled with healing agent. Cement and Concrete Composites. 2011;**33**(4):497-505. DOI: 10.1016/j.cemconcomp.2011.01.004

[92] Jefferson A, Joseph C, Lark R, Isaacs B, Dunn S, Weager B. A new system for crack closure of cementitious materials using shrinkable polymers. Cement and Concrete Research. 2010;**40**(5):795-801. DOI: 10.1016/j.cemconres.2010.01.004

[93] Qian SZ, Zhou J, Schlangen E. Influence of curing condition and precracking time on the self-healing behavior of engineered cementitious composites. Cement and Concrete Composites. 2010;**32**(9):686-693. DOI: 10.1016/j.cemconcomp.2010.07.015

[94] Sahmaran M, Yildirim G, Erdem TK. Self-healing capability of cementitious composites incorporating different supplementary cementitious materials. Cement and Concrete Composites. 2013;**35**(1):89-101. DOI: 10.1016/j.cemconcomp.2012.08.013

[95] Sisomphon K, Copuroglu O, Koenders EAB. Effect of exposure conditions on self healing behavior of strain hardening cementitious composites incorporating various cementitious materials. Construction and Building Materials. 2013;**42**:217-224. DOI: 10.1016/j.conbuildmat.2013.01.012

[96] Li W, Jiang Z, Yang Z, Zhao N, Yuan W. Self-healing efficiency of cementitious materials containing microcapsules filled with healing adhesive: Mechanical restoration and healing process monitored by water absorption. PLoS One. 2013;**8**(11):e81616. DOI: 10.1371/journal.pone.0081616

[97] Ariffin NF, Hussin MW, Sam MAR, Lee HS, Khalid NHA, Lim NHAS, et al. Mechanical properties and self-healing mechanism of epoxy mortar. Jurnal Teknologi. 2015;**12**:37-44

[98] Luhar S, Gourav S. A review paper on self healing concrete. Journal of Civil Engineering Research and Practice. 2015;**5**(3):53-58. DOI: 10.5923/j.jce.20150503.01

[99] Cao QY, Hao TY, Su B. Crack self-healing properties of concrete with adhesive. Advances in Materials Research. 1880-1884;**2014**:919-921. DOI: 10.4028/www.scientific.net/AMR.919-921.1880

[100] Granger S, Pijaudier-Cabot G, Loukili A. Mechanical behavior of self-healed ultra high performance concrete: From experimental evidence to modeling. In: Proceedings of 3rd International Conference on Construction Materials: Performance, Innovations and Structural Implications (ConMat'05), 22-24th August 2005. Canada: Vancouver; 2005

[101] Van TK, Tsangouri E, Van HD, Belie ND. The efficiency of self-healing concrete using alternative manufacturing procedures and more realistic crack patterns. Cement and Concrete Composites. 2015;**57**:142-152. DOI: 10.1016/j.cemconcomp.2014.12.002

[102] Van TK, Gruyaert E, Rahier H, Belie ND. Influence of mix composition on the extent of autogenous crack healing by continued hydration or calcium carbonate formation. Construction and Building Materials. 2012;**37**:349-359. DOI: 10.1016/j.conbuildmat.2012.07.026

[103] Elmoaty A. Self-healing of polymer modified concrete. Alexandria Engineering Journal. 2011;**50**(2):171-178

[104] Zhong W, Yao W. Influence of damage degree on self-healing of concrete. Construction and Building Materials. 2008;**22**(6):1137-1142. DOI: 10.1016/j.conbuildmat.2007.02.006

[105] Yang Z, Hollar J, He X, Shi X. A self-healing cementitious composite using oil core/silica gel shell microcapsules. Cement and Concrete Composites. 2011b;**33**(4):506-512. DOI: 10.1016/j.cemconcomp.2011.01.010

[106] Huang H, Ye G. Self-healing of cracks in cement paste affected by additional Ca^{2+} ions in the healing agent. Journal of Intelligent Material Systems and Structures. 2014;**26**(3):309. DOI: 10.1177/1045389X14525490

[107] Xu J, Yao W. Multiscale mechanical quantification of self-healing concrete incorporating non-ureolytic bacteria-based healing agent. Cement and Concrete Research. 2014;**64**:1-10. DOI: 10.1016/j.cemconres.2014.06.003

[108] Sauvant-Moynot IV, Gonzalez S, Kittel J. Self-healing coatings: An alternative route for anticoorosion protection. Progress in Organic Coatings. 2008;**63**(3):307-315. DOI: 10.1016/j.porgcoat.2008.03.004

[109] Gonzalez-Garcia Y, Garcia SJ, Mol JMC. Electrochemical techniques for the study of self healing coatings. In: Hughes AE, Mol JMC, Zheludkevich ML, Buchheit RG, editors. Active Protective Coatings: New-Generation Coating for Metals. Vol. 233. Springer Series in Materials Science; 2016. pp. 203-239. DOI: 10.1007/978-94-017-7540-3

[110] Aparicio M, Mosa J. Electrochemical characterization of sol–gel coatings for corrosion protection of metal substrates. Journal of Sol-Gel Science and Technology. 2018;**88**:77-89. DOI: 10.1007/s10971-018-4785-9

[111] Aramaki K. An attempt to prepare nonchromate, self-healing protective films containing molybdate on iron. Corrosion. 1999;**55**(11):1020-1030. DOI: 10.5006/1.3283939

[112] Hughes AE, Yang S, Oezkaya B, Ozcan O, Grundmeier G. Physico-chemical characterization of protective coatings and self healing processes. In: Hughes AE, Mol JMC, Zheludkevich ML, Buchheit RG, editors. Active Protective Coatings: New-Generation Coating for Metals. Vol. 233. Springer Series in Materials Science; 2016. pp. 203-239. DOI: 10.1007/978-94-017-7540-3

[113] Lucca D, Hermann K, Klopfstein M. Nanoindentation: Measuring methods and applications. CIRP Annals - Manufacturing Technology. 2010;**59**(2):803-819. DOI: 10.1016/j.cirp.2010.05.009

Section 4

Advanced Functional Materials for Textile

Chapter 14

Smart E-Textile Materials

Dilan Canan Çelikel

Abstract

Smart textiles are one of the areas that provides added value to textile materials. It is a sector that has been developed with new technologies, new fibers, and textile materials. The production of smart or intelligent textiles cooperate with other branches of science like nanotechnology, materials science, design, electronics, and computer engineering, etc. Smart textiles are classified into three groups as passive smart textiles, active smart textiles and ultra smart textiles according to their performance characteristics. Passive smart textiles are the first generation of smart textiles and sense the external conditions; for instance, UV protecting clothing, conductive fibers, etc. As active smart textiles respond to external conditions, ultra smart textiles sense, react, and adopt themselves to conditions. Shape memory materials, chromic materials, heat storage, and thermo-regulated fabrics are the typical applications of active smart textiles.

Keywords: smart textiles, E-textiles, intelligent textiles, color-changing materials, phase-changing materials

1. Introduction

Textiles, with the basic characteristics of clothing, protection, and esthetics, are the indispensable part of our lives, but in recent years with the development of technology and the variation of requirements, the demand to smart materials and intelligent textiles grows increasingly all over the world. In other words, technology has also taken control of textile industry. Smart textiles have superior performance and functionalities for the applications ranging from simple to more complicated uses such as military, healthcare, sportswear, etc. Smart or intelligent textiles can also be called as the next-generation textiles.

Many classifications related to smart textiles are available in the literature. In this chapter, the classifications based on the esthetic and performance functions are mentioned as two categories. Esthetic smart textiles use the technology for fashion design, because of their ability to light up and change color. Light-emitting clothes and luminous dresses are the typical and commercial examples for esthetic smart textiles. As for the performance, smart textiles are classified into three categories as passive smart textiles, active smart textiles, and ultra smart textiles.

Passive smart textiles can only sense the environment, as they are just sensors. UV protecting clothing, conductive fibers, plasma-treated clothing, and waterproof fabrics are the typical examples of passive smart textiles. Active smart textiles can sense the stimuli from the environment and also react to them; besides the sensor function, they also have an actuator function. Phase change materials, shape memory materials, and heat sensitive dyes are active smart textile applications.

IntechOpen

	Sensing external conditions	Reacting	Responding and adopting
Passive smart textiles	√		
Active smart textiles	√	√	
Ultra smart textiles	√	√	√

Table 1.
Classification of smart textiles.

Ultra smart textiles take a step further. Ultra smart textiles are materials that sense, react, monitor, and adopt themselves according to the stimuli or environmental conditions, such as thermal, mechanical, chemical, magnetic, or other sources. An ultra smart or intelligent textile essentially consists of a unit which works like a brain, with cognition, reasoning, and activating capacities. For instance, spacesuits, musical jackets, and wearable computers are ultra smart materials [1, 2] (**Table 1**).

In the mid-1970s, with the development of personal computers, a technological explosion was recorded in all the areas of human activity for any purposes. In the early 1990s, the benefits of smart textiles became apparent. Many researchers have studied on smart materials and textiles. Chan Vili studied the use of shape memory materials in developing high performance smart textiles, taking into consideration the ways for enhancing the esthetics of woven interior textiles. Dunne et al. provided an overview on textile integration strategies and component attachments. Choi and Jiang presented a system intended for cardiorespiratory measurement to monitor sleep condition. Mattmann et al. analyzed a yarn sensor that is nearly hysteresis-free while measuring elongation along body parts, for example, the back. Paradiso et al. presented a smart garment that can be used as wearable healthcare system. Cho et al. compared different conductive textiles and their performance for measuring joint angles.

The market for smart textile is growing with a high potential globally. The rise in demand for smart textile products is causing the existing market to expand, leading the way to new players to enter the smart textile market. In the emerging economies, the market share of smart textile consumed relative to conventional textile products is increasing. The global smart textile market size is expected to reach $5369 million by 2022 from $943 million in 2015, with a CAGR of 28.4% from 2016 to 2022. The global smart textile market is thriving and witnessing significant growth owing to the numerous applications in various industries.

2. Functions of smart textiles

Smart textiles are smart systems that can both perceive or communicate the environmental conditions and can detect and process the wearer's condition. They can use electrical, heat, mechanical, chemical, magnetic, and other detection systems to detect them. Smart garments are separated from wearable computing systems by revealing the importance of the garment on which they are integrated. Wearable computing systems are formed by the traditional systems being attached to the garment in some way. The equipment used is placed in non-textile ways without being integrated. Although some electronic materials have been reduced to be used in garments, the actual smart garments should use materials made entirely by textile production. The electronic materials to be placed must not impair the comfort of the standard textile material garment. Therefore, providing this combination is vital for wearability in smart garment and textile manufacturing. It is clear that smart textiles are simple computer systems and have five functions basically

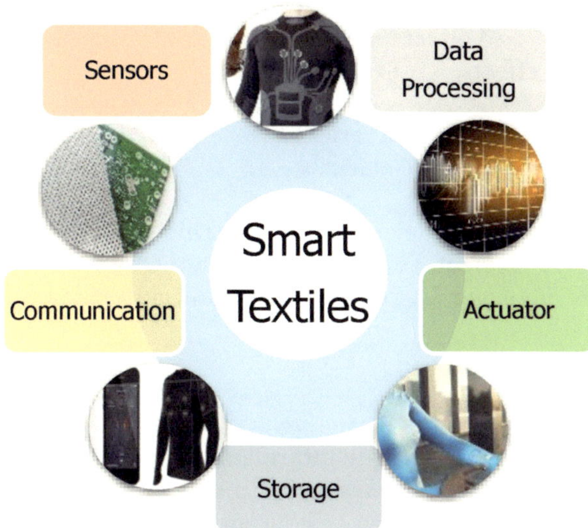

Figure 1.
Functions of smart textiles.

as sensors, data processing, actuators, storage, and communication (**Figure 1**). But it must be compatible with the function of clothing such as comfort, durability, resistant to regular textile maintenance processes, and so on [1, 2].

2.1 Sensors

Sensors are the components that transform one type of signal into another type of signal. There are already systems in the textiles that measure heart, breath rate, temperature, movement, and moisture, but these systems work with the installation of traditional sensors in textiles. At the present stage of intelligent textiles, the sensors are produced from real textile material, and the heart, breath, and movement sensitive sensors are already produced with satisfactory results. There are also different materials and structures that have the capacity of transforming signals:

- **Thermal sensors:** a thermal sensor detects thermal change, for example, a thermistor that changes resistance due to thermal change. Another example is the stimuli-responsive hydrogels that swell in response to a thermal change.

- **Light sensors:** these sensors that convert light energy into voltage output, for example, photoresistors.

- **Sound sensors:** these converts sound into an electrical signal, for example, piezoelectric materials.

- **Humidity sensors:** these sensors measure absolute or relative humidity. An example that can be interesting for textile use is the capacitive device that changes dielectric properties with the absorption of moisture.

- **Pressure sensors:** these sensors convert pressure to an electrical signal. A pressure sensor can be based on simple operations such as opening or closing a circuit. But they may also be based on more sophisticated forms like capacitive or piezoelectric phenomena.

- **Strain sensors:** these sensors convert strain into an electrical signal. Strain sensors may be based on semiconducting materials, strain sensing structures, or piezoelectric effects.

- **Chemical sensors:** these are a series of sensors that detect presence and/or concentration of chemical/chemicals.

- **Biosensor:** it is a sensing device that contains biological elements which is the primary sensing element. This element responds with a property change to an input analyte, for example, the sensing of blood glucose levels [3–5].

2.2 Data processing

Data processing is one of the components that is required only when active processing is necessary. According to information theory, it is necessary to process every collected information and data and obtain the desired output. Therefore, in order to obtain the desired output by processing the parameters collected by the sensors, a processor suitable for the relevant purpose is required in smart textiles. The information processing element is only needed when the textile is actively processing information. Textile sensors can provide information to a large extent, but the main problem lies in how the information is evaluated and the processor component comes to the fore. Variation of signals and analysis of signals are main problems for data processors. Furthermore, the energy required for the processor is another problem encountered today. Since the electronic components required for energy do not have sufficient smallness and flexibility, they differ from the structure of the textile. The waterproofing requirement of these energy units and other electronic units is another problem. However, these problems are generally seen more in the garment-type smart textiles. In the case of vehicles, this is not a problem; the information processing elements can be mounted inside the vehicle [4, 6].

2.3 Actuators

Actuators are the devices designed to perform the necessary action according to signals from the sensor or processor. These devices are also called actuators. Actuators act by an effect sent from the sensor and possibly by first passing this effect through an information processor to perform objects such as moving objects, releasing materials, and making noise. Shape memory materials are the best examples in this field. Shape memory alloys can be formed in the form of lattice. Its responsiveness to heat changes enables shape memory materials to be used as an actuator and meets the requirements of intelligent textiles very well. Another type of actuator is the materials that are capable of releasing certain chemicals under certain conditions, which can be trapped in protective microcapsules or chemically bound to the fiber polymer. Such secretory materials have various commercial applications. Odors, skin protectors, antimicrobial products, and so on. Application studies have been started with active secretion methods and some simple projects have been implemented yet. It is contemplated that the release will be effected by triggering other environment variables such as temperature, pH, humidity, chemical substance, and the like. In one view, a system capable of actively controlling drug secretion would integrate the body with a smart suit capable of receiving simple health findings. For this reason, it is expected that the actuators will have some technological and mechanical components and will bring problems in both fields [6, 7].

2.4 Storage

Storage is another component of smart textiles. Although not a fundamental goal, smart suits are expected to need a storage capacity to operate on their own. While the information to be stored in smart textiles is usually information or energy, examples such as textiles that inject or emit drugs or odors indicate that this storage unit will also serve different areas. Detection, computing, actuators, and communication units generally require energy, especially electrical energy. Efficient energy management is achieved by combining the energy source and storage in an appropriate manner. Examples of the energy sources that can be used in clothing are body temperature, mechanical movement (the energy generated by movement resulting from the elasticity of fabrics or kinetic energy from body movement), radiation (solar energy), and so on. The energy source required for the operation of sensors, processors, and moving systems in smart textiles should be combined with energy storage capability. Nowadays, very small and light batteries are available, and this solution of this energy requirement is a method that comes to mind in the first place. Even if the flexible ones are manufactured, they are not sufficient in performance and are still under development. On the other hand, the situation is easier and the energy requirement can be achieved by direct contact with clothing or by wireless connection [8, 9].

2.5 Communication

One of the components of smart textiles is the communication component, which is shaped according to the type and need of communication. There are many types of communication within smart textiles. Some of the basic situations in which smart textiles are contacted are as follows: in one element of the garment itself; can be mounted between two different elements of the garment; and in order to command the garment by the wearer, contact is made to inform the wearer or his surroundings. In today's prototypes, communication within the garment is provided by optical fibers or by conductive fine wires. They are naturally woven and can be placed in textiles without the use of stitches. A specific communication protocol is followed to communicate with the wearer. The outlines of this protocol can be provided by the technologies described below. Optical fibers are used in the creation of optical screens, and France Telecom has managed to produce several prototypes by producing a sweater and a backpack. On the other hand, since it requires more than one fiber for a pixel, it appears that the present situation needs further consideration. Another communication protocol in smart textiles is pressure-transmission systems. Information can be provided to the garment with pressure-sensitive textile materials, and a data processing element needs to process these entered orders. In some intelligent textile applications, communication with wider environmental elements is important. For example, there are many situations in which the suit is required to interact with the vehicle when 41 drivers are handled. The first thing that comes to mind in the communication of the dress with the vehicle is the seat in which it is in direct contact. A wireless connection can be achieved by integrating an antenna into the suit. This antenna is integrated into the clothes. The major advantage of this integration of antennas to the garment is that a large area can be used for communication without the user even being aware of it. In the summer of 2002, a prototype was produced in Philips Research Laboratories. With regard to road safety or driving comfort, a lot of data about the driver, such as heat comfort, concentration etc., can be obtained using intelligent textiles for a better and safer driving quality. For risky human profiles, for example, heart patients can anticipate the problems of clothing and provide information to stop the vehicle or even call for help if necessary.

Although these developments are considered unquestionable, it is thought that these studies can only be implemented in the future with more advanced technology due to the limited information that can be obtained from the human body and the lack of materials and concepts for the systems to process this information [8, 9].

3. Applications of smart textiles

3.1 Shape memory textiles

Materials capable of remembering the original shape are called shape memory materials. Materials are shaped out of its original shape as the temperature change returns to its original shape with a chemical, mechanical, magnetic, or electrical external effect. There are many classes of shape memory materials such as alloys, polymers, gels, and ceramics. Shape memory alloys and shape memory polymers are the types of shape memory materials with applications in textiles. The important point in these applications is that the material used exhibits the shape memory effect at temperatures close to body temperature.

Shape memory alloys are composed of a combination of two or more elements with the properties of hardness and elasticity that vary considerably at certain temperatures. An example of the application of shape memory alloys in textiles is a nickel-titanium alloy, which is used in protective clothing against fire and high temperatures and provides different levels of protection according to temperature.

Under the degree of activation, the easily deformable alloy becomes more rigid at the degree of activation, taking its original shape. Alloy applied to the fabric in the form of a flat surface takes the form of spring with the effect of temperature, increasing the air gap in the fabric, thus increasing the protection of the garment and the formation of second-degree burns under the same conditions.

The degree of activation can be adjusted by changing the ratio of nickel to titanium in the alloy. T-shirts developed by an Italian company Corpo Nove shortens the sleeves with the increase in temperature and do not require ironing, which is another example of the application of shape memory alloys in textiles.

Shape memory polymers can be used in fiber production or can be applied to the fabric by finishing, coating, or lamination processes. Polymers have different water vapor permeability, air permeability, modulus of elasticity, refractive index, and expansion properties below and above the glass transition temperature (Tg).

The shape-memory polymer, placed between two layers of fabric, has a tight structure below a certain temperature and prevents heat, water and wind circulation around the body. By the increase in temperature it starts the molecules motion and becomes a porous structure resulted with the expulsion in body heat. This flexible barrier function makes it possible to adjust the insulation properties of the garment to temperature changes and to provide optimum comfort in any environment.

The crystal structure of a material at a given temperature determines its many physical properties. During the phase change, besides microscopic changes, macroscopic changes such as modulus of elasticity, coefficient of friction, electrical conductivity, and hardness occur. One of the important applications using these changes is surgical yarns.

Self-tangled surgical threads are designed for endoscopic surgery, and implants that are small in normal ambient conditions are designed for use in endoscopic surgery. Thus, it will be possible to perform operations with small incisions, shortening the healing time and reducing the risk of infection. Shape memory textiles can also be used for esthetic and decorative purposes. Textile materials that deform with the stimulating effect acquire a third dimension [1, 8, 9].

3.2 Color-changing textiles

They are intelligent textile materials that have the ability to change color with an external stimulus effect. They are obtained by incorporating color-changing materials into the structure of textile materials. Color-changing materials are chromic materials or chameleon materials. There are many different color-changing mechanisms, but mostly the electron density or molecular structure of the material changes due to the external stimulus effect and the color change occurs; when the stimulus effect disappears, they return to their initial state where they are more stable and get their first color.

Color changing materials are specified according to the effect mechanism. Light, heat, pH change, solution, friction and pressure are basic effect parameters. They are also called by the effect type as photochromic (light effected), thermochromic (heat effected), electrochromic (electric effected), solventchromic (solution effected), halochromic (pH effected), tribochromic (friction effected), mechanochromic (pressure effected).

The application of chromic materials to textile materials can be done by different methods at different stages. For example, a chromic dyestuff can be used for dyeing fibers by conventional dyeing methods; the fibers can be added to the fiber structure at the polymer stage; color-changing fibers can be obtained by melt spinning or wet spinning; they can be mixed with resin and coated onto the fabric surface, thereby using them for fabric printing or dyeing.

Smart textiles change colors depending on environmental factors; they are important because of their esthetic advantages. It is thought that the use of color-changing textiles will become more widespread in the future in the field of fashion and will change the color depending on many other effects besides the existing ones. Photochromic, thermochromic, electrochromic, and solventchromic textile applications can be seen in fashion and decoration. They are available for T-shirts, bags, and hats.

The reversible color-changing property of thermochromic dyes indirectly changes the heat absorbing property of the textile material. While light reflection increases, darker colors increase heat absorption. Because of these properties, thermochromic dyestuffs are used to coat the uniforms of firefighters who turn white under very high temperatures and reflect the heat in this way and also in building coatings. The fact that thermochromic dyes accelerate the dimensional change of fibers provides another thermoregulation effect. At high temperatures, fibers containing thermochromic dyestuff shorten. The pores of the fabric are enlarged so that a large amount of air is introduced in and consequently the body temperature decreases. At low temperatures, the fibers are elongated, the pores are closed, and the fabric maintains the body's temperature [1, 2, 10, 11].

3.3 Phase-changing textiles

Phase change materials, with a textile substrate, are basicly thermo regulating materials. When the melting temperature of the material is reached during the heating process, the transition from solid state to liquid, that is, a phase change occurs, during which the phase change material absorbs and stores a large amount of heat. The temperature of the phase-changing material remains virtually constant during the entire phase change. During cooling of the same material, the stored heat is transferred to the medium and the transition from liquid to solid state takes place. Again, the temperature of the material remains constant throughout the phase change process. If the temperature change continues except for phase change, the temperature of the material also changes.

By using two or more phase-changing materials together, the temperature range at which the phase change occurs can be adjusted so that it can be used in specific applications. Textile materials in which the phase-changing materials are applied have a cooling effect, heating effect, or thermoregulation effect caused by the absorption or dissipation of heat depending on the ambient temperature conditions. The degree of all effects depends on the type of material used, its thermal capacity, and the amount of application. Of course, in order to obtain the desired effect efficiently, the temperature values in which the material changes phase must match the temperature values to be encountered during use.

The application of phase-changing materials to textile materials can be done in different ways. Microencapsulated phase-changing materials can be added to the structure of synthetic fibers during fiber drawing, can be added to the nonwoven structure, or can be coated on textile surfaces. Product design is also very important in all these applications. For example, when changing from a warm indoor environment to a cold outdoor environment, it was seen that the garment containing the phase-changing material showed heating effect on average 12–15 min depending on the phase-changing material content and outdoor conditions.

If the structure of the garment is not designed well, it is also possible to dissipate heat from the phase-changing material. When we look at their usage, they have commercialized usage in hospital beds and pillows. With the effects of thermoregulation, they keep the temperature at levels that do not disturb the patient and ensure that the patient does not sweat and thus contribute to the healing process of the patient. There are also studies on heating or cooling plasters and heating blankets for use in the medical field. In long-term operations, it is possible to provide thermal comfort by preventing surgeons' sweats by providing a coating on the inside of their garments with a phase-changing material. There are commercialized everyday garments, underwear, shoes, and sportswear where phase-changing materials are applied [2, 10–12].

3.4 Wearable smart electronic textiles

Wearable smart electronic textiles make lifes more reliable, healthy and comfortable in many areas. Wearable smart electronic textiles; temperature change, light, moisture, such as environmental stimuli can detect, react to these stimuli, can change itself according to external conditions, store data, these data are used to produce information and communication purposes. In this sense, they are perceived as intelligent technologies that will have the qualities to support the vital activities of human beings such as sensation, movement, communication, taking action, and adapting to environmental conditions.

The four basic elements of wearable smart electronic textiles are conductivity, sensors, wireless communication module, and power supply. Depending on the nature of these components, the degree to which they can be integrated into the textile material varies. As a first method, existing electronic devices can be integrated into the textile material. An example of a life belt to which the existing sensors are attached is an example. The biggest advantage is that the process is very easy. However, the disadvantages of the large, inflexible electronic components used are disturbing the user and washing problems. The second method is the production of electronic components using textile materials and textile manufacturing techniques (textronics) and their use as part of the garment. The disadvantage of this method is that integration processes can be carried out easily, but that a limited number of electronic components can be produced by textile materials and methods. The third method is to produce and use fibers to provide some electronic functions (fibertronics).

Clothing equipped with sensors that monitor vital functions such as breathing, heart rate, and body temperature increases the mobility of the patients while providing the confidence of being constantly monitored and increasing the standard of living for chronic patients and the disabled. High-performance active sportswear provides a performance increase by following the athletes' body functions such as pulse, breath, body temperature, and activity-related values such as speed, distance, time, and calories.

Various applications are available in the field of medicine, sportswear, and protective clothing. Lifeshirt is an example of the protective use of intelligent electronic garments designed for pioneers, hazardous workers, firefighters, and industrial cleaning workers. Lifeshirt is a belt that contains sensors that detect indicators related to vital activities such as respiratory rate, heart rate, and body temperature and can transmit this information to a remote monitor via a modem. Through this belts, the health status of the wearer can be monitored continuously, and strategic decisions can be made by evaluating the general situation of the team.

The Cyberia Smart Coverall with wearable technology is designed to be worn in polar areas. The project was started with the aim of developing a garment displaying the health data of the wearer. The garment displaying health data also includes a global positioning system (GPS) for use in the event of a loss and a GSM module that can automatically send the coordinates and health information to a predetermined number in the event of an abnormal condition [2, 5, 13].

4. Innovations in smart textiles

4.1 Laser-printed waterproof and stretchable e-textiles

The next generation of waterproof smart fabrics will be laser printed and made in minutes. That is the future imagined by the researchers behind new e-textile technology. Scientists from RMIT University in Melbourne, Australia, have developed a cost-efficient and scalable method for rapidly fabricating textiles that are embedded with energy storage devices. In just 3 min, the method can produce a 10 × 10 cm smart textile patch that is waterproof, stretchable, and readily integrated with energy harvesting technologies. The technology enables graphene supercapacitors—powerful and long-lasting energy storage devices that are easily combined with solar or other sources of power—to be laser printed directly onto textiles (**Figure 2**).

4.2 Conductive textiles

A conductive textile can be defined as a fabric which is made from the strands of a metal that are woven, blended, or coated during the creation of the textile. Conductive metals such as silver, titanium, gold, nickel, and carbon are utilized by the textile. Conductive textiles inhabit the property that it can conduct electricity and thus is used in several applications by different end-use industries. The primary function of the conductive textile is controlling the static electricity and protecting from the electromagnetic interference. Based on type, the woven textile segment has significant growth during the forecast period. Woven textiles are widely utilized by various end-use industries such as military and defense, healthcare, and sports and fitness. As these textiles offer high standard performance in shielding and conductivity, they are considered to be the preferred type of conductive textiles utilized across the globe, thereby boosting the growth of the woven textile segment [14].

Figure 2.
Laser-printed waterproof and stretchable e-fabric.

4.3 Medical smart textile as a cardiac supporting device

Knitted and woven fabrics are being used as a cardiac supporting device. An innovative medical device has been made by using the knitted and woven fabrics, which corrects the life-threatening conditions of the heart and vascular system. Implantation of the new devices requires less invasive surgical procedures and involves less risk than traditional procedures, while also causing fewer complications in hospital days. Heart failure is a chronic syndrome that occurs when the heart is not getting enough amount of blood. Generally, valve leakage is reliable for this. The treatment of heart failure is only the drug therapies and surgical, but they are temporary treatment. The only permanent treatment is a heart transplant. But most of the patients cannot qualify for heart transplantation. So they have to do the surgical treatment or drug therapy.

So the scientists have developed a device named as a cardiac support device (CSD), which is intended to halt the progression of heart failure. The cardiac support device (CSD) research was conducted to determine the best material, yarn configuration, knit pattern, and processing to use to produce CSD fabric. It is a mesh-like warp knitted fabric. The fabric is fabricated from the multifilament texture (**Figure 3**).

Polyester fabric is used for it. Polyester fabric has biological tissue response and it has the compatibility for the epicardial surface of the heart. The polyester yarns are warp knitted into a mesh configuration using a variation of an atlas stitch. After knitting, the fabric is conditioned to ease its handling during the processing to manufacture the CSD [9].

4.4 Smart clothing with improved comfort and safety for firefighters

Temperature is a major challenge in numerous professions—thermal comfort and occupational safety. For example, in emergency missions of fire and rescue services as well as in mines and construction sites, the working conditions often cause extreme physical strain (**Figure 4**).

Working in hot situations without wearing appropriate protective clothing and equipment often causes high heat stress. It will be perfect if the amount of such stress could be monitored in real time during the performance of different work tasks. To solve this, researchers and companies together developed a wearable technology solution for firefighters. It allows the real-time monitoring of heat stress, thus improving the occupational health and safety in challenging temperatures.

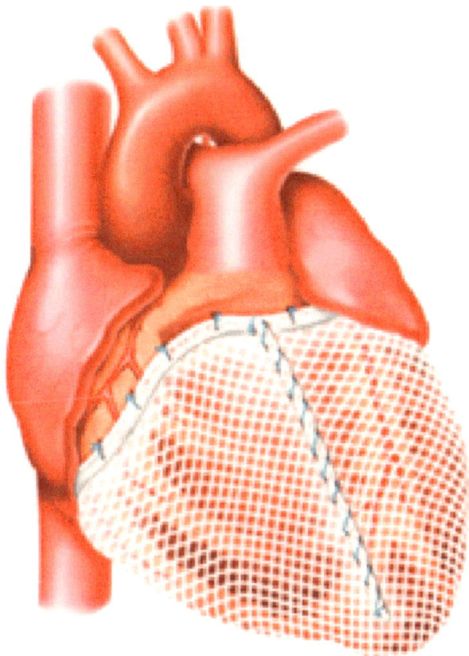

Figure 3.
Medical textiles working as a cardiac support device.

The new method has been tested at the Finnish Institute of Occupational Health in Oulu and at the Emergency Services College in Kuopio. Based on the first tests, it would seem to offer a very promising tool for commanding rescue missions and enhancing the occupational health and safety of firefighters [10–12].

4.5 Graphene-based smart textiles

Graphene has already made a huge blast in the next step of wearable technology. Due to the thermal conductive properties of graphene, the warmth produced by the human body is preserved and distributed evenly in cold climates and allows an even body temperature during physical activity.

A renowned company Directa Plus, a producer and supplier of graphene-based products, teamed up with Colmar, the high-end sportswear company, has launched a new collection of SKI jackets containing graphene-based products.

Figure 4.
Smart clothing, adjusting the heating control autonomously.

The new technology SKI jacket contains graphene Plus (G+) and is worn by the French national SKI team for multiple successful tournaments. It was explained that the key benefit of incorporating G+ is that it enables the fabric to act as a filter between the body and the external environment, ensuring the ideal temperature for the wearer (**Figure 5**).

A Chinese company called Shanghai Kyorene New Material Technology has also developed a graphene fiber that has been used to produce clothes, sportswear, and underwear products.

Recently, researchers have designed a low-cost, sustainable, and environmentally friendly method for making conductive cotton fabrics using graphene. These fabrics could lead to smart textiles and interactive clothes that will find applications in healthcare, wearable, and more. Functionalization of these conductive cotton fabrics was done by thermal reduction of graphene oxide (GO) adsorbed on cotton. Besides, researchers have created two ways to apply thin graphene sheets that either make the fabric super-hydrophobic or super-hydrophilic.

A team of scientists in Korea also announced the successful development of a technology to make a washable, flexible, and highly sensitive textile-type gas sensor. This technology is based on coating graphene using molecular adhesives to fiber like nylon, cotton, or polyester so that the fabric can check whether or not gas exists in the air.

Graphene has also strong cytotoxicity toward bacteria. So, this can be highlighted for maternity clothes to create coatings that prevent the growth of bacteria on the surface of the fabrics, thus protecting the pregnant against the possible diseases transmitted by bacteria. This type of protection will be very useful in gynecologists, nurses, and midwives clothing who assist the birthing woman in order to avoid the spreading of bacterial infections in newborns [6].

4.6 Smart denim jacket

The smart denim jacket designed by Levi's turns a portion of the fabric on the sleeve into a touch-sensitive remote control for phones to be helpful in everyday life. This is a second version of their Jacquard smart jacket first introduced in 2017.

Figure 5.
Graphene-based jacket.

The iconic jacket merges style with innovative Jacquard technology and allows the wearer to answer calls, play music, and take photos right from the sleeve. With the Jacquard technology, the jacket lets you access digital services right from your cuff, wherever you go. Get updates about your day, take a remote selfie, get notified if you leave your phone or jacket behind, and more, so you can stay focused on what is important (**Figure 6**).

The technology allows to use touch gestures, like swiping and tapping, on the left cuff of the jacket to issue commands. The new and improved Jacquard Tag wirelessly connects your Trucker jacket to your smartphone. Jacquard also provides you helpful alerts, like when you have left your phone behind, using lights on the Tag and vibrations in the cuff to get your attention [5].

4.7 Smart film fabric

DuPont Intexar is a revolutionary electronic ink and film that seamlessly transforms fabric into smart clothing for multiple applications. The technology is embedded directly onto fabric using standard apparel manufacturing processes, offering both ease of integration and ease of design. It is currently leveraged for three applications: fitness, heat, and shealth (**Figure 7**).

The technology for fitness and health function similarly with key components that monitor and transmit biometric signals. A thin layer of carbon or silver serves as a sensor, sensing electrical signals, while a conductor, made of a layer of silver, transmits currents throughout. Other films are integrated onto the textiles to shield the technology from water and additional exposure. The data received is captured and monitored via a third-party app. The heat application utilizes a battery-powered technology that includes a resistor, a thin layer of carbon that radiates heat, a conductor, a thin layer of silver that transmits the electrical currents, and additional films for protection.

Intexar is engineered and tested to perform as designed each and every time, with durability to outlast any alternative and offer unmatched comfort with its seamless stretchability. Intexar also offers a powered heating solution in a thin and safe application. The battery-operated technology enables clothing to generate heat, creating actively controlled on-body warming. This technology is particularly well-suited for outdoor activity and industry professionals within the utility,

Figure 6.
Smart denim jacket.

Figure 7.
Smart film fabric.

construction, military, forestry, mining, and infrastructure industries, among others. This technology also delivers advanced wearable health care through the sense and transmission of biometric signals. Primary uses include monitoring of pregnancy, telemetry and respiratory disorders as well as heat and electro-stimulation therapies.

5. Conclusion

Current developments in textile technologies, new materials, nanotechnology and miniature electronics, and wearable makes systems more convenient, but the most important parameter for users to accept wearable devices is comfort is sufficient. This is recognized as a challenging environment for the human body and the environment, mechanics resistance, and durability. In addition, the circuit design of the development of intelligent textiles, the knowledge of intelligent materials, microelectronics, and chemistry is basically integrated with a deep understanding of textile production. It requires a multidisciplinary approach.

DOI: http://dx.doi.org/10.5772/intechopen.92439

Author details

Dilan Canan Çelikel
Technical Sciences Vocational School, Gaziantep University, Gaziantep, Turkey

*Address all correspondence to: celikel@gantep.edu.tr

References

[1] Cherenack K, van Pieterson L. Smart textiles: Challenges and opportunities. Journal of Applied Physics. 2012;**112**(9): 091301 (1-14)

[2] Castano LM, Flatau AB. Smart fabric sensors and e-textile technologies: A review. Smart Materials and Structures. 2014;**23**(5):1-27

[3] Kallmayer C, Simon E. Large area sensor integration in textiles. In: International Multi-Conference on Systems, Signals and Devices (SSD), 20-23 March 2012, Chemnitz, Germany. 2012. p. 5

[4] Sergio M, Manaresi N, Campi F, Canegallo R, Tartagni M, Guerrieri R. A dynamically reconfigurable monolithic CMOS pressure sensor for smart fabric. IEEE Journal Solid-State Circuits. 2003;**38**(6):966-975

[5] Cho G, Jeong K, Paik MJ, Kwun Y, Sung M. Performance evaluation of textile-based electrodes and motion sensors for smart clothing. IEEE Sensors Journal. 2011;**11**(12):3183-3193

[6] Kim H, Kim Y, Kim B, Yoo HJ. A wearable fabric computer by planar-fashionable circuit board technique. In: 6th International Workshop on Wearable and Implantable Body Sensor Networks, 3-5 June 2009, Berkeley, CA. 2009. pp. 282-285

[7] Hasegawa Y, Shikida M, Ogura D, Sato K. Novel type of fabric tactile sensor made from artificial hollow fiber. In: Proceedings of the 20th International Conference on Micro Electro Mechanical Systems, 21-25 January 2007, Hyogo, Japan. 2007. pp. 603-606

[8] Engin M, Demirel A, Engin EZ, Fedakar M. Recent developments and trends in biomedical sensors. Measurement. 2005;**37**(2):173-188

[9] Meyer J, Lukowicz P, Tröster G. Textile pressure sensor for muscle activity and motion detection. In: Proceedings 10th IEEE International Symposium on Wearable Computers, 11-14 October 2006, Montreux, Switzerland. 2006. pp. 69-72

[10] Zhang RQ, Li JQ, Li DJ, Xu JJ. Study of the structural design and capacitance characteristics of fabric sensor. Advanced Materials Research. 2011;**194-196**:1489-1495

[11] Avloni J, Lau R, Ouyang M, Florio L, Henn AR, Sparavigna A. Polypyrrolecoated nonwovens for electromagnetic shielding. Journal of Industrial Textiles. 2008;**38**(1):55-68

[12] Holleczek T, Rüegg A, Harms H, Tröster G. Textile pressure sensors for sports applications. In: 9th IEEE Sensors Conference, 1-4 November 2010, Kona, HI. 2010. pp. 732-737

[13] Hui Z, Ming TX, Xi YT, Sheng LX. Pressure sensing fabric. In: Proceedings of MRS, 920, 0920-S05-05. 2006

[14] Wallace GG et al. Conductive Electroactive Polymers: Intelligent Materials Systems. 2nd ed. Boca Raton, FL: CRC Press; 2002

Section 5

Advanced Functional Materials for Glasses

Chapter 15

Tellurite Glass and Its Application in Lasers

Pengfei Wang, Shijie Jia, Xiaosong Lu, Yuxuan Jiang, Jibo Yu, Xin Wang, Shunbin Wang and Elfed Lewis

Abstract

This chapter provides expert coverage of the physical properties of new noncrystalline solids—tellurite glass and the latest laser applications of the material—offering insights into innovative applications for laser and sensing devices, among others. In particular, there is a focus on specialty optical fibers, supercontinuum generation and laser devices, and luminescence properties for laser applications. This chapter also addresses the fabrication and optical properties and uses of tellurite glasses in optical fibers and optical microcavities, the significance of from near infrared (NIR) to mid-infrared (MIR) emissions and the development of tellurite glass-based microcavity lasers. The important attributes of these tellurite glasses and their applications in lasers were discussed in this chapter.

Keywords: tellurite glass, fiber lasers, supercontinuum sources, specialty fibers, microcavity lasers

1. Introduction of tellurite glass

Tellurite glasses are noncrystalline solids with many applications in photonics, appear in a wide range of compositions, and can be operated over a large temperature range [1–4]. Tellurite glasses have been studied for more than 150 years [5], but more recent versions have been produced with purities exceeding 98.5% [6]. They are characterized by a low melting point and the absence of hygroscopic properties, and hence tellurite glasses have limited the application of phosphate and borate glasses and aroused widespread interest in the field of photonics and associated technologies. Moreover, they have high density and a low transition temperature [7, 8]. Their optical properties include relatively high refractive index, high nonlinear refractive index, high dielectric constant, as well as good chemical stability and a wide infrared (IR) transmission range (1–6 μm) [9–11].

In 1952, Stanworth [1] conducted preliminary research on the formation and structure of tellurite glass. The main raw material is TeO_2 and at that time this was relatively expensive, and hence tellurite glass was considered to be of low practical value and had not been further studied. Since the late 1980s to the mid of 1990s [12, 13], considerable progress had been made in the advancement and understanding of the optical and physical properties of new tellurite glasses, including their molecular structure and bonding properties.

Research in tellurite glass-based broadband fiber amplifiers was initially concentrated around erbium-doped tellurite fibers. This was primarily due to its relatively broadband gain spectrum, which led to it attracting a great deal of research attention which has persisted up to the present. Currently, many world-class university-based research institutions and industrial companies have investigated the potential of tellurite glass for use in fibers, and this has resulted in rapid progress. In this section, the composition, structure, and thermal stability of tellurite glasses will be considered.

1.1 Composition of tellurite glass

The selection of tellurite glass components when used in binary combinations with other materials is very important. It directly affects the glass-forming ability, thermal stability, refractive index, rare earth ion doping concentration, and spectral characteristics. **Table 1** lists the range of TeO_2 glass formation in several binary tellurite glass systems. **Table 1** shows that TeO_2 exhibits the largest glass formation range in the case of the three binary systems TeO_2-ZnO (100–52 mol%), TeO_2-WO_3 (94.7–61.3 mol%), and TeO_2-TiO_2 (100–52 mol%).

The structure of tellurite glass is always generally based on binary systems. The ternary and multivariate tellurite glass systems have been generally used as rare earth-doped substrates for the investigation of the waveguide spectral properties. The diversity of components has helped to improve the chemical and thermal stability of tellurite glass-based devices. **Table 2** [14] includes the composition of the tellurite glass systems that have been reported in recent years. In all cases TeO_2 was used as the glass-forming material, and its content was generally higher than 50 mol% (as shown in **Table 1**). Other oxides were generally used as modified bodies. From **Table 2** [14], it can be seen that the research objects of the binary system were more diversified. In addition to the common monovalent alkali metal oxides and divalent alkaline earth metal oxides, many other oxide components were involved, including CeO_2, SmO_2, V_2O_5, etc. It should be pointed out that in the case of the ternary tellurite glass systems, the TeO_2-ZnO-R_mO_n and TeO_2-WO_3-R_mO_n glass systems were the most widely investigated, because these two systems possessed a wide range of glass formation regions and a wide range of adjustable components.

1.2 The structure of tellurite glasses

Early research was reported to suggest that the molecular structure of pure tellurite glass molecules comprised TeO_4 double triangular bipyramids (tbp's) [36].

Composition	Glass formation range TeO_2 mol%	Composition	Glass formation range TeO_2 mol%
Cs_2O	98.0–87.5	ZnO	100–52.5
Rb_2O	96.5–73.0	CdO	60.0–48.0
K_2O	95.5–77.0	PbO	60.0–48.0
Na_2O	91.5–59.5	Bi_2O_3	66–60
Li_2O	87.0–69.5	WO_3	94.7–61.3
BaO	93.0–80.0	Nb_2O_5	100–73.2
TiO_2	100–52.5		

Table 1.
The formation range of binary system tellurite glass.

Binary system	Ternary system	Multicomponent glass system
TeO_2-R_2O (R = Li, Na, K, Rb, Cs, Ti) [15]	TeO_2-ZnO-R_2O (R = Li, Na, K) [16]	TeO_2-ZnO-B_2O_3-K_2O [10]
TeO_2-MO (M = Zn [17], Ba, Pb) [18]	TeO_2-ZnO-RO (R = Ba, Mg, Sr) [19]	TeO_2-ZnO-GeO_2-Na_2O [20]
TeO_2-M_3O_4 (M = Co) [21]	TeO_2-WO_3-R_2O (R = Li, Na, K) [22]	TeO_2-ZnO-B_2O_3-GeO_2-Na_2O [23]
TeO_2-M_2O_3 (M = Sm, La) [21]	TeO_2-WO_3-BaO [15]	TeO_2-ZnO-Na_2O-Bi_2O_3 [24]
TeO_2-CeO_2 [21]	TeO_2-WO_3-Bi_2O_3 [15]	TeO_2-ZnO-WO_3-TiO_2-Na_2O [25]
TeO_2-M_2O_5 (M = P [21], V [26], Nb [27])	TeO_2-WO_3-Nb_2O_5 [28]	TeO_2-ZnO-Nb_2O_5-Nb_2O_3 [29]
TeO_2-MO_3 (M = W [30], Mo [31])	TeO_2-B_2O_3-M_2O_3 (M = Al, Ga, Sc, La, Bi) [16]	TeO_2-Li_2O-Nb_2O_5-K_2O [32]
TeO_2-PbF_2 [33]	TeO_2-K_2O-La_2O_3 [34]	TeO_2-ZnO-Nb_2O_5-Gd_2O_3 [35]

Table 2.
Tellurite glass systems [14].

In this polyhedron, one Te atom is surrounded by four oxygen atoms, of which two oxygen atoms O_{eq} are at the equator position and the other two oxygen atoms O_{ax} at the axial position. The Te atoms are linked by O_{ax} or O_{eq} into Te—O—Te; an apex of the tetrahedron on the equatorial plane remains unoccupied by oxygen atoms and is occupied by Te's lone electron pair [16]. This special polyhedral structure and the chemical bonding were different from the traditional glass-forming bodies (B_2O_3, SiO_2, GeO_2, and P_2O_5), which determined the specificity of the tellurite glass structure.

Some scholars used various testing methods to conduct research and analysis on tellurite glasses, especially binary system tellurite glass. Jha et al. [37] considered that the main structural units of tellurite glass were TeO_4 double triangular bipyramids (tbp's) and TeO_3 bipyramids (bp's) triangular pyramids. In 1995, Neov et al. [36] were the first to perform neutron diffraction analysis on lithium tellurite glass and pointed out that in addition to the TeO_4 structural unit, a deformed double triangular pyramid TeO_{3+1} existed in the glass network. One of the Te—O bonds was significantly longer than the other three. Due to the short-range similarity between the glass and crystal structures, the structure of tellurite glasses can be studied and analyzed based on the structure of tellurite crystals with the same composition. Sakida et al. [32] compared the Raman spectra of alkali tellurite crystals, pure tellurite glasses, and alkali tellurite glasses. The resulting Raman spectra were considered to correspond to structural elements in the glass. TeO_4 (tbp's) double triangular bipyramids were finally transformed into a TeO_3 (bp's) triangular pyramid by TeO_{3+1}. Tatsumisago et al. [38] studied the change of tellurite glass structure with temperature using Raman spectroscopy. Throughout the above research, the general laws could be classified as follows:

1. It was generally considered that there were two kinds of structural units that form a tellurite glass network. One was TeO_4 (tbp's) double triangular pyramid in which the Te atoms were arranged as a four ligand, and the other

was TeO_3 (bp's) triangular pyramid in which the Te atoms were in a triple coordination. It was considered that there were generally five kinds of structural units in alkali tellurite crystals, as shown in **Figure 1(a–e)**. $Q^n{}_m$ can be used to represent the structural unit in **Figure 1**, where n is the number of bridge oxygen molecules in the $[TeO_4]$ group and m represents the number of covalent bonds. Research on the distribution of various structural units (a–e) in tellurite glass has become a significant focus of research in this field.

2. When an alkali metal oxide or alkaline earth metal oxide was introduced into tellurite glass as a network modifier, the original glass network structure was destroyed. TeO_4 (tbp's) double triangular pyramid was finally transformed into TeO_3 (bp's) triangular pyramid by TeO_{3+1}. Sekiya [39] investigated the TeO_2-$MO_{1/2}$ binary system and considered that when the alkali metal oxide content was low, the glass was composed of TeO_4 (tbp's) double triangular pyramid and TeO_{3+1} polyhedron. When the alkali content was less than 20 mol %, the number of TeO_{3+1} polyhedra increased with the increase of the alkali metal oxide content. When the alkali content was between 20 and 30 mol%, TeO_3 (bp's) triangular pyramids with non-bridged oxygen bonds appeared in the glass network structure, and the numbers of TeO_4 (tbp's) and TeO_{3+1} decreased accordingly. When the alkali metal oxide content exceeded 30 mol %, the $Te_2O_5^{2-}$ polyhedron was formed in the network structure. When the alkali metal oxide content was greater than 50 mol%, it was considered that the glass network structure at this time was composed of TeO_3 (bp's) polyhedrons, TeO_{3+1} polyhedrons, and independent $Te_2O_5^{2-}$ and TeO_3^{2-}. At this time, the number of TeO_4 in the glass was very small, and the glass structure had become extremely complex.

3. Temperature also affects the structure of tellurite glass. For example, when the glass temperature was gradually increased and exceeded the melting temperature, the TeO_4 (tbp's) double triangular pyramid would also be transformed into a TeO_3 (bp's) triangular pyramid. This is mainly due to the fact that Te-O_{ax} is caused by fracture with increasing temperature, and its structural transformation process is shown in **Figure 2**.

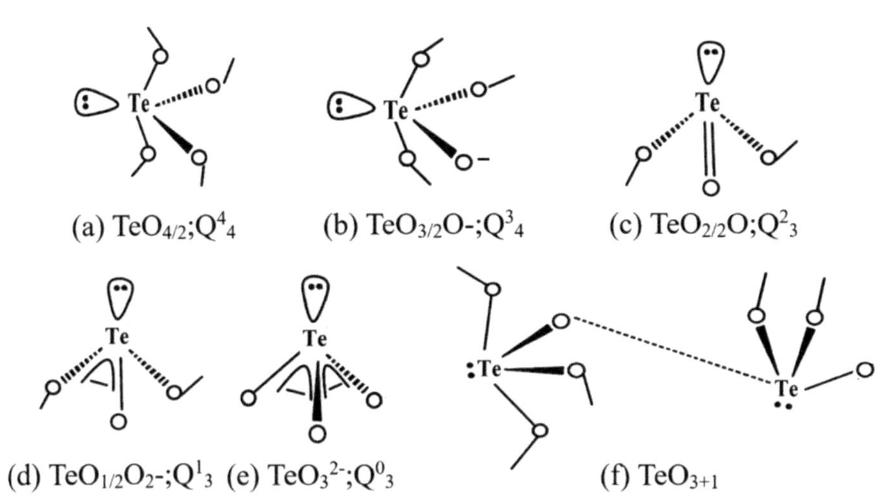

Figure 1.
(a–e) Five basic structural units in alkali tellurite crystals. (f) Deformed bitriangular cone TeO_{3+1}.

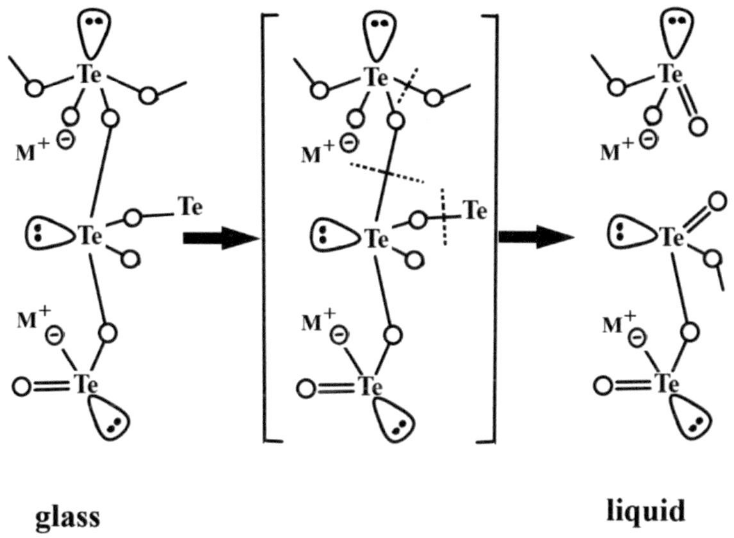

Figure 2.
Transformation of glass structure during heating.

1.3 Thermal properties of tellurite glass

The thermal stability of tellurite glass is primarily dictated by composition and the doping concentration of rare earth ions. The characteristic glass temperature values include glass transition temperature T_g, incipient crystallization temperature T_x, peak crystallization temperature T_c, and glass-melting temperature T_m.

The thermal stability of glass is usually expressed by ΔT, which is the differential value between T_x and T_g. A higher value of ΔT generally means that the glass has good thermal stability. If the value of the T_x is close to T_f, it will lead to crystallization during a fiber drawing process which leads to an increase in the loss (attenuation) of the resulting glass fiber. **Table 3** includes a listing of several kinds of tellurite glass with good thermal stability together with their characterized glass temperatures (Tg, Tx, and ΔT). In the case of TeO_2-R_2O (R = Li, Na, K, or other alkali metal) tellurite glass systems, as the content of the alkali metal oxide increases, T_g gradually increases, while T_x remains almost unchanged. Consequently, the ΔT increases correspondingly, and the resistance against crystallization of the tellurite glass also increases.

In addition, the introduction of rare earth ions also has an influence on the thermal stability of tellurite glasses. For example, 1 wt% Pr_2O_3 introduced to a

Glass component	T_g (°C)	T_x (°C)	$T_x - T_g$ (°C)
$85TeO_2$-$15Na_2O$	277	447	170
$70TeO_2$-$10ZnO$-$20Li_2O$	265	392	127
$70TeO_2$-$20ZnO$-$10BaO$ [40]	339	495	156
$82.5TeO_2$–$7.5WO_3$-$10Nb_2O_5$ [41]	391	562	171
$80TeO_2$-$10WO_3$-$10Nb_2O_5$-$1Yb_2O_3$ [42]	404	566	162
$60TeO_2$-$20ZnO$-$7.5B_2O_3$-$7.5GeO_2$-$5K_2O$	200 ± 5	378 ± 2	178

Table 3.
Characteristic temperature of tellurite glasses.

$75TeO_2$-$20ZnO$-$5Na_2O$ [43] system increases the ΔT value from 118 to 150°C, while 1 wt% Er_2O_3 introduced to $90TeO_2$-$10P_2O_5$ [44] system decreases the ΔT value from 147 to 101°C. Furthermore, the concentration of the rare earth ion has a significant influences on the ΔT value of the $75TeO_2$-$20ZnO$-$5Na_2O$ [45] glass system.

2. Fiber lasers and fiber amplifier based on tellurite glass fibers

2.1 Tellurite glass-based fiber lasers

Since the discovery of the first ruby laser (Maiman in 1960), the laser has attracted worldwide attention for its excellent collimation, high brightness, and monochromaticity [46]. Since then, the development of the laser has accelerated. In 1961, Javan et al. developed the helium-neon gas laser [47], and in 1962, Hall et al. created the GaAs coherent light emission [48]. In 1963, Koester et al. first proposed the idea of fiber lasers and amplifiers [49]. However, due to the shortcomings of the optical fiber at that time, the development of the optical fiber laser was slow during this period. In 1966, Gao et al. proposed the basic concept of optical fiber communication [50]. Subsequently, optical fiber communication underwent a major research and development stage (1966–1976), a practical application stage (1976–1986), and a large-scale optical fiber communication infrastructure construction stage after 1986. With the rapid development of optical communication, optical fiber manufacturing technology and semiconductor laser production technology have matured, which formed the foundation for the subsequent development of doped fibers, optical fiber lasers, and fiber amplifiers.

The tellurite fiber laser is based on a tellurite glass fiber which acts as a gain medium. The first significant characterization of the optical properties of tellurite glass in fiber form was reported in 1994 [12]. In 1997, Mori et al. realized that Er^{3+}-doped tellurite glass fiber could be used for broadband optical amplifiers [13]. Further research on tellurite fiber was initiated worldwide driven by the development of the communication industry. Over the next few years, Japan's NTT led the research in this field. They fabricated tellurite fiber with a loss of 0.02 dB/m and developed the first erbium-doped tellurite fiber amplifier (EDTFA) module for use in commercial WDM systems [51]. Further subsequent significant contributions to tellurite fiber laser development have been used by several university groups as well as industry-based research institutions including American Corning corporation, Fujitsu and Nippon of Japan, Korea's ETRI, etc.

Tellurite glass has a broad transmission window in the infrared wavelength range which extends up to 6 μm, a relatively low phonon energy of about 700 cm^{-1}, and high solubility of rare earth ions. It is therefore an excellent host material for constructing single and high repetition frequency fiber lasers. Yao et al. measured the transmission spectrum of 2-mm-thick glass samples after they were immersed in deionized water for 12 days [52], and the results are shown in **Figure 3**. There was no obvious change in the transmission spectra, and no hydrated layer was formed at the end face of the tellurite glass, which proves its great resistance to water. Several researchers have studied the reduction of water molecules and hydroxyl groups, in order to further improve the performance of tellurite glass materials used in mid-infrared fiber lasers. Specific test procedures included melting the glass in a dry atmosphere, raw material dehydration, and the use of fluoride [52] or chloride raw materials [53], as eliminating hydroxyl groups diminishes loss (at specific wavelengths) and is therefore favorable for the commercialization of tellurite glass fibers.

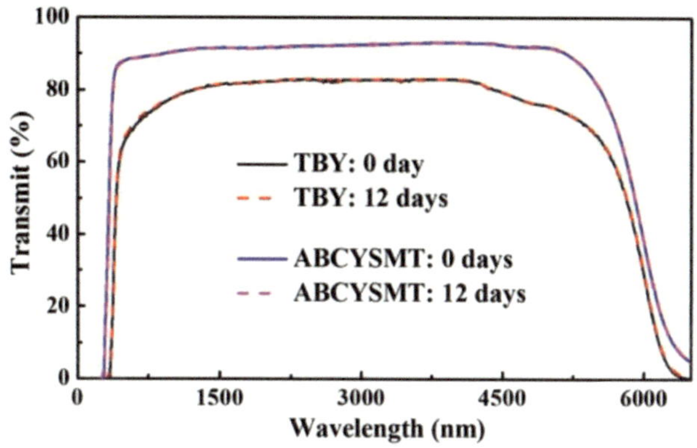

Figure 3.
Transmission spectra of two kinds of tellurite glasses before and after dipping in water [52].

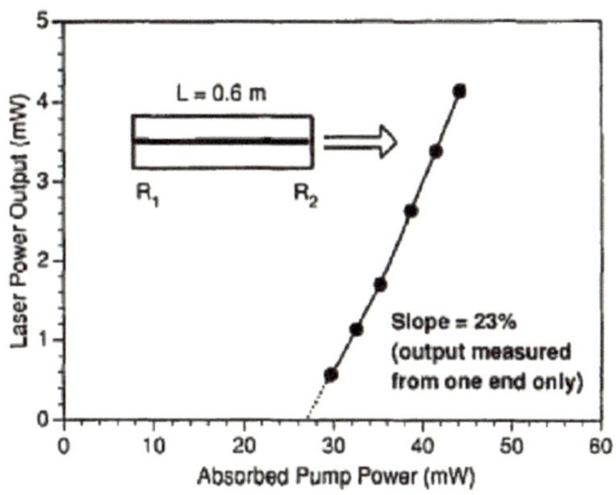

Figure 4.
The relationship between the laser output power and the pump power ($R_1 = R_2 = 11.9\%$) in Nd^{3+}-doped tellurite fiber. Considering that the laser output power at both ends of the fiber is the same, the total slope efficiency should be 46% [12].

With the above advantages coupled with excellent thermal stability, tellurite glass preforms could be handled with relative ease for casting [24, 53], drilling [54], and extrusion techniques [55], providing precursors for tellurite fiber-based nonlinear optical processing [56] and fiber lasers.

In 1994, Wang et al. successfully prepared a Nd^{3+}-doped tellurite single-mode fiber for the first time. The numerical aperture of the fiber was determined as 0.21. The laser resonator was formed as a consequence of multiple Fresnel reflection (~11.9%) from the end surfaces of the fiber. A laser with a wavelength of 0.818 μm was used as the pump source. A laser output with a wavelength of 1.061 μm was obtained from a 0.6 m long fiber, with a laser threshold of ~27 mW. When only single ended output is considered, the slope efficiency of the laser was 23%, as shown in **Figure 4** [12].

In 1998, Ohishi et al. used a 0.9-m-long Er^{3+}-doped tellurite fiber as the gain medium to construct a ring laser cavity. When the pump power was 300 mW, a continuous tunable laser output covering 1529–1623 nm was obtained using a tunable filter. A 2.4-m-long Er^{3+}-doped tellurite fiber was used as the gain medium to obtain laser output at ~1624.5 nm, with a slope efficiency of 3.6% as shown in **Figure 5** [57].

In 2011, Dong et al. demonstrated a high-performance Er^{3+}/Ce^{3+} co-doped tellurite fiber amplifier and tunable fiber laser using a dual-pumping scheme.

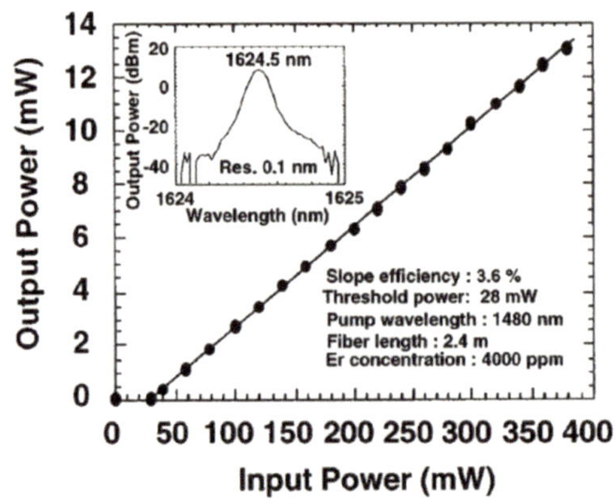

Figure 5.
Laser characteristics of a tellurite-based fiber laser operating in the 1625 nm band. The inset shows the lasing spectrum of the fiber laser [57].

The short 22 cm fiber exhibits a net gain of 28 dB at 1558 μm, a wide positive net gain bandwidth of 122 μm, and a noise figure of 4.1 dB. As shown in **Figure 6**, a widely tunable Er^{3+}/Ce^{3+} co-doped tellurite fiber ring laser with a tuning range of 83 μm was demonstrated [58].

In 2012, M. Oermann et al. fabricated Er^{3+}-doped tellurite microstructured fibers with three air holes using an extruding method. The resulting fibers are shown in cross section in **Figure 7**. The core diameter of the fiber was about 1.5 μm, the loss was 1.3 dB/m, and the doping concentration of Er^{3+} was 0.022 mol %. A 2.2-m-long Er^{3+}-doped tellurite microstructure fiber was used as the gain medium to construct the laser cavity and was pumped using a 976 nm laser source. As shown in **Figure 8**, its threshold power is only 1.5 mW, and its slope efficiency reaches 13% [59].

In the same year, Chillcce et al. fabricated an Er^{3+}-doped tellurite microstructured fiber using the stack-and-draw technique. They demonstrated laser emission using a simple double-pump configuration with two sources at 980 μm. The fiber core had a hexagonal structure as shown in **Figure 9**, the Er_2O_3 doping concentration was 7500 ppm, and the background loss of the resulting microstructured fiber was 0.2 dB/cm at ~1117 μm. Two short segments of fiber of 5 and 12 cm generated laser emissions at 1532.3, 1536.3, and 1558.5 μm, as shown in **Figure 10**. The maximum optical signal-to-noise ratio (OSNR) obtained was 21.2 dB [60].

In 2014, Yao et al. fabricated microstructured fibers consisting of a solid core surrounded by six air holes using a rod-in-tube method. A maximum unsaturated power of 9 mW laser operating at ~1872 μm was obtained in a Tm^{3+}-doped 2.8 cm

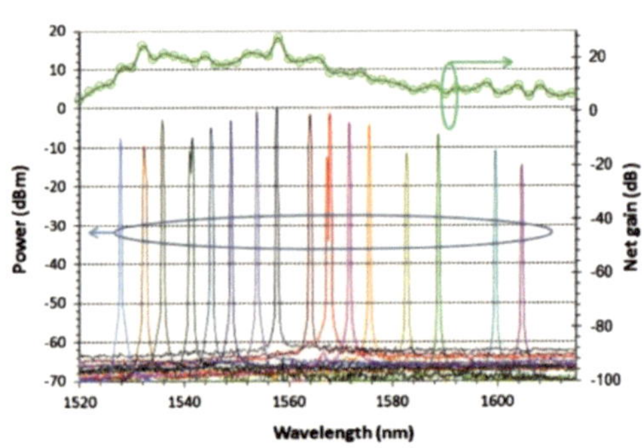

Figure 6.
Output spectra of the E Er^{3+}/Ce^{3+} co-doped tellurite fiber ring laser [58].

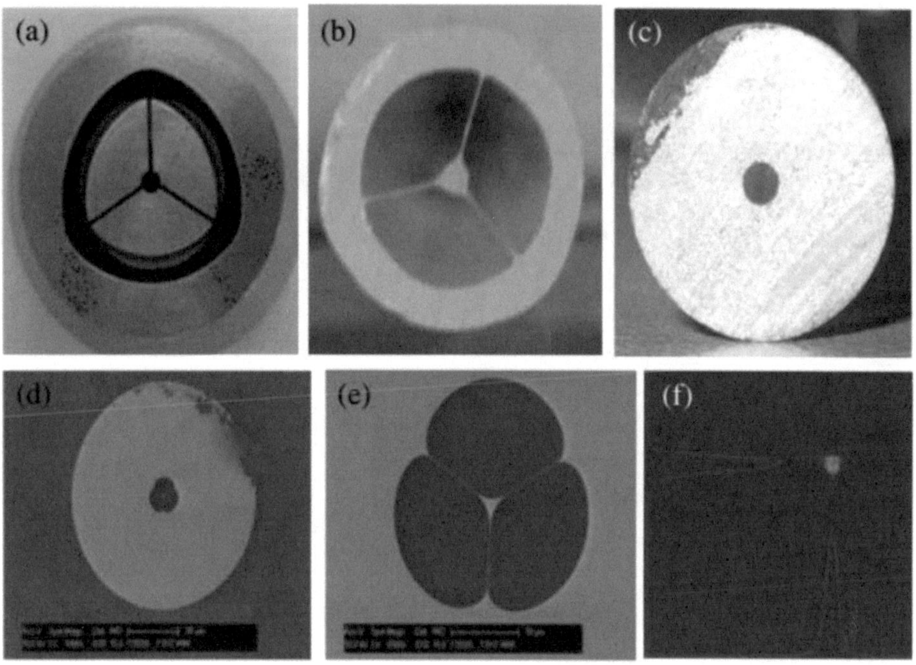

Figure 7.
Photographs of (a) stainless steel die exit used for the extrusion of the structured preform and (b and c) the extruded structured and jacket preforms, respectively. SEM images of the (d) fabricated fiber cross section, (e) enlarged SEM image of the fiber's core and cladding, and (f) beam profile of the laser mode emitted from the output of the fiber [59].

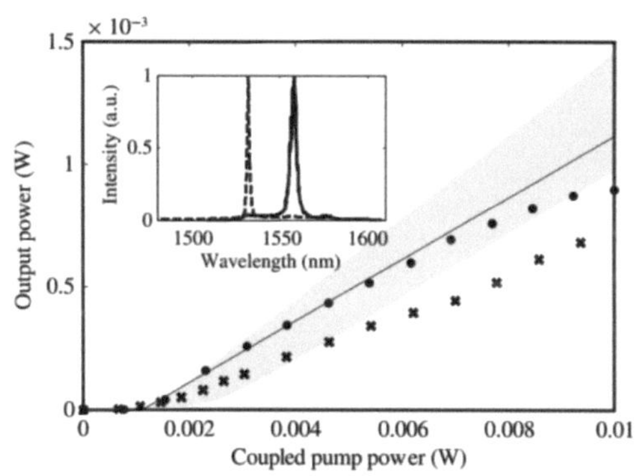

Figure 8.
Fiber laser output plotted against the coupled pump power for a fiber length of 2.2 m (circles). The figure inset is a plot of the laser output spectrum for 5 mW of coupled pump power into the 1 m (dashed) and 2.2 m (solid) lengths of fiber [59].

long microstructure fiber with a slope efficiency of ~ 6.53% and a threshold power of ~200 mW. The results shown in **Figure 11** indicate that the Tm^{3+}-doped tellurite microstructure fiber is a promising material for achieving a compact 2 μm output fiber laser [61].

In 2015, Meng et al. used a 22-cm-long Tm^{3+}/Ho^{3+} co-doped tellurite fiber to obtain a continuous laser output with a maximum output power of 8.34 mW and a wavelength of 2065 nm when the pump power was 507 mW, as shown in **Figure 12**. The slope efficiency was 2.97% [62].

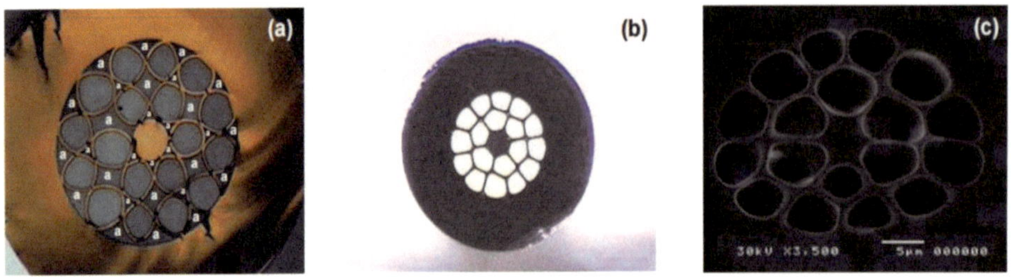

Figure 9.
(a) Preform with the first clad before eliminating the air trapped. The air regions are indicted with a white "a". (b) Preform without the air trapped. (c) Scanning electron microscope image of the microstructure fiber [60].

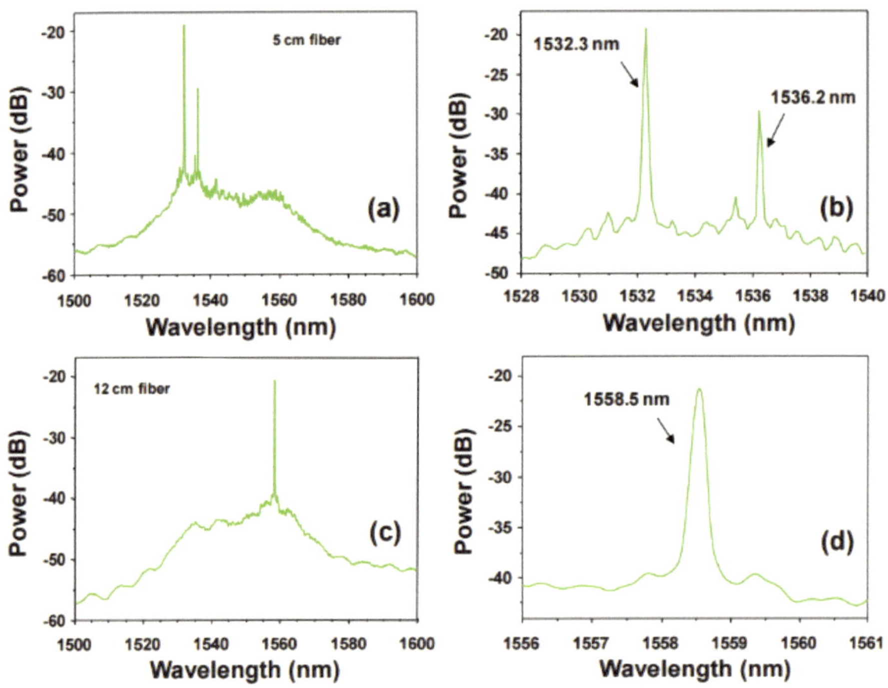

Figure 10.
Laser emission spectra. (a) 5 cm fiber segment. (b) A zoom of the emission region observed in (a). (c) 12 cm fiber segment. (d) A zoom of the emission region observed in (c) [60].

2.2 Tellurite fiber-based supercontinuum light source

The supercontinuum (SC) light source is defined as a broadband laser source whose output spectrum is greatly broadened through the interaction of nonlinear effects and dispersion when a high peak power pulsed laser output (e.g., a soliton pulse) propagates in nonlinear optical medium. The SC spectra generated in transparent materials do not usually originate from a single nonlinear process—typically the initiated self-phase modulation (SPM) modulates the phase of the input laser, and then other nonlinear effects including cross-phase modulation (XPM), stimulated Raman scattering (SRS), four-wave mixing (FWM), soliton self-frequency shifting (SSFS), etc. broaden the output frequency (wavelength) spectrum [63]. The first observation and application of SC spectra were obtained in solids and liquids [64–66], but recent investigations and applications of SC light sources have utilized optical fibers including single-mode and microstructured fibers [67]. The latter is a widely used medium due to its unique geometry and low transmission loss

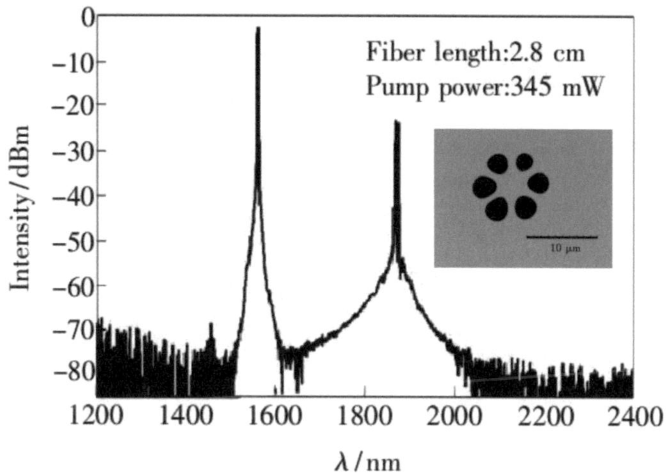

Figure 11.
Laser spectrum of fiber laser pumped by 1560 μm band fiber laser. The figure inset shows cross section of Tm^{3+}-doped TZNB microstructure fiber [61].

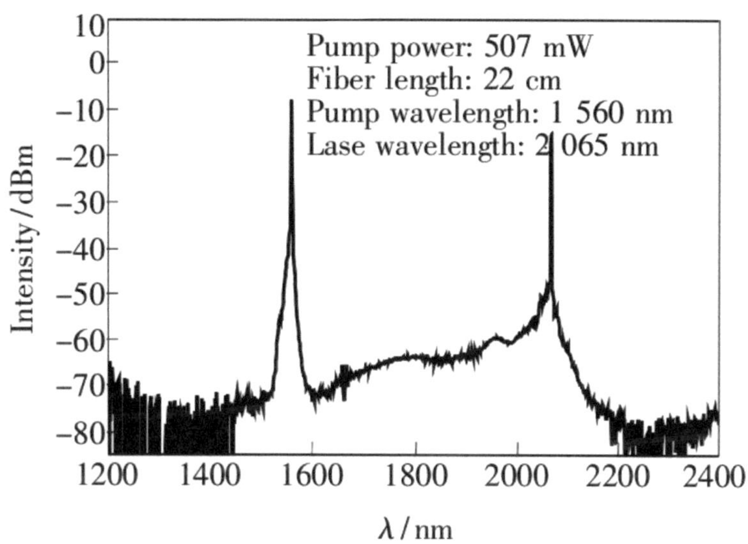

Figure 12.
Spectrum of the fiber laser pumped by 1560 μm band fiber laser [62].

that can accumulate the power intensity of pump sources and provide adequate interaction length to facilitate the occurrence of the nonlinear processes. In 2005, half the Nobel Prize in Physics were awarded for the development of optical frequency combs that was generated from the SC coherent light source employing microstructured silica fiber. SC light sources based on silica microstructured fiber with outputs spanning from the ultraviolet to the near infrared spectral regions have been widely commercialized by major optics firms, such as American Corning corporation, Fujitsu and Nippon of Japan, Korea's ETRI and so on.

The 2–5-μm-mid-infrared region is the typical wavelength range corresponding to the "atmospheric optics window," the "molecular fingerprint region," and "strong absorption band of hydroxyl and amino groups." Therefore, SC light sources in this region offer great possibilities for optical telecommunication, remote sensing, atmospheric pollution monitoring, molecular spectroscopy, medical diagnosis, hyperspectral imaging, laser surgery, and IR opto-electric countermeasures [63, 68], all of which greatly attracted intense worldwide research interest over the past two decades [69–71]. There are several requirements of nonlinear fibers used for 2–5 μm SC light sources, e.g., they must be transparent within the 2–5 μm window, they must have a relatively high laser damage threshold for potentially high-power light transmission, they should have a high nonlinear refractive index, and they need to be fabricated based on mature processing technology. Silica is not a candidate material for generating SC spectra at wavelengths longer than 2.2 μm, due to its high intrinsic loss and relatively low nonlinear parameters. Alternatively, soft glass fibers, mainly including fluoride, tellurite, and chalcogenide glass fibers, are being investigated to develop SC light sources in the 2–5 μm spectral region and have achieved remarkable progress to date with their broad IR transparency range as well as prominent optical nonlinearity.

Among the soft glass materials investigated, tellurite glass provides many several attractive features for use in high-power SC light sources. These include a broad IR transmission window (0.3–7 μm) that can be matched with fluoride glass while possessing lower intrinsic losses than chalcogenide glass and possessing the highest optical damage threshold than other soft glass materials. Moreover, with outstanding thermal and chemical stability, tellurite glass can be drawn as microstructured fiber from a preform constructed using the rod-in-tube method or extrusion technique. The dispersion profile and nonlinearity of the fabricated fiber can be readily optimized. In the past two decades, much effort has been concentrated on fabricating a microstructured tellurite fiber for SC generation.

Kumar et al. prepared low-loss tellurite microstructured fiber for the first time using an extrusion and rod-in-tube method, whose minimum loss was 2.3 dB/m at 1055 nm [72]. Photographs of the fiber are shown in **Figure 13**. In such a microstructured fiber with 1.02 m length, they studied the stimulated Raman scattering generation pumped using a 1064 nm pulsed laser.

In 2008, Domachuk et al. generated a SC spectrum with a broad bandwidth covering the spectral range 789–4870 nm in tellurite microstructured fiber pumped using a1550 nm pulsed laser [73]. As shown in **Figure 14**, the fiber core was surrounded by six large diameter air holes to achieve strong light confinement, and the calculated nonlinear waveguide coefficient at 1550 nm was 596 $km^{-1}\ W^{-1}$,

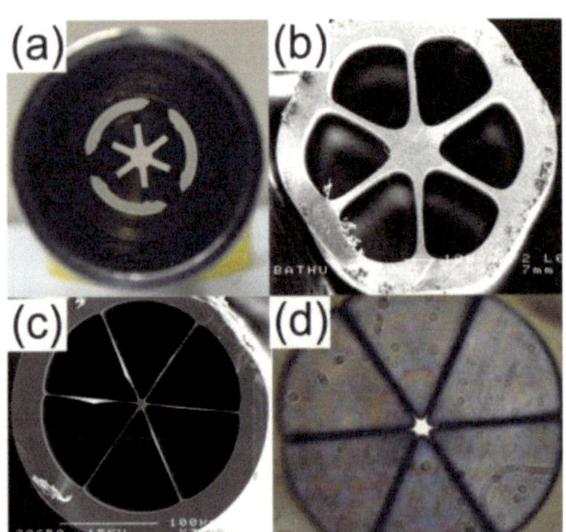

Figure 13.
(a) The cross section of the die used for extrusion. (b) Electron micrograph of an extruded tellurite preform, with outer diameter 1 mm. (c) Electron micrograph of tellurite PCF. (d) Transmission view of a tellurite PCF as seen in microscope [72].

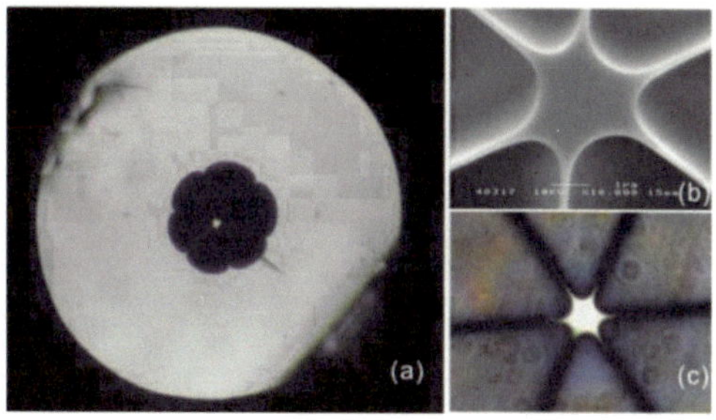

Figure 14.
Picture as seen in optical microscopy (a and c) and cross section profile of the tellurite PCF in electron microscopy (b). Scale bar in (b) is 1 μm [73].

which broadened the SC spectrum spanning two and a half optical octaves in the fiber having only a length 0.8. Such a short fiber length results in flatter SC spectra, lower dispersion, and reduced material absorption at longer wavelengths.

In 2008, Feng et al. fabricated a large-mode-area tellurite holey fiber from an extruded preform, with a core diameter of ~80 μm, attenuation of 2.9 dB/m at 1.55 μm, and zero-dispersion wavelength (ZDW) at 2.15 μm (**Figure 15**) [74]. Using such microstructured fiber with a 9 cm length, a broadband SC spanning of 0.9–2.5 μm was achieved.

In 2009, Liao et al. fabricated the hexagonal core fiber (**Figure 16**) for the first time [75]. They studied the SC generation in such a fiber of 6 cm length pumped by a 1557 nm femtosecond laser and with a 30-cm-long fiber pumped using a 1064 nm picosecond fiber laser. Additionally, they demonstrated that the holey region has an important influence on the dispersion, nonlinear coefficient, and SC generation.

In 2010, a 36-cm-length tellurite microstructured fiber with four holes [76] was used to generate a flattened SC spectrum spanning from 900 to 2800 nm (**Figure 17**) and was pumped using a 1550 nm pulsed laser. The calculated nonlinear coefficient at 1550 nm was 539 $km^{-1} W^{-1}$ [76].

In 2012, Savelii et al. prepared a low-loss suspended-core tellurite fiber, from which they generated a 0.75–2.8 μm SC spectrum when pumped at 1745 nm [77]. And in 2015, Belal et al. generated SC spectra extending to 3 μm in a suspended-core tellurite fiber. Their numerical study show that the structure of the fiber can have a significant impact on the dispersion profile and hence the nonlinear processes and SC broadening [78].

In 2013, Klimczak et al. reported a breakthrough in the design of optical fiber transverse structure. They produced a novel, regular hexagonal-lattice tellurite photonic crystal fiber (PCF) as shown in **Figure 18** [79]. Pumping the 2-cm-long

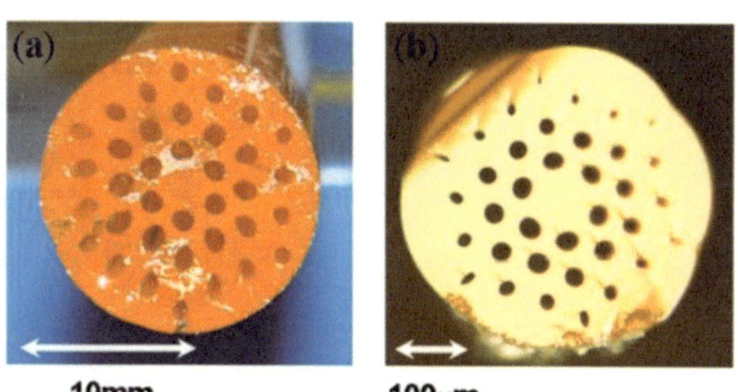

Figure 15.
Optical photographs of the cross-sectional views of (a) the extruded tellurite preform and (b) the resulting tellurite holey fiber with 410 μm outer diameter [74].

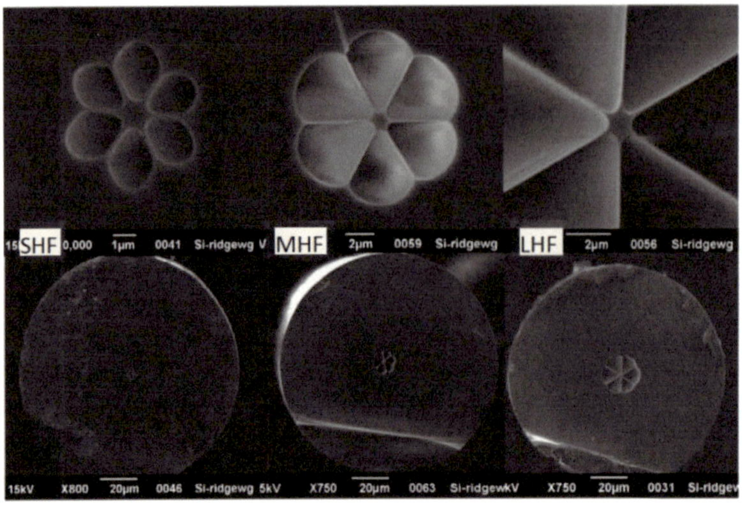

Figure 16.
Scanning electron microscope (SEM) images of the fibers [75].

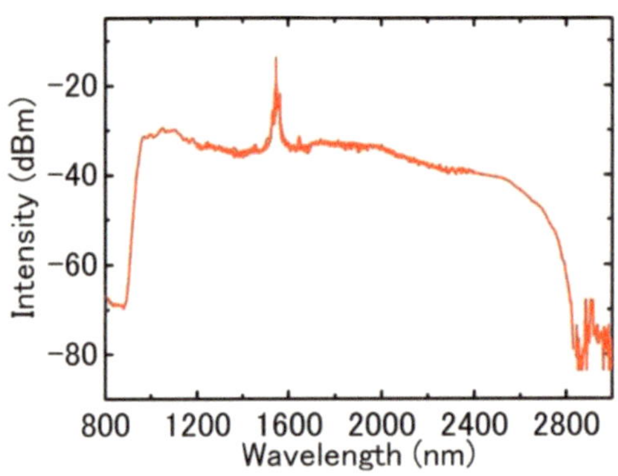

Figure 17.
SC spectrum generated from the tellurite fiber when the peak power of the pump laser is fixed at 3.9 kW [76].

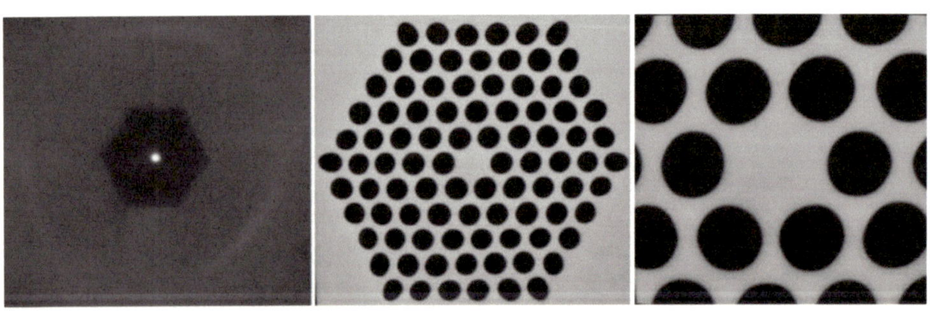

Figure 18.
Microstructure of tellurite PCF: close-up picture of the structure with propagating mode as seen in a CCD camera, and SEM images of photonic structure [79].

PCF with 150 fs/36 nJ/1580 nm pulses, they achieved an output of 800–2500 nm SC spectrum that is comparable to that generated in suspended-core tellurite PCF pumped at wavelengths over 1800 nm.

Yao et al. proposed a novel fluorotellurite fiber with the composition $65TeO_2$-$25BaF_2$-$10Y_2O_3$ (TBY) [80]; the authors claimed further improvement in the performance of tellurite fiber-based MIR laser sources. BaF_2 was included for the purpose of reducing the OH— content, and the introduction of Y_2O_3 was for better thermal stability in fiber drawing process as well as providing a higher glass transition temperature raised by the high melting temperature of Y_2O_3. In 2016, Wang et al. achieved SC generation extending from 0.47 μm to 2.77 μm (zero-dispersion wavelength at 1730 nm) using a tapered TeO_2-BaF_2-Y_2O_3 (TBY)-based microstructured fiber whose core was surrounded by six air holes [81].

Tellurite microstructured fibers with dispersion modification and nonlinear coefficient enhancement have been widely studied and applied for SC generation, and significant progress has been achieved over the last 10 years. However, the air holes present in microstructured fibers readily accommodate moisture and dust particles from the atmosphere, which lead to incremental losses, which act to deteriorate the SC output. In addition, the performance of tellurite microstructured fiber for high-power output in the mid-IR SC is not satisfactory, because the thermal conductivity of the air holes and the core of microstructure fiber greatly differ, and hence means that heat dissipation is a significant problem of high-power light transmission in the fiber. Solid-state tellurite fibers (comprising a solid core and cladding with no air holes) have therefore become the nonlinear medium of choice for high-power mid-infrared SC light sources [82].

Hydroxyl ions have deleterious broad absorption peaks centered at ~3.3 μm and ~4.3 μm, which hinder the tellurite fiber from extending its spectrum to the multiphonon edge (5 μm). In 2013, Thapa et al. of NP Photonics Incorporation developed ultra-low-loss solid-core tellurite fibers which eliminate almost all molecular species, especially hydroxyl ions [83]. Using a 1922-nm all-fiber-based mode-locked fiber laser oscillator, a 1–5 μm SC spectrum shown in **Figure 19b** was generated in a tellurite fiber with a W-type (**Figure 19a**) index profile for strong light confinement, and the ZDW shifted from 2.5 μm to ~1.9 μm. It was argued that the broadened anti-Stokes wavelength portion originated from self-phase modulation (SPM) and the long wavelength portion with increased power originated from the generation of a Raman soliton because of the self-frequency Raman shift. In the same year, Savelii et al. reported SC generation extending from 840 nm to 3000 nm in a low-loss suspended-core tellurite fiber with different lengths (**Figure 20**), pumped at its anomalous dispersion regime at 1745 nm [54]. It was found that the introduction of fluoride ions into the tellurite glass reduced the OH— content and resulted in a fiber that was still transparent at 4.1 μm.

The W-type index profile makes it possible to tailor the ZDW, and this fiber can be fusion spliced to robust step-index silica fiber with relative ease. In 2016

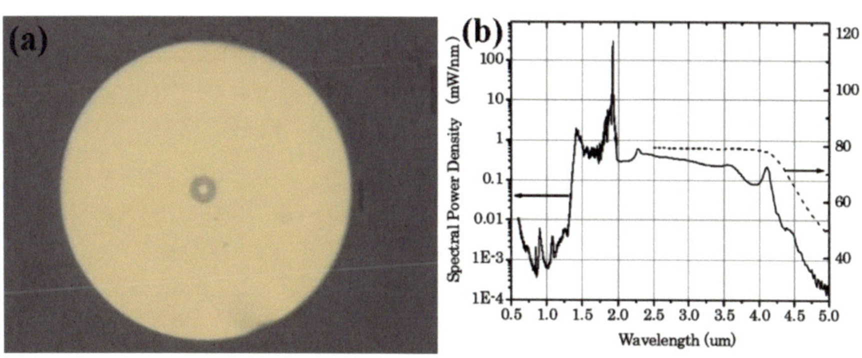

Figure 19.
(a) Cross section of the W-type tellurite fiber. (b) SC spectra in W-type proprietary tellurite fiber pumped by 3 W of ~20 ps pulses from a 32 MHz repetition rate amplified mode-locked laser at 1.92 μm. (Note: dotted line is the transmission measured in the corresponding 1-cm-thick tellurite glass sample.) [83].

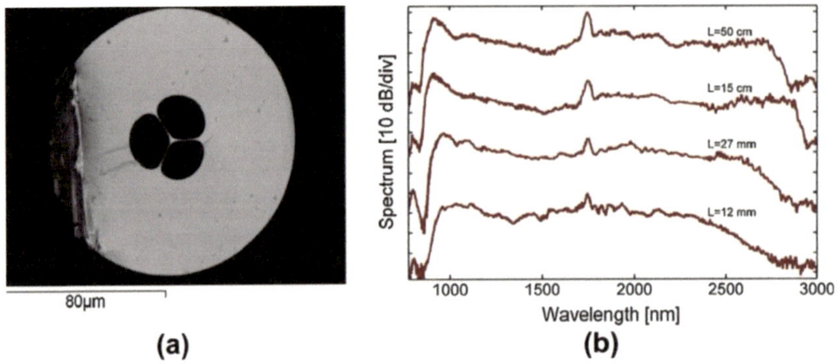

Figure 20.
(a) SEM picture of the cross section of the fiber. (b) SC spectra generated from the suspended-core tellurite fiber with different lengths [54].

Kedenburg et al. studied SC generation spanning 2.6–4.6 μm in low-loss W-type index tellurite fiber with a length of 15 cm [84]. Additionally, they studied the variation of spectral bandwidth with core diameter, pump wavelength, length of fibers, and pump power. In 2017, Kedenburg et al. studied the effects of the core size, pump wavelength, and fiber length on SC generation in a robust step-index tellurite fiber, and they achieved broadband SC generation spanning 1.3–5.3 μm in the fiber with a length of 9 cm and a core diameter of 3.5 μm, when pumped using a 2.4 μm femtosecond pulsed laser [85].

In 2016, Shi et al. prepared a solid-state tellurite optical fiber with a numerical aperture (NA) of 0.21 and a core diameter of 12 μm [86]. **Figure 21(a)** shows a micrograph of the end face of the fiber. They studied the SC generation in the fiber which was 0.8 m-in length. **Figure 21(b)** shows the spectrum of the pump laser and SC output in the fiber when different pump powers were used. When the pump power was 9.8 W, the power spectral density of the SC spectrum in the wavelength range of 1975–3000 nm is above 5 dBm/nm. In this investigation, the maximum output power of the SC light source was 5.1 W, and the power of the spectrum at wavelengths longer than 2.5 μm was about 2.1 W.

In 2017 Jia et al. obtained a stable 4.5 W SC output spanning 1017–3438 nm, using a TBY-based 60-cm-long all-solid fluorotellurite fiber fabricated using the rod-in-tube method. The fiber was pumped using a 2 μm femtosecond fiber 10.48 W output power laser and thus demonstrated the capability of all-solid fluorotellurite fibers for

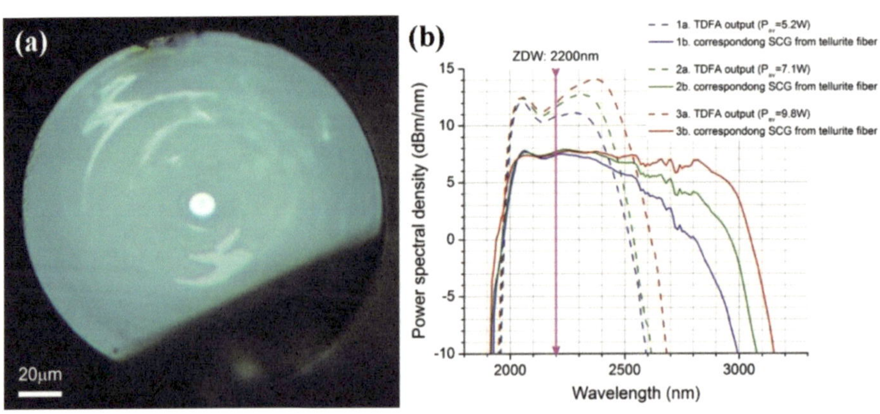

Figure 21.
(a) Photograph of the tellurite fiber. (b) Spectrum of thulium-doped fiber amplifier (TDFA) and the SC spectrum generated from the tellurite fiber, pumped by various power: 5.2 W, 7.1 W, and 9.8 W [86].

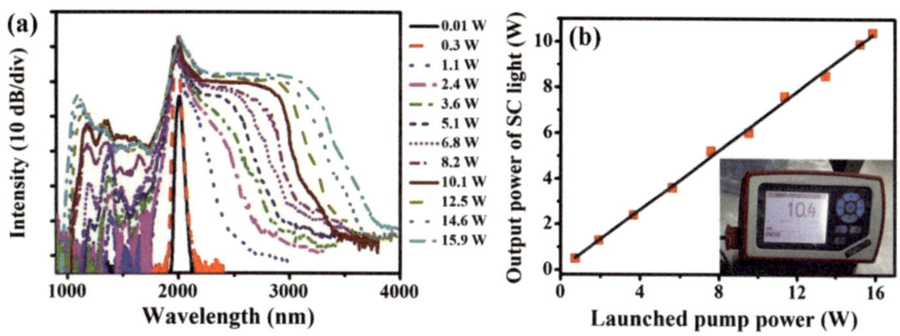

Figure 22.
(a) Dependence of the measured SC spectra generated from 60-cm-long fluorotellurite fiber on the average power of the 1980 nm femtosecond fiber laser. (b) The dependence of the SC average power on the pump power. Inset: photograph of the power meter when the mid-IR SC laser source is operating at the output power of 10.4 W [52].

use as high-power mid-IR SC light sources [87]. The same authors used a tapered all-solid fluorotellurite fiber with ultra-high NA to generate an SC output spectrum covering the entire 0.4–5 μm transmission window and pumped using a 1560-nm mode-locked fiber laser [88]. Yao et al. achieved stable 10.4 W SC generation in the wavelength ranging from 947 to 3934 nm from a TBY-based all-solid fluorotellurite fiber when pumped using a high-power 1980 nm femtosecond fiber laser [52]; when the average pump power was increased to 1.1 W, large spectral broadening occurred as shown in **Figure 22(a)**. Because the fiber was pumped at anomalous dispersion regime, the spectral broadening for a pump power of ≥1.1 W originated from the SPM, the formation of higher-order soliton, soliton fission, soliton self-frequency shift (SSFS), and the generation of blue-shifted dispersive waves. The average output power of the SC laser source increases linearly with the average pump power (**Figure 22(b)**), and the corresponding optical-to-optical conversion efficiency was measured to be as high as 65%. The successful achievement of a 10 W output power level represented a significant breakthrough in all-solid fluorotellurite fiber, demonstrating its bright future for high-power MIR SC light sources.

3. Tellurite glass-based microcavity lasers

3.1 Experimental preparation of tellurite glass microcavities

Over the past few decades, research interest in microsphere resonators has grown rapidly. For a microsphere resonator, the pump light can be coupled into the microsphere through a tapered optical fiber or via free space. Most current microsphere resonators are fabricated from the silica optical fiber, but it is also possible to fabricate microsphere resonators from compound glass materials (such as tellurite glass) other than silica. At present, the principal method used for making microsphere cavities is based on melting of the glass materials, which uses the surface tension of molten glass to form the microsphere when suspended at the tip of a fiber. There are two common methods for the preparation of tellurite glass microsphere cavities, one is to melt glass fiber and the other is a powder floating method.

3.1.1 Melting glass fiber method

Most glass microsphere cavities are prepared using a CO_2 laser, arc discharge, or high temperature ceramics to melt glass fibers. These methods have also been widely applied for fabricating tellurite glass microsphere cavities and other compound glass microsphere cavities resulting in a good shape and high Q factor. The fabrication method using a CO_2 laser is described here.

A schematic diagram of the experimental setup for manufacturing a microsphere resonator is shown in **Figure 23**. The main positioning and alignment instrument used in the experiment was a precision three-dimensional (3D) translation stage, a continuous CO_2 laser with an output wavelength of 10.6 μm, and a ZnSe lens for focusing. The experimental step for fabricating the tellurite microsphere resonator was divided into three stages. In the first step, the tellurite fiber was mounted vertically on the 3D translation stage and a weight hung at the end of the fiber. The ZnSe lens was used to focus the laser beam on the tip of the tellurite fiber, causing it to absorb the incident light which resulted in a temperature rise. The glass softened and gradually stretched into a tapered fiber under the influence of the weight. The heating was terminated when the waist diameter of the tapered fiber reached around 30 μm. In the second step, the tapered fiber was accurately cleaved at the waist region to obtain a half-tapered fiber. In the third step, using a ZnSe lens once more to focus the laser beam on the end of the half-tapered fiber, the tellurite microsphere was formed at the fiber end due to the surface tension acting

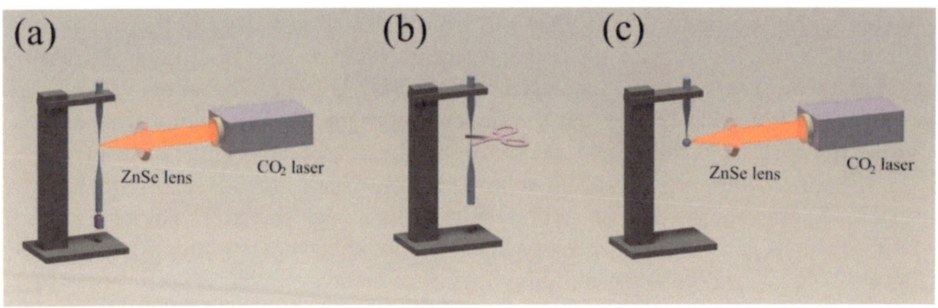

Figure 23.
Schematic diagram of the experimental setup for making a tellurite glass microsphere. (a) A ZnSe lens is used to focus a CO_2 laser beam on the tellurite fiber. (b) The waist region of tapered fiber is cleaved. (c) A tellurite microsphere is obtained by focusing a CO_2 laser beam on the end of the cleaved tapered [92].

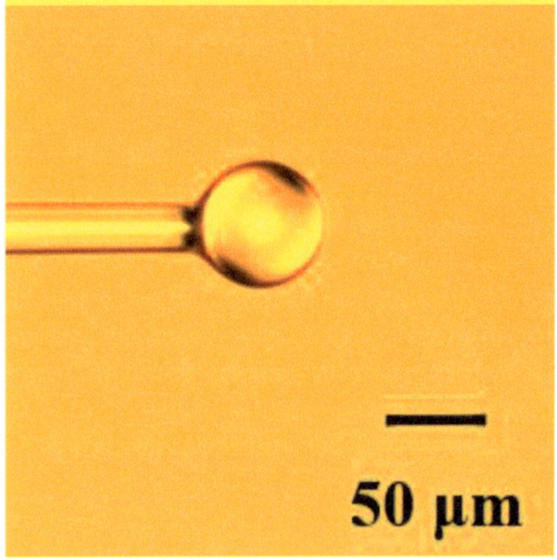

Figure 24.
Microscope image of tellurite microsphere made with CO_2 laser.

on the molten glass. A microscope image of a tellurite microspheres fabricated in this manner is shown in **Figure 24**.

Although the resulting microsphere cavity made of molten glass fiber includes a glass fiber attached at a pole of the sphere, the light field is mainly concentrated around the equatorial plane of the microsphere cavity, and hence the loss induced by the fiber to the whispering gallery mode (WGM) resonance is negligible [89, 90]. In general, the fiber rods are only used to hold the microspheres in place and to facilitate light coupling. However, in recent years, some other uses of fiber rods have attracted increasing attention. In 2017, Murphy et al. [91] reported an alternative method for precise coupling control using the fiber rod. In the experiment, 980 nm laser light was input into the fiber rod, and the coupling distance between the microsphere cavity and the tapered fiber was precisely controlled by heating the connection between the microsphere and the fiber rod using the 980 nm laser. The adjustment range of microsphere cavity position was from 0.61 ± 0.13 μm to 3.49 ± 0.13 μm.

3.1.2 Powder floating method

The previously mentioned method for fabricating tellurite glass microspheres was only capable of producing one microsphere at a time, and the powder floating represents an alternative method for preparing glass microspheres in large quantities. In this method, the tellurite glass was ground into powder and poured into a high temperature furnace, which was placed vertically with a proper protective gas (nitrogen or inert gas) from the bottom to top. The tellurite glass powder melts and forms into microspheres due to surface tension at high temperature. In addition, the protective gas reduced the falling speed of the glass powder and increased the exposure time of the powder to the high temperature in the furnace. Additionally, the method isolates the glass powder from the atmosphere [93]. For some glass materials with less stringent experimental requirements, the microspheres can be formed without the use of protective gas [94].

Tellurite microspheres prepared using this method have no attached fiber rods, which is different from melting glass fiber or sol-gel methods. Using the powdered method, a large number of microspheres with different diameters can be prepared simultaneously, which is beneficial to the selection of experimental size and enables the integration and commercialization of the microsphere laser on a mass produced basis. **Figure 25** is a schematic diagram of fabrication of microspheres, and **Figure 26** is a microscope image of the microspheres fabricated using the powder floating method.

3.2 Tellurite glass microsphere lasers

Tellurite glass has emerged as a promising material for use in microsphere resonators in the near infrared wavelength region and tellurite glass microsphere lasers have been widely reported. In 2002, Sasagawa et al. reported continuous-wave oscillation in an Nd^{3+}-doped tellurite glass microsphere laser at 1.06 μm for the first time [96]. Tellurite glass microspheres with diameters in the range of 50 μm to a few hundred micrometers were fabricated by melting using a Kanthal wire heater. Resonances were excited in the microsphere pumped using a 800 nm laser, the threshold of the output laser was measured as 81 mW, and the emission spectrum is shown in **Figure 27**.

Later in 2005, an Er^{3+}-doped tellurite glass microsphere laser was reported by Peng et al. [97]. The threshold of 1561 nm microsphere laser with 0.5 wt% Er_2O_3 doped was measured to be as low as 1.4 mW, and the maximum output power

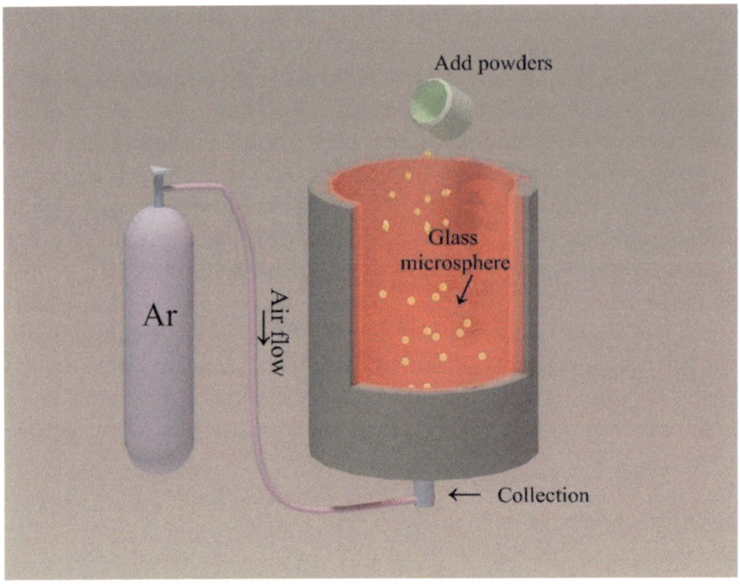

Figure 25.
Schematic diagram of fabrication of microsphere by powder floating method [95].

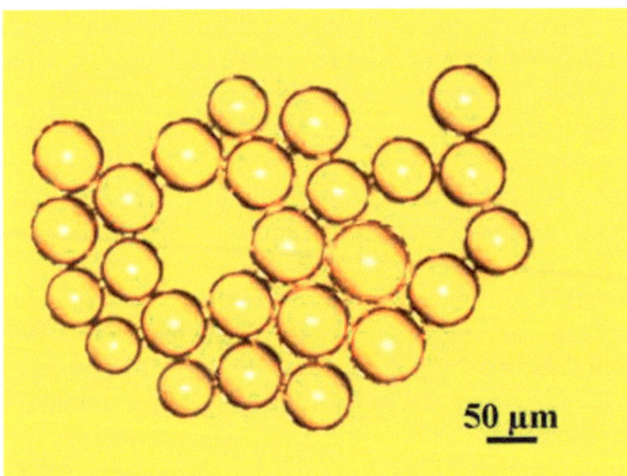

Figure 26.
Microspheres fabricated by powder floating method.

achieves 124.5 μW. **Figure 28** shows the relationship between the output laser power and the 1480 nm pump power.

The output wavelength of the laser around 1.9 μm, and 1.47 μm band is generated from the transition of Tm^{3+} ions: $^3F_4 \rightarrow {}^3H_6$ and $^3H_4 \rightarrow {}^3F_4$ [98]. Wu et al. proposed a microcavity laser based on a Tm^{3+}-doped tellurite glass microsphere at 1.9 μm [99]. However, there are two problems in realizing a laser at the wavelength 1.47 μm. Firstly, the lifetime of the $^3H_4$ level is shorter than that of the $^3F_4$ level in Tm^{3+} ions, so the transition is sometimes described as self-terminating [100]. Secondly, the glass host material should have very low phonon energy, as in the case of silica and phosphate glass lasers, and amplification is essential. Tellurite and other heavy metal fluoride glasses have been considered as key materials for thulium-doped fiber amplifier operation in the S band, mainly due to their lower phonon energies (~580 cm^{-1}) [12]. In 2004, Sasagawa et al. solved the population inversion problem in Tm^{3+} ions and realized a cascade laser with output wavelengths in the

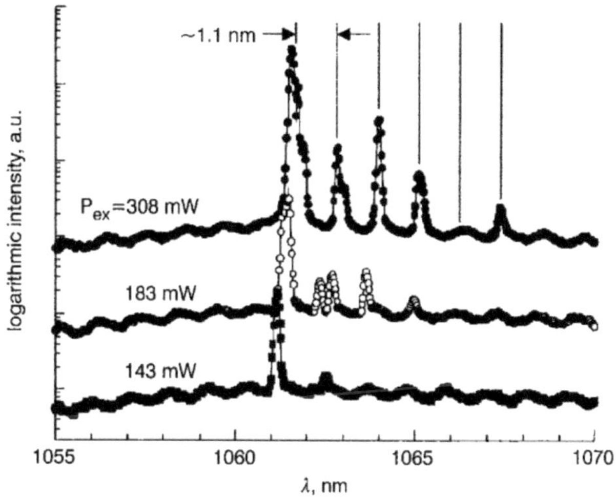

Figure 27.
Emission spectra for $^4F_{3/2} \rightarrow {}^4I_{11/2}$ transition of Nd^{3+} ions in tellurite glass microsphere at various pumping powers [96].

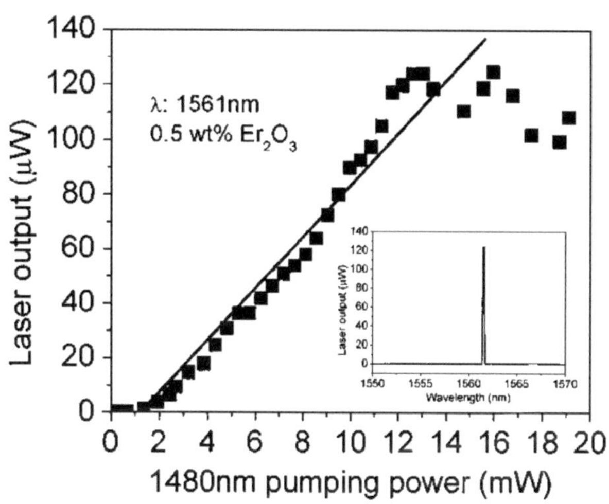

Figure 28.
The microsphere laser pumped by a 1480 nm laser. The Er_2O_3 doping concentration is 0.5 wt%, and the diameter of the microsphere is 32 μm. The maximum output power is 124.5 μW. The inset shows the single-mode profile of this L-band microsphere laser [97].

1.47 μm and 1.9 μm bands using a tellurite glass microsphere [101]. The output spectrum of the Tm^{3+}-doped tellurite glass laser is shown in **Figure 29**, which shows the laser emission in the S band and at 1.9 μm. The average output power is plotted as a function of the average pump power in **Figure 30**. The threshold of the laser in the S band is 4.6 mW, while the thresholds measured for 1.9 μm are 3.0 mW and 4.8 mW, respectively. The differential quantum efficiency in the S band and at 1.9 μm were calculated as 1.4% and 1.1% for bidirectional lasing.

In 2019 [3], Li et al. fabricated Tm^{3+}-Ho^{3+} co-doped tellurite glass samples to solve the problem of the population inversion and obtained a 1.47 μm output using a tellurite glass microsphere laser. **Figure 31(a)** shows the output spectrum at 1.47 μm of Tm^{3+}-Ho^{3+} co-doped tellurite glass microspheres when pumped using a 802 nm laser source. It is clear from **Figure 31(b)** that the lifetime of $^3F_4$ energy level is attenuated through the energy transfer process in Tm^{3+}-Ho^{3+} co-doped tellurite

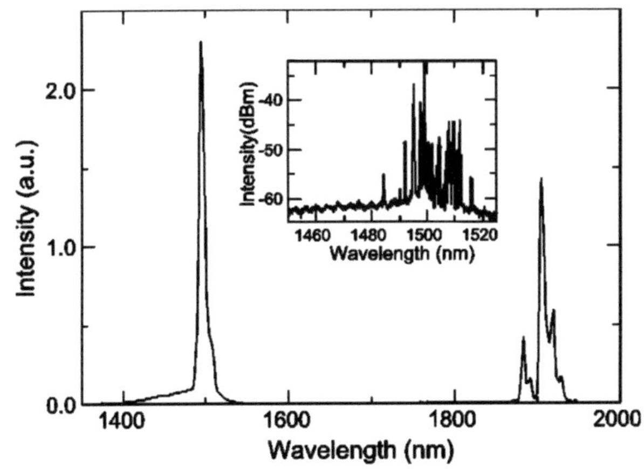

Figure 29.
Emission spectrum of a Tm^{3+}-doped tellurite microsphere laser with diameter of 104 μm. (Inset) OSA emission spectrum [101].

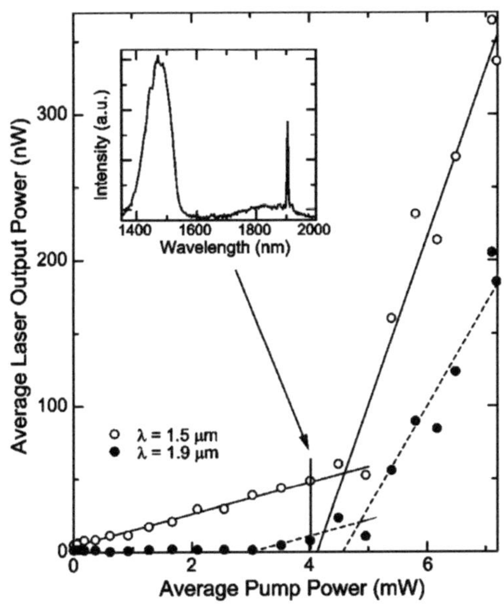

Figure 30.
Average laser output power against average pump power for a Tm^{3+}-doped tellurite microsphere laser. (Inset) Laser emission spectrum at average pump power of 4.0 mW [101].

glass. The Tm^{3+} ions are excited from $^3F_4$ to the $^3H_4$ energy level by the 802 nm pump laser, and the lifetime of Tm^{3+}-doped and Tm^{3+}-Ho^{3+} co-doped material are shown in **Figure 32**. The emission process originates from the Tm^{3+}: $^3H_4 \rightarrow {}^3F_4$ transition, and the energy transfer efficiency of the Tm^{3+}: $^3F_4$ level to Ho^{3+}: $^5I_7$ level is 34.9% in Tm^{3+}-Ho^{3+} co-doped tellurite glass sample.

3.3 Summary

The last two decades have witnessed significant progress of tellurite fiber-based SC light sources, whose original progress was primarily implemented through the development of microstructured and all-solid fiber devices. The microstructured fiber demonstrated greater flexibility in tailoring the dispersion profile than the

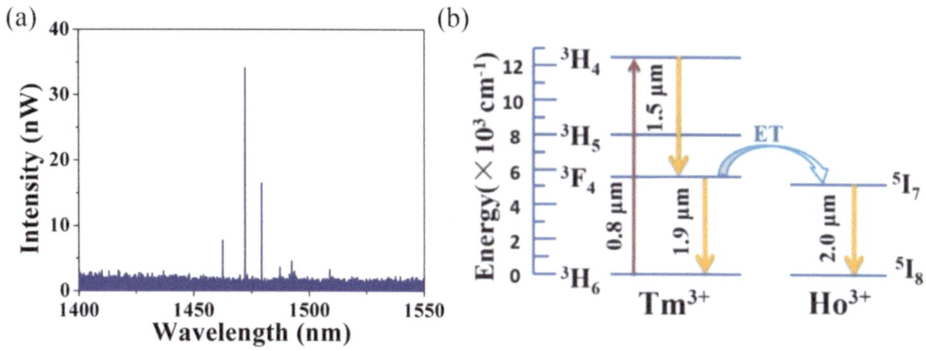

Figure 31.
(a) Laser emission spectrum from the Tm^{3+}-Ho^{3+} co-doped microsphere when the pump power was set to 2.5 mW. (b) Energy level diagram and energy transfer model in tellurite glass [3].

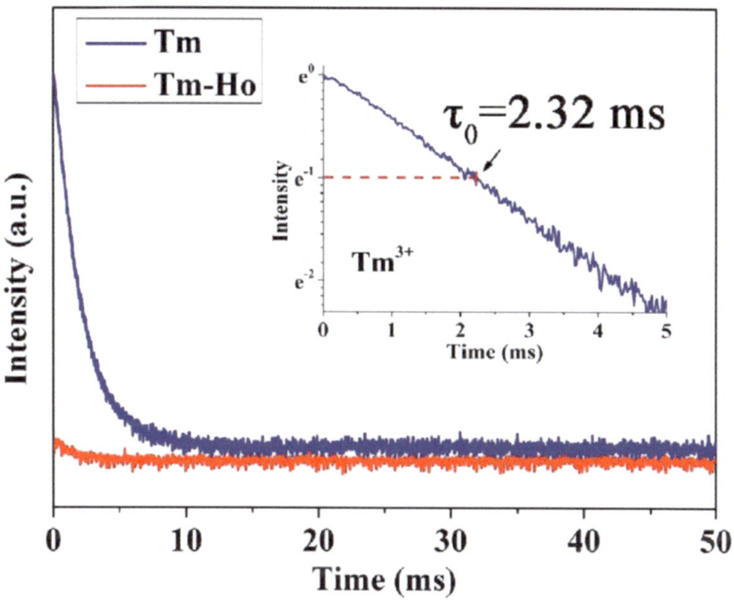

Figure 32.
Fluorescence decay curves of Tm^{3+}-Ho^{3+} co-doped and Tm^{3+}-doped tellurite glass samples at 1.9 μm. The inset figure shows that the lifetime of Tm^{3+} at 1.9 μm is 2.32 ms in Tm^{3+}-doped tellurite glass samples [3].

all-solid version, which provided greater options for using different pump sources, producing higher coherence. In the case of high-power output and stable MIR SC generation, the all-solid tellurite fiber performed much better than the microstructured one, and hence the fluorotellurite fiber is a promising candidate for high-power Mid-IR laser emission. Potentially this technology could be expected to reach the hundred-watt output level even after losses with careful design for heat management, fiber structure, and pump parameter optimization.

Tellurite glass microsphere resonators have overcome the limitations associated with traditional resonators in terms of glass materials. In the future, it is envisaged that tellurite glass microsphere resonators will have wide-ranging applications in photonics, having a high Q value and fast response. Meanwhile, doping rare earth ions in different host materials is expected to achieve higher-power output and more efficient lasers accessing different wavelength ranges, most notably in the infrared band.

Author details

Pengfei Wang[1,2]*, Shijie Jia[1], Xiaosong Lu[1], Yuxuan Jiang[1], Jibo Yu[1], Xin Wang[1], Shunbin Wang[1] and Elfed Lewis[3]

1 Key Laboratory of In-Fiber Integrated Optics of Ministry of Education, College of Physics and Optoelectronic Engineering, Harbin Engineering University, Harbin, China

2 Key Laboratory of Optoelectronic Devices and Systems of Ministry of Education and Guangdong Province, College of Optoelectronic Engineering, Shenzhen University, Shenzhen, China

3 Department of Electronic and Computer Engineering, Optical Fibre Sensors Research Centre, University of Limerick, Limerick, Ireland

*Address all correspondence to: pengfei.wang@tudublin.ie

References

[1] Stanworth JE. Tellurite glasses. Nature. 1952;**169**:581-582

[2] Xing Z, Liu X, Gao S, Zhang Y, Liao M. ~3 μm fluorescence behavior of Ho^{3+} doped transparent tellurite glass ceramics. Journal of Luminescence. 2019;**215**:116562

[3] Li A, Li W, Zhang M, Zhang Y, Wang S, Yang A, et al. Tm^{3+}-Ho^{3+} codoped tellurite glass microsphere laser in the 1.47 μm wavelength region. Optics Letters. 2019;**44**:511-513

[4] Li A, Dong Y, Wang S, Jia S, Brambilla G, Wang P. Infrared-laser and upconversion luminescence in Ho^{3+}-Yb^{3+} codoped tellurite glass microsphere. Journal of Luminescence. 2020;**218**: 116826

[5] Berzelius J. Tellurite glasses. Annal. Physike Chemie. 1834;**32**:577

[6] Lambson E, Saunders G, Bridge B, El-Mallawany R. The elastic behaviour of TeO_2 glass under uniaxial and hydrostatic pressure. Journal of Non-Crystalline Solids. 1984;**69**:117-133

[7] El-Mallawany R. The optical properties of tellurite glasses. Journal of Applied Physics. 1992;**72**:1774-1777

[8] Westman A, Crowther J. Constitution of soluble phosphate glasses. Journal of the American Ceramic Society. 1954;**37**:420-427

[9] El-Mallawany R. Tellurite glasses part 1. Elastic properties. Materials Chemistry & Physics. 1998;**53**:93-120

[10] El-Mallawany R. Structural and vibrational investigations of thermal properties of tellurite glasses. Journal of Materials Research. 1992;**7**:224-228

[11] Liu J, Xiao Y, Huang S, Mao L, Wang W, Zhang Q. The glass-forming region and 2.7 μm emission of Er^{3+}-doped $TeO_2-Ta_2O_5-ZnO$ tellurite glass. Journal of Non-Crystalline Solids. 2019; **522**:119564

[12] Wang J, Vogel E, Snitzer E. Tellurite glass: A new candidate for fiber devices. Optical Materials. 1994;**3**:187-203

[13] Mori A, Ohishi Y, Sudo S. Erbium-doped tellurite glass fibre laser and amplifier. Electronics Letters. 1997;**33**: 863-864

[14] El-Mallawany RA. Tellurite Glasses Handbook: Physical Properties and Data. Boca Raton, Florida, USA: Taylor & Francis; 2011

[15] Yakhkind A. Tellurite glasses. Journal of the American Ceramic Society. 1966;**49**:670-675

[16] Jha A, Richards B, Jose G, Teddy-Fernandez T, Joshi P, Jiang X, et al. Rare-earth ion doped TeO_2 and GeO_2 glasses as laser materials. Progress in Materials Science. 2012;**57**:1426-1491

[17] Vetrone F, Boyer J-C, Capobianco JA, Speghini A, Bettinelli M. 980 nm excited upconversion in an Er-doped $ZnO–TeO_2$ glass. Applied Physics Letters. 2002;**80**:1752-1754

[18] Weber MJ, Myers JD, Blackburn DH. Optical properties of Nd^{3+} in tellurite and phosphotellurite glasses. Journal of Applied Physics. 1981;**52**: 2944-2949

[19] Manikandan N, Ryasnyanskiy A, Toulouse J. Thermal and optical properties of TeO_2–ZnO–BaO glasses. Journal of Non-Crystalline Solids. 2012; **358**:947-951

[20] Froidevaux P, Lemière A, Kibler B, Désévédavy F, Mathey P, Gadret G, et al. Specialty Optical Fibers.

Zurich, Switzerland: Optical Society of America; 2018. p. SoTh3H. 4

[21] Yong D, Shibin J, Bor-Chyuan H. Spectral properties of erbium-doped lead haloellurite glasses for 1.5 μm broadband amplification. Optical Materials. 2000;**15**:123

[22] Savelii I. Fibres optiques à coeur supendu en verre d'oxyde de tellure et génération d'effets non linéaires dans l'infrarouge au-delà de 2 microns. Autre. Université de Bourgogne; 2012 [Français]

[23] Lei Y, Ren F, Mei Y, Gao C, Zhu L, Lu A. Judd-Ofelt analysis and improvement in thermal stability and optical properties of Er^{3+} doped TeO_2–ZnO–Na_2O–B_2O_3–GeO_2 glasses. Materials Research Innovations. 2014; **18**:259-266

[24] Moiseev A, Dorofeev V, Chilyasov A, Kraev I, Churbanov M, Kotereva T, et al. Production and properties of high purity TeO_2–ZnO–Na_2O–Bi_2O_3 and TeO_2–WO_3–La_2O_3–MoO_3 glasses. Optical Materials. 2011;**33**:1858-1861

[25] Lakshminarayana G, Qiu J, Brik M, Kumar G, Kityk I. Spectral analysis of Er^{3+}-, Er^{3+}/Yb^{3+}- and $Er^{3+}/Tm^{3+}/Yb^{3+}$-doped TeO_2–ZnO–WO_3–TiO_2–Na_2O glasses. Journal of Physics: Condensed Matter. 2008;**20**:375101

[26] Mazurier F, Levy M, Souquet J. Reversible and irreversible electrical switching in TeO_2-V_2O_5 based glasses. Le Journal de Physique IV. 1992;**2**: C2-185-C182-188

[27] Dai S, Wu J, Zhang J, Wang G, Jiang Z. The spectroscopic properties of Er^{3+}-doped TeO_2-Nb_2O_5 glasses with high mechanical strength performance. Spectrochimica Acta Part A: Molecular Biomolecular Spectroscopy. 2005;**62**: 431-437

[28] Kashchieva E, Pankova M, Dimitriev Y. Liquid phase separation in the systems TeO_2-B_2O_3-M_2O_3 (M_2O_3 = Al_2O_3, Ga_2O_3, Sc_2O_3, La_2O_3, Bi_2O_3). Ceramics-Silikáty. 2001;**45**:111-114

[29] Kamalaker V, Upender G, Ramesh C, Mouli VC. Raman spectroscopy, thermal and optical properties of TeO_2–ZnO–Nb_2O_5–Nd_2O_3 glasses. Spectrochimica Acta Part A: Molecular Biomolecular Spectroscopy. 2012;**89**: 149-154

[30] Shen S, Naftaly M, Jha A. Tungsten–tellurite—A host glass for broadband EDFA. Optics Communications. 2002; **205**:101-105

[31] Dimitriev Y, Dimitrov V, Arnaudov M. IR spectra and structures of tellurite glasses. Journal of Materials Science. 1983;**18**:1353-1358

[32] Sakida S, Hayakawa S, Yoko T. Part 2. ^{125}TeNMR study of M_2O–TeO_2 (M = Li, Na, K, Rb and Cs) glasses. Journal of Non-Crystalline Solids. 1999;**243**:1-12

[33] Sennaroglu A, Kabalci I, Kurt A, Demirbas U, Ozen G. Spectroscopic properties of Tm^{3+}: TeO_2–PbF_2 glasses. Journal of Luminescence. 2006;**116**:79-86

[34] Zhou D, Wang R, Yang Z, Song Z, Yin Z, Qiu J. Spectroscopic properties of Tm^{3+} doped TeO_2-R_2O-La_2O_3 glasses for 1.47 μm optical amplifiers. Journal of Non-Crystalline Solids. 2011;**357**: 2409-2412

[35] Elkhoshkhany N, Abbas R, El-Mallawany R, Fraih A. Optical properties of quaternary TeO_2–ZnO–Nb_2O_5–Gd_2O_3 glasses. Ceramics International. 2014;**40**:14477-14481

[36] Neov S, Kozhukharov V, Gerasimova I, Krezhov K, Sidzhimov B. A model for structural recombination in tellurite glasses. Journal of Physics C: Solid State Physics. 1979;**12**:2475

[37] Jha A, Richards BDO, Jose G, Toney Fernandez T, Hill CJ, Lousteau J, et al.

Review on structural, thermal, optical and spectroscopic properties of tellurium oxide based glasses for fibre optic and waveguide applications. International Materials Reviews. 2012; **57**:357-382

[38] Tatsumisago M, Lee S-K, Minami T, Kowada Y. Raman spectra of TeO_2-based glasses and glassy liquids: Local structure change with temperature in relation to fragility of liquid. Journal of Non-Crystalline Solids. 1994;**177**:0-163

[39] Sekiya T, Mochida N, Ohtsuka A, Tonokawa M. Normal vibrations of two polymorphic forms of TeO_2 crystals and assignments of Raman peaks of pure TeO_2 glass. Journal of the Ceramic Society of Japan. 1989;**97**:1435-1440

[40] Zhou B, Rapp CF, Driver JK, Myers MJ, Myers JD, Goldstein J, et al. Oxide-based Materials and Devices IV. Vol. 8626. San Francisco, California, USA: International Society for Optics and Photonics; 2013. p. 86261F

[41] Kaur A, Khanna A, Sathe VG, Gonzalez F, Ortiz B. Optical, thermal, and structural properties of Nb_2O_5–TeO_2 and WO_3–TeO_2 glasses. Phase Transitions. 2013;**86**:598-619

[42] Souri D, Salehizadeh SA. Glass transition, fragility, and structural features of amorphous nickel–tellurate–vanadate samples. Journal of Thermal Analysis Calorimetry. 2013;**112**:689-695

[43] Aida K, Komatsu T, Dimitrov V. Thermal stability, electronic polarisability and optical basicity of ternary tellurite glasses. Physics Chemistry of Glasses. 2001;**42**:103-111

[44] Ohishi Y. Optical Components and Materials IV. Vol. 6469. International Society for Optics and Photonics; 2007. p. 646908

[45] Mohamed E, Ahmad F, Aly K. Effect of lithium addition on thermal and optical properties of zinc–tellurite glass. Journal of Alloys & Compounds. 2012;**538**:230-236

[46] Maiman TH. The Laser Inventor: Memoirs of Theodore H. Maiman. Cham: Springer International Publishing; 2018. pp. 279-280

[47] Javan A, Bennett WR, Herriott DR. Population inversion and continuous optical maser oscillation in a gas discharge containing a He-Ne mixture. Physical Review Letters. 1961;**6**:106-110

[48] Hall RN, Fenner GE, Kingsley JD, Soltys TJ, Carlson RO. Coherent light emission from GaAs junctions. Physical Review Letters. 1962;**9**:366-368

[49] Koester CJ, Snitzer E. Amplification in a fiber laser. Applied Optics. 1964;**3**: 1182-1186

[50] Kao KC, Hockham GA. In: Brown J, editor. Electromagnetic Wave Theory. Pergamon; 1967. pp. 441-444

[51] Kawanishi S, Takara H, Uchiyama K, Shake I, Mori K. 3 Tbit/s (160 Gbit/s× 19 ch) OTDM/WDM transmission experiment. In: Optical Fiber Communication Conference (OFC); San Diego, California United States. 1999

[52] Yao C, Jia Z, Li Z, Jia S, Zhao Z, Zhang L, et al. High-power mid-infrared supercontinuum laser source using fluorotellurite fiber. Optica. 2018;**5**: 1264-1270

[53] Feng X, Shi J, Segura M, White NM, Kannan P, Loh WH, et al. Halo-tellurite glass fiber with low OH content for 2-5μm mid-infrared nonlinear applications. Optics Express. 2013;**21**: 18949

[54] Savelii I, Desevedavy F, Jules J-C, Gadret G, Fatome J, Kibler B, et al. Management of OH absorption in tellurite optical fibers and related supercontinuum generation. Optical Materials. 2013;**35**:1595-1599

[55] Ebendorff-Heidepriem H, Kuan K, Oermann MR, Knight K, Monro TM. Extruded tellurite glass and fibers with low OH content for mid-infrared applications. Optical Materials Express. 2012;**2**:432

[56] Wei C, Zhu X, Norwood RA, Song F, Peyghambarian N. Numerical investigation on high power mid-infrared supercontinuum fiber lasers pumped at 3 μm. Optics Express. 2013; **21**:29488

[57] Ohishi Y, Mori A, Yamada M, Ono H, Nishida Y, Oikawa K. Gain characteristics of tellurite-based erbium-doped fiber amplifiers for 1.5-μm broadband amplification. Optics Letters. 1998;**23**:274-276

[58] Dong J, Wei YQ, Wonfor A, Penty RV, White IH, Lousteau J, et al. Dual-pumped tellurite fiber amplifier and tunable laser using Er/Ce codoping scheme. IEEE Photonics Technology Letters. 2011;**23**:736-738

[59] Oermann MR, Ebendorff-Heidepriem H, Ottaway DJ, Lancaster DG, Monro TM. Extruded microstructured fiber lasers. IEEE Photonics Technology Letters. 2012;**24**: 578-580

[60] Chillcce EF, Narro-García R, Menezes JW, Rodriguez E, Marconi JD, Fragnito HL, et al. National Fiber Optic Engineers Conference. Los Angeles, California: Optical Society of America; 2012. p. JW2A.27

[61] Chuan-fei Y, Zhi-xu J, Shun-bin W. Tm^{3+} doped tellurite microstructure fiber laser. Chinese Journal of Luminescence. 2014;**35**:1109-1113

[62] Xiang-wei M, Chuan-fei Y, Shan-de W. Tm^{3+}/Ho^{3+} co-doped tellurite microstructure fiber lasers. Chinese Journal of Luminescence. 2015;**36**:94-98

[63] Alfano RR. The Supercontinuum Laser Source: The Ultimate White Light. Springer; 2016

[64] Alfano RR, Shapiro SL. Emission in the region 4000 to 7000 Å via four-photon coupling in glass. Physical Review Letters. 1970;**24**:584-587

[65] Alfano RR, Shapiro SL. Observation of self-phase modulation and small scale filaments in crystals and glasses. Physical Review Letters. 1970;**24**:592-594

[66] Alfano RR, Shapiro SL. Picosecond spectroscopy using the inverse Raman effect. Chemical Physics Letters. 1971;**8**: 631-633

[67] Dudley JM, Genty G, Coen S. Supercontinuum generation in photonic crystal fiber. Reviews of Modern Physics. 2006;**78**:1135-1184

[68] Dudley JM, Taylor JR. Cambridge CB2 1TN, UK: Cambridge University Press; 2010

[69] Huang D, Swanson EA, Lin CP, Schuman JS, Stinson WG, Chang W, et al. Optical coherence tomography. Science. 1991;**254**:1178

[70] Hartl I, Li XD, Chudoba C, Ghanta RK, Ko TH, Fujimoto JG, et al. Ultrahigh-resolution optical coherence tomography using continuum generation in an air–silica microstructure optical fiber. Optics Letters. 2001;**26**:608-610

[71] Brown DM, Shi K, Liu Z, Philbrick CR. Long-path supercontinuum absorption spectroscopy for measurement of atmospheric constituents. Optics Express. 2008;**16**: 8457-8471

[72] Kumar VVRK, George AK, Knight JC, Russell PSJ. Tellurite photonic crystal fiber. Optics Express. 2003;**11**: 2641-2645

[73] Domachuk P, Wolchover NA, Cronin-Golomb M, Wang A, George AK, Cordeiro CMB, et al. Over 4000 nm bandwidth of mid-IR supercontinuum generation in sub-centimeter segments of highly nonlinear tellurite PCFs. Optics Express. 2008;**16**:7161-7168

[74] Feng X, Loh WH, Flanagan JC, Camerlingo A, Dasgupta S, Petropoulos P, et al. Single-mode tellurite glass holey fiber with extremely large mode area for infrared nonlinear applications. Optics Express. 2008;**16**:13651-13656

[75] Liao M, Chaudhari C, Qin G, Yan X, Suzuki T, Ohishi Y. Tellurite microstructure fibers with small hexagonal core for supercontinuum generation. Optics Express. 2009;**17**: 12174-12182

[76] Qin G, Yan X, Kito C, Liao M, Suzuki T, Mori A, et al. Highly nonlinear tellurite microstructured fibers for broadband wavelength conversion and flattened supercontinuum generation. Journal of Applied Physics. 2010;**107**: 043108

[77] Savelli I, Mouawad O, Fatome J, Kibler B, Désévédavy F, Gadret G, et al. Mid-infrared 2000-nm bandwidth supercontinuum generation in suspended-core microstructured. Optics Express. 2012;**20**:27083-27093

[78] Belal M, Xu L, Horak P, Shen L, Feng X, Ettabib M, et al. Sulfide and tellurite optical fibers. Optics Letters. 2015;**40**:2237-2240

[79] Klimczak M, Stepniewski G, Bookey H, Szolno A, Stepien R, Pysz D, et al. Mid-infrared supercontinuum generation in suspended core tellurite microstructured optical fibers. Optics Letters. 2013;**38**:4679-4682

[80] Yao C, He C, Jia Z, Wang S, Qin G, Ohishi Y, et al. Holmium-doped fluorotellurite microstructured fibers for 2.1 μm lasing. Optics Letters. 2015; **40**:4695-4698

[81] Wang F, Wang K, Yao C, Jia Z, Wang S, Wu C, et al. Tapered fluorotellurite microstructured fibers for broadband supercontinuum generation. Optics Letters. 2016;**41**: 634-637

[82] Price JHV, Feng X, Heidt AM, Brambilla G, Horak P, Poletti F, et al. Supercontinuum generation in non-silica fibers. Optical Fiber Technology. 2012;**18**:327-344

[83] Thapa R, Rhonehouse D, Nguyen D, Wiersma K, Smith C, Zong J, et al. SPIE. 2013;**8898**:889808:1-8

[84] Kedenburg S, Steinle T, Mörz F, Steinmann A, Nguyen D, Rhonehouse D, et al. Solitonic supercontinuum of femtosecond mid-IR pulses in W-type index tellurite fibers with two zero dispersion wavelengths. APL Photonics. 2016;**1**:086101

[85] Kedenburg S, Strutynski C, Kibler B, Froidevaux P, Désévédavy F, Gadret G, et al. High repetition rate mid-infrared supercontinuum generation from 1.3 to 5.3 μm in robust step-index tellurite fibers. Journal of the Optical Society of America B. 2017;**34**:601-607

[86] Shi H, Feng X, Tan F, Wang P, Wang P. Multi-watt mid-infrared supercontinuum generated from a dehydrated large-core tellurite glass fiber. Optical Materials Express. 2016;**6**: 3967-3976

[87] Jia Z, Yao C, Jia S, Wang F, Wang S, Zhao Z, et al. 4.5 W supercontinuum generation from 1017 to 3438 nm in an all-solid fluorotellurite fiber. Applied Physics Letters. 2017;**110**:261106

[88] Jia ZX, Yao CF, Jia SJ, Wang F, Wang SB, Zhao ZP, et al. Supercontinuum generation covering

the entire 0.4–5 μm transmission window in a tapered ultra-high numerical aperture all-solid fluorotellurite fiber. Laser Physics Letters. 2018;**15**:025102

[89] Zou C-L, Shu F-J, Sun F-W, Gong Z-J, Han Z-F, Guo G-C. Theory of free space coupling to high-Q whispering gallery modes. Optics Express. 2013;**21**: 9982-9995

[90] Little BE, Laine J-P, Haus HA. Analytic theory of coupling from tapered fibers and half-blocks into microsphere resonators. Journal of Lightwave Technology. 1999;**17**:704

[91] Murphy RM, Lei F, Ward JM, Yang Y, Chormaic SN. Bandpass transmission spectra of a whispering gallery microcavity coupled to an ultrathin fiber. Optics Express. 2017;**25**:13101-13106

[92] Yu J, Lewis E, Farrell G, Wang P. Compound glass microsphere resonator devices. Micromachines. 2018;**9**:356

[93] Yang Z, Wu Y, Zhang X, Zhang W, Xu P, Dai S. Low temperature fabrication of chalcogenide microsphere resonators for thermal sensing. IEEE Photonics Technology Letters. 2016;**29**: 66-69

[94] Yang Z, Wu Y, Yang K, Xu P, Zhang W, Dai S, et al. Fabrication and characterization of Tm^{3+}-Ho^{3+} co-doped tellurite glass microsphere lasers operating at ~2.1 μm. Optical Materials. 2017;**72**:524-528

[95] Yu J, Lewis E, Brambilla G, Wang P. Temperature sensing performance of microsphere resonators. Sensors. 2018; **18**:2515

[96] Sasagawa K, Kusawake K, Ohta J, Nunoshita M. Nd-doped tellurite glass microsphere laser. Electronics Letters. 2002;**38**:1355-1357

[97] Peng X, Song F, Gonokami MK, Jiang S, Peyghambarian NN. Er^{3+}-doped tellurite glass microsphere laser: Optical properties, coupling scheme, and lasing characteristics. Optical Engineering. 2005;**44**:034202

[98] Zhao H, Yi Y, Wang X, Li A, Yang A, Yang Z, et al. Triple-wavelength lasing at 1.50 μm, 1.84 μm and 2.08 μm in a Ho^{3+}/Tm^{3+} co-doped fluorozirconate glass microsphere. Journal of Luminescence. 2019:116889

[99] Wu J, Jiang S, Qiu T, Morrell M, Schulzgen A, Peyghambarian N. Optical Components and Materials II. Vol. 38. International Society for Optics and Photonics; 2015. pp. 1355-1357

[100] Percival R, Szebesta D, Williams J, Lauder R, Tropper A, Hanna D. Diode pumped operation of thulium doped fluoride fibre amplifier suitable for first window systems. Electronics Letters. 1994;**30**:1598-1599

[101] Sasagawa K, Yonezawa Z-O, Iwai R, Ohta J, Nunoshita M. S-band Tm^{3+}-doped tellurite glass microsphere laser via a cascade process. Applied Physics Letters. 2004;**85**:4325-4327

Chapter 16

Global Trend of Glass Bonding for Appliance Industry Assemblies

Chulsoo Woo, Bin Liang and Jonghyuk Oh

Abstract

Due to the ever-increasing need for production efficiency and reliability as well as cost saving, assembly industries have been looking for a better solution compared to current methodologies. It is critically important to know that there is a solution this industry had not paid much attention and can benefit a lot due to not only historical reason but also limited knowledge management or awareness. This chapter examines and qualifies the effect of a couple of special adhesive bonding solutions on various glass plate bonding applications at appliance industry using a total solution such as dispensing system and robot. The result clearly shows its benefit over current methodologies, and also as industry trend moves toward more exterior decoration for high-end image products, this chapter should contribute on glass bonding industry not only for faster production, better efficiency, less production space, and better reliability but also for lower manufacturing cost using special adhesive bonding solutions.

Keywords: decorative glass, bonding strength, polyurethane hot melt, 2 component silicone

1. Introduction

Ever since the appliance industry started its debut a several decades ago, appliances have evolved through various changes of performance both with their original functions and energy consumption and better high-end appearance while other parameters have come to almost achievable limit, the demand for diverse functions and appearance images that still draw hearts of multiple consumers. Now appliances have been inevitable items for every house for its convenience yet and improvement of life.

As customer's interests for high-end image products are ever increasing, and more simplified manufacturing processes with a high quality at lower cost have to be met, assembly industries have been looking for various solutions. In the beginning, they have adopted various mechanical assembly methods, one of which has been a tape method. For example, the front side design of previous major home appliances has been plastic, painted metal, or printed film on top of bare metal frame. It was difficult to achieve a high-end image with such simple design. With the use of printed glass plate, the whole image of appliances has been upgraded and met the demands of consumers who aspired better appearance of appliances; it has given more freedom to make diverse front images using all different printings at the

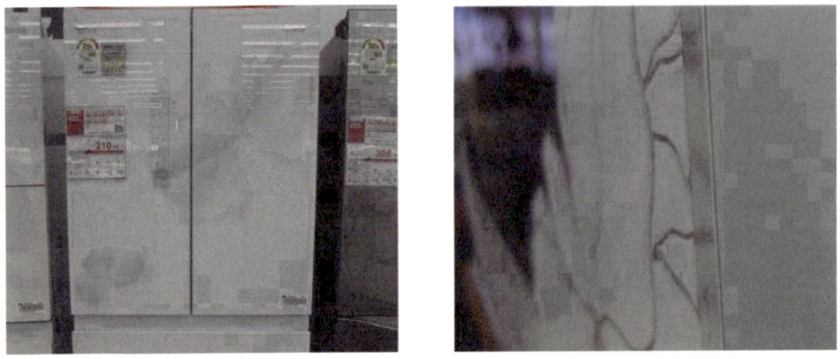

Figure 1.
Glass bonding for refrigerator and side view.

Figure 2.
Glass bonding for top loading washing machine.

Figure 3.
Glass bonding for front door of microwave.

back side of glass plate. A few examples can be shown in **Figures 1–8**. The latest solution was a special structural adhesive that is cured by a chemical reaction. In this chapter, we would like to introduce two structural adhesive technologies, such as polyurethane hot melt technology and 2 component silicone technology, evaluate their performance and compare and analyze its difference from the current tape solution and 1 component silicone + tape solution and mechanical way on process efficiency, reliability as well as cost saving. Those appliances (or white goods) have been used at almost all the housings, and those appliances that have adopted glass plate bonding are refrigerator, washing machine, microwave, cook top, air

Figure 4.
Glass bonding of ceramic glass for microwave.

Figure 5.
Glass bonding for wall type air conditioner.

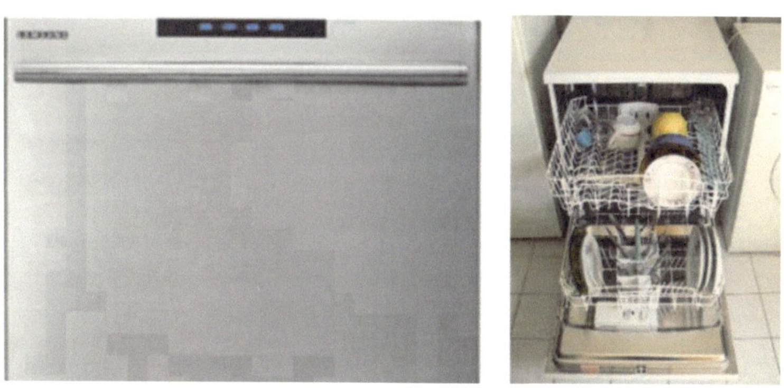

Figure 6.
Glass bonding for stand-alone type dish washer.

conditioner, dish washer, kitchen ventilation hood, styler, etc. Each appliance is explained further on the area of glass plate bonding applications. In the literature, there is only a few research on glass bonding with conventional methods, but there is not any research on glass bonding with adhesives which we focused on in this report. In this study, double-sided tape and actual test substrates were received from appliance customers, and PUR hot melt and 2C silicone adhesives were received from Henkel laboratory.

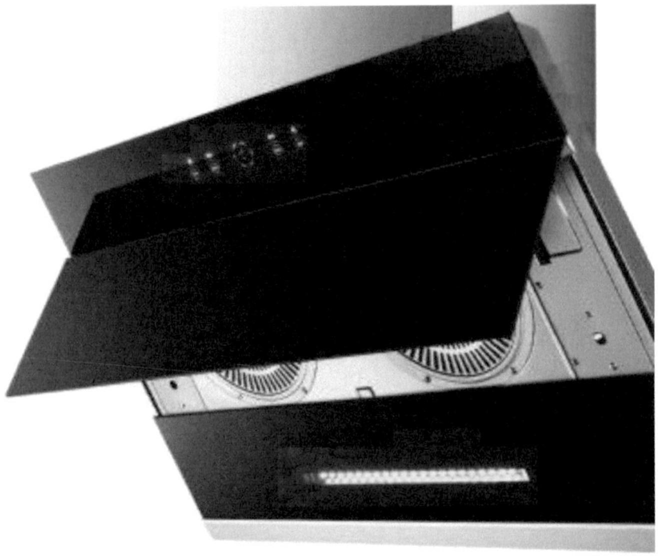

Figure 7.
Glass bonding for kitchen ventilation hood.

Figure 8.
Glass bonding for styler.

2. Major appliances

Those requiring glass bonding are described as follows.

2.1 Refrigerator

Refrigerator is one of the items that can adopt glass bonding applications for various and different designs in terms of shape and function such as one door, two door, triple door, double door, French door, chest type, upright type, freezer, and kimchi refrigerator. They all can adopt glass plate bonding applications such as all front doors, inclined area between double and two triple doors, and water dispenser area on a front door and control panel. The most popular glass bonding application is a front glass plate bonding to the plastic door chassis. The below **Figure 1** shows a glass plate bonded to front door and a glass plate bonded from side view [1].

2.2 Washing machine

Washing machine has a few application areas that can adopt glass plate bonding applications in terms of shape and function such as top loading, front loading, and combined top and front loading, and they all can adopt a printed glass plate bonding application including control panel area. The most popular application is a glass bonding of top cover to plastic cover frame for top loading washing machine. Front loading type has a door assembled with glass and plastic frame by screws. This structure can be assembled by using a special structural adhesive eliminating screws and its related extra process to use screws. The combined front loading (on top) and top loading (on the bottom) has a function of washing clothe at the same time. The top loading type washing machine in the bottom part slides out front and do the washing. Other application is for bonding of a glass door frame to a glass door for the function of opening and throwing in a forgotten laundry while the washing machine is already running. The below **Figure 2** shows a glass bonding of top cover to plastic frame for top loading washing machine [2].

2.3 Microwave

Microwave has a few glass plate bonding applications. The most popular application is for bonding of front glass plate to outside front plastic door. Other applications are for bonding of glass plate to inner front metal door frame. One more application for a high-end microwave is for bonding of glass plate to bottom body frame. The below **Figure 3** shows a front door glass bonding to plastic door frame [3].

2.4 Cook top

Cook top uses an electrical coil with ceramic glass placed on top where they put pots and pans to cook food instead of using gas. The coil generates heat that transfer through a ceramic glass to food. It is a much more high-end image with a sleek design for easier cleaning and better health, and it has a safety advantage during cooking as people can wipe out contamination easily and do not need to inhale the burning gas. The most popular application is for bonding of ceramic glass to ceramic-coated metal frame. Other application is for bonding of ceramic glass to inner metal chassis. The below **Figure 4** shows a ceramic glass that is bonded to metal ceramic coating surface and to metal chassis [3].

2.5 Air conditioner

Air conditioner has a different type of design that can adopt glass plate bonding in terms of shape and function such as window type, wall type, standalone vertical type, and ceiling type. The most popular application is for bonding of a glass plate to plastic frame for wall type air conditioner. This type of application needs a reliable bonding solution as the glass might fall off from air conditioner and cause a safety problem for people if an improper solution is used. The below **Figure 5** shows a glass bonding to plastic frame of wall type air conditioner [4].

2.6 Dish washer

Dish washer has a different type in terms of shape and function that can use a glass bonding application such as table top, built in stand-alone type, and drawer type that can adopt glass bonding application in the front side. The most popular application is for bonding of glass plate to plastic front door for stand-alone type. The below **Figure 6** shows a glass bonding for stand-alone type washing machine [5].

2.7 Kitchen ventilation hood

Kitchen ventilation hood can have three glass plate bonding application for its function of easy cleaning of contamination from cooking and high-end image and appearance. This type of design has the glass plate in the middle which is in a closed state. When it is turned on, the glass plate in the middle opens up and the fan inside sucks those fumes, smokes, and microparticles generated during cooking. For a situation where the glass plate is contaminated with oil due to splash of boiling oil during cooking, it can be wiped off easily for clean surface. The most popular application is for bonding of glass plate to moving chassis. Other applications are for bonding of glass to control panel chassis and bonding of glass to other chassis. The below **Figure 7** shows an overall glass bonding area of kitchen ventilation hood.

2.8 Styler

Styler has a glass plate bonding application such as bonding of glass plate to plastic frame on the front side of it. It is a stand-alone type and has a function of eliminating dust by vibrating the clothe, deodorization by vacuum fan, sterilization by steam heating, and ironing of the clothe by heat pressing overnight inside it and next day they can take it out of the styler and put it on. The below **Figure 8** shows a glass bonding in front of styler.

3. Assembly solutions

In this section, we are going to list up and review major solutions that have been used in the industry and new solutions that can be used in this industry. Conventional methods consist of mainly double-sided tape, 1 component silicone, and mechanical fasteners, and new special adhesive solutions are polyurethane hot melt technology and 2 component silicone technology. We will review characteristics of each solution.

3.1 Double-sided tape

This technology has been used widely in the assembly industry. It has had a wide range of awareness in the market place. It can fix the parts within 1 min and no chemical reaction is required. But at the same time, it requires a high labor-intensive pre-work such as priming the surface with a primer to improve a bonding durability. It also requires a peel-off of release paper on one side and place the sticky part on the application area mostly manually. It is very difficult to re-adjust the location once it is placed on the substrate, and very high labor cost is required if it has to be re-adjusted. It has to be pressed for a certain amount of time; otherwise, there is no good affinity achieved between the sticky surface of tape and the substrate. The de-bonding happens when bonding a surface not completely flat. Tape solution also generates a bulky waste of release paper. Tape solution tends to release the bonding property under aging condition; therefore, a gap is generated in the glass bonding area as time goes causing a repair. The adhesion of tape is the surface temperature dependent; therefore, if the temperature of assembly line, tape, and substrate was not properly controlled, especially in the winter time due to the effect of cold temperature on the bonding durability of tape, it generates a poor adhesion to substrate and consequently a bonding failure, scrap, and liability issue [6].

3.2 Mechanical fastener

This technology has been available with a history of mankind for a very long time. This can provide a very fast mechanical fixture and consequent assembly, but at the same time, this requires mechanical fasteners and drilled thread holes on both parts, and operator has to assemble them by using fastening tools manually. The assembly leaves a screw mark that is not good in appearance and it is also difficult to make an automation assembly line, too [7].

3.3 1 component RTV silicone

This technology is a room temperature vulcanized (RTV) silicone that reacts with moisture in the air and substrate, and generates some by-products such as oxime, alcohol, acetic acid, and acetone. This technology in general has a high viscosity with a slow curing property that takes a few days for functional bonding and more than 1 week for a complete curing. Normally this technology requires mechanical clips or double-sided tapes for initial fixture time and a stacking space and lengthy curing time before it goes to the next process. Some process adopts 1 component RTV silicone or double-sided tape which requires more manual process and manpower cost [8].

3.4 Polyurethane hot melt

This technology is at first a thermoplastic (remeltable) material that contains some free isocyanate groups, and then, it is similar with 1 component polyurethane technology that reacts with moisture to form a tough thermoset (not meltable) plastic material. Its initial thermoplastic property can melt easily and provide a very excellent high position tack capability that can hold the two substrates in place by losing an energy from molten liquid state, and also its thermoset property provides a high temperature heat and creep resistance with a good durability. Polyurethane hot melt is available with a wide range of open time. This technology has a good adhesion and is compatible with a wide variety of substrates. It has better bonding strength performance and environment resistance compared to normal hot melt due its cross-linked chemical reaction property. It has a relatively lower operation temperature around 120°C; therefore, a dispensing at lower energy is possible. It also has longer open time 1–4 min; therefore, it has a wide bonding window for assembly. This technology is just like other 1 component and 2 component polyurethane adhesives, and during the reaction, generates some carbon dioxide (CO_2) to evaporate.

Additional characteristics are excellent toughness, flexibility, high elongation, excellent gap filling, paintable when cured, and excellent chemical resistance.

As shown in **Figure 9**, isocyanate NCO reacts with moisture in air and substrate and becomes an oligomer with by-product carbon dioxide evaporating; then, these oligomers react each other to form a cured polyurethane [9].

As shown in **Figure 10**, polyurethane hot melt is in a solid state in the sealed aluminum foil packaging to prevent the ingress of moisture in the solid product that can react to become a thermoset plastic. After the product becomes a molten state at elevated temperature by melting machine, it is dispensed onto the substrate, and substrates are mated at “a: joining time”, as it loses the energy, and it solidifies just

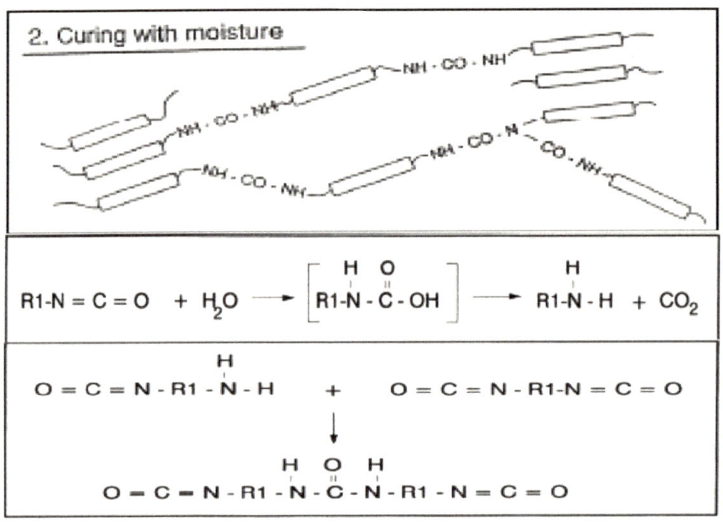

Figure 9.
PUR HM curing mechanism with moisture.

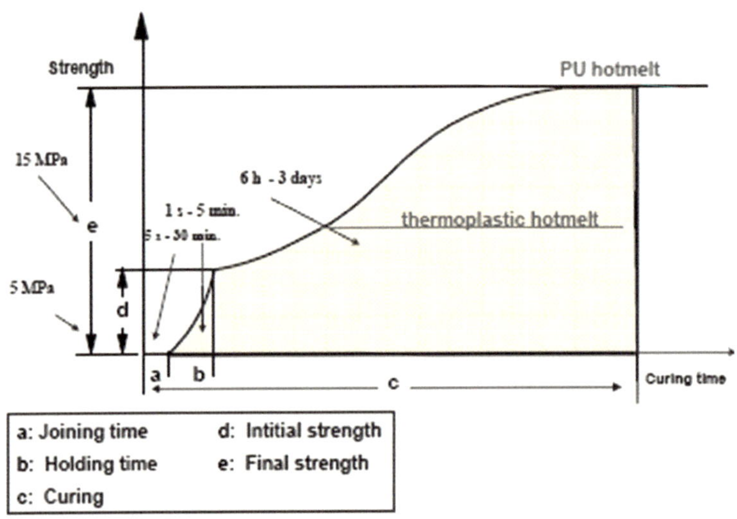

Figure 10.
PUR HM curing strength vs. curing time.

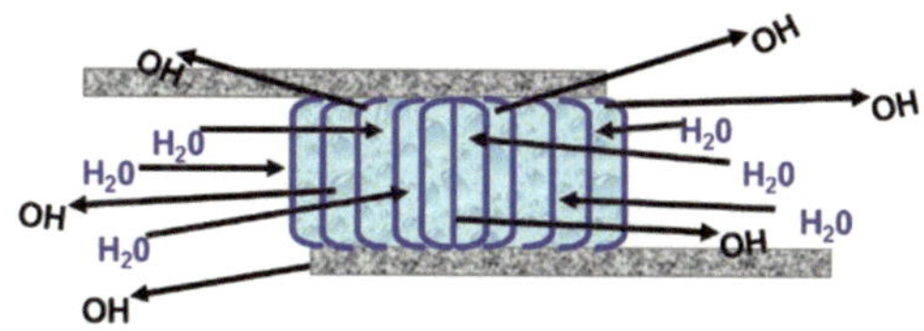

Figure 11.
2C silicone curing mechanism.

like normal hot melt and starts to build an initial bonding strength "d" at "b: holding time." Once it solidifies by losing energy, it starts to react with moisture and cures chemically to form a much higher bonding strength at "curing time c" and finally comes to a final bonding strength "e." Therefore, a very fast fixturing and reliable bonding can be achieved with this technology. Polyurethane hot melt is dispensed by melting machine and robot or by roll coating machine; in this chapter, for glass bonding application, melting machine and robot are used mostly [9].

3.5 2 component silicone

When part A and part B are mixed in a proper mixing ratio, a chemical reaction occurs which is initiated by the catalyst reacting with water. This chemical reaction occurs not only at the surface but also inside the mixed product. Therefore, it has a very fast fixturing and curing time for faster fixture and assembly, and it allows a structural bonding assembly in more compact assembly line and automated assembly with less manpower. This technology can replace various assemblies that have used double-sided tape or RTV silicone that requires a manual and lengthy assembly line and time with much manpower. During the chemical reaction, moisture ("H_2O") comes in and alcohol ("OH") that was generated during chemical reaction as a by-product comes out and evaporates (see **Figure 11**) [8].

Two-component silicone characteristics are fast tack free, fixture and curing time, and cure speed not affected by moisture, and cure speed less affected by bond line thickness, and excellent thermal resistance ("up to 270°C"), and flexible, tough, high modulus, and agency UL 746°C listed for high temperature applications for appliance, and excellent gap filling, and black, gray and almond colors are available [8].

4. Refrigerator

First of all, in this section, we are going to compare polyurethane hot melt adhesive solution vs. tape solution based on parameters such as assembly process, environment reliability, tensile shear strength against open time, cross pull strength against fixture time, design benefit, air pockets on bond line, bond line thickness, creep resistance, dispensing equipment, and overall value analysis. The application is for bonding of glass to plastic frame for various refrigerator doors [1].

4.1 Assembly process

Tape solution is to apply primer to plastic frame by 1 worker (5 s) → apply tape to plastic frame and cut manually by 2 workers (20–40 s) → peel off release paper manually by 2 workers (5 s) → place glass plate onto the plastic frame manually by 1 worker (10 s) → compress the glass plate by pressing (5–10 s) → next process. This solution takes a total of 4 workers and 70 s.

Polyurethane hot melt adhesive solution is to apply adhesive to plastic frame by machine (10 s) → place glass plate onto the plastic frame manually by 1 worker or by machine (0–10 s) → compress glass plate by pressing (5–10 s) → next process.

This polyurethane hot melt adhesive solution takes only a total of 1 or no workers and 30 s, which means it requires almost no worker and a half process time compared to tape solution. This adhesive bonding solution can save 5 or 6 workers and can use them in other more important work places, can save overall assembly time in half, and can reduce assembly line space more compact and eliminate waste coming from release paper from tape solution. Tape solution requires a primer process for more reliable bonding as its adhesion decreases when the ambient temperature was relatively low; then, the complete wetting of tape onto the substrate was not fully established resulting in bonding reliability issue down the road, while adhesive bonding solution wets all the mating surfaces and penetrates all the surface imperfections and mechanically grip the surface after cure contributing more reliable bonding, and also much reduced process steps allow more reduction of quality check points, too [1].

4.2 Tensile shear strength vs. aging conditions

Testing has been carried out to measure a tensile shear strength to compare the bonding reliability before and after aging condition between adhesive solution and tape solution. Substrate was a high temperature ink printed glass and a plastic combined with PC and ABS. Testing conditions were at room temperature (23°C), low temperature (−40°C for 7 days), high temperature (80°C for 7 days), high temperature, and high humidity (50°C, 95%RH for 8 h, 5°C 95%RH for 8 h as 1 cycle for 1000 h) and thermal cycle tests (60°C for 4 h to −30°C for 4 h as 1 cycle for 1000 h). As shown in **Figure 12**, adhesive solution has at least 10 times higher bonding strength for all aging conditions. It was shown that both adhesive and tape have increased their bonding strength a bit, but adhesive solution still has maintained at least 10 times higher bonding strength than the tape solution. If we review the thermal cycle test, strength has increased for adhesive solution, but it has decreased for tape; consequently, adhesive solution has 13 times of higher bonding strength than the tape solution. Overall, it is very clear that as far as environmental durability is concerned, adhesive solution provides a far more reliable bonding than the tape solution.

4.3 Tensile shear strength vs. open time

Testing has been conducted to measure a tensile shear strength per different open time which means the time after adhesive dispensing until mating the other

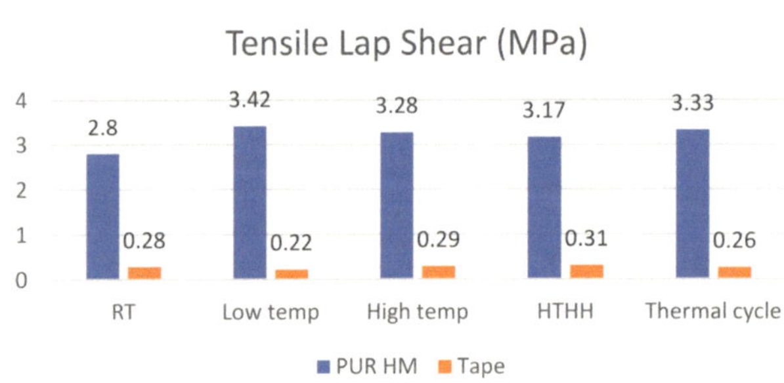

Figure 12.
Tensile shear strength vs. different aging conditions.

substrate. A variation of tensile shear strength between adhesive and tape was checked and compared in **Figure 13**. As it is shown below, adhesive solution has 10 times of higher bonding strength than the tape solution from open time of 15–30 s to 2 min; and the adhesive solution still maintains 7–8 times of higher tensile shear strength than the tape solution from open time of 4 min to max 15 min. It explains that adhesive solution has a lot of assembly process freedom as same as tape solution, yet it achieves a much higher and reliable bonding strength. The realistic handling time that manufacturer allows is within 2 min after adhesive dispensing as the assemblies need to move to the next station as early as possible. Therefore, the quality of bonding strength with adhesive solution can be achieved in most appliance assembly process conditions.

Tape solution maintains the same level of tensile shear strength but in a much lower figure.

4.4 Cross pull strength vs. fixture time

Testing has been carried out to measure an initial fixture strength per holding time between adhesive and tape solution. An initial fixture time means a bonding strength developed after dispensing and assembly. As shown in the below **Figure 14**, tape solution shows a higher initial fixture strength at 1 min holding time; however, as time goes by adhesive solution achieves higher strength than tape, and as time goes further, adhesive solution reacts with moisture. The holding time in **Figure 14** is the one adhesive solution that still maintains a solidified hot melt status which is not chemically reacted, yet shows higher strength after 2 min of holding time. As time goes further, the adhesive reacts with moisture and achieves a chemically reacted bonding strength which is at least 10 times higher than the tape solution as seen in **Figure 12**. The holding time of less than 2 min is not realistic condition as most of manufacturers use more than 2 min of holding time due to handling other parts until the next process. The substrate tested was glass bonded to combined PC and ABS.

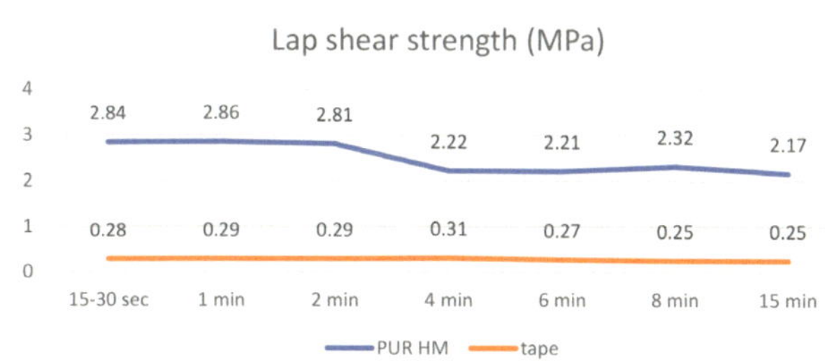

Figure 13.
Tensile shear strength vs. open time.

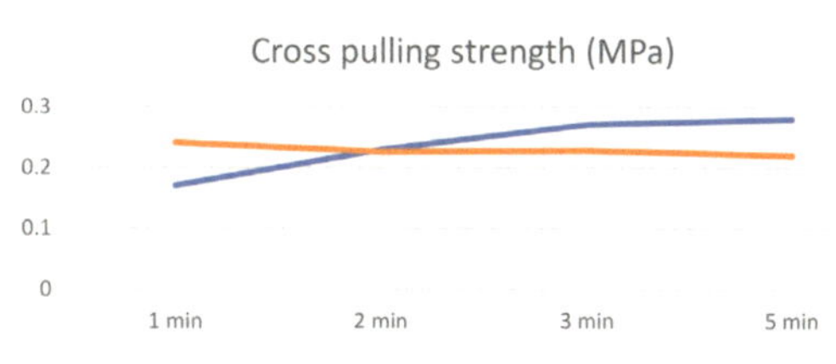

Figure 14.
Cross pull strength vs. fixture time.

4.5 Better design vs. design complexity

Tape solution has a certain thickness (1 mm) of tape and once it is pressed, it becomes to 0.8 mm thickness. When adopting the glass plate bonding design, its own tape thickness puts some limitations on the freedom of design and causes additional unnecessary and complicated design and assembly process due to its design. For instance, 0.8 mm of gap between glass pate and body frame is clearly and easily visible to the eyes of consumers as the application of glass bonding is always external and it affects the image of their product on consumer's preference and expectation on high-end image appliances. Manufacturers who used a tape solution have adopted additional side frame to cover the gap and not to let consumers realize that there was a gap clearly visible, or designed a special threshold on the outer end of body frame so that the gap is not visible outside from the eyes of consumer as shown in **Figure 15**; ultimately, it has caused more cost, more process complexity, and more manpower as well as poor high-end image as it is clearly and vividly different if they compare the image of refrigerator between the one that has a side cover on the glass (with tape solution) and the one that has no side cover, just bonding of glass plate to body frame (with adhesive solution).

4.6 Air pockets on the bondline

Tape solution has an inheritable structural limitation to achieve a complete contact to substrate as it is a solid form, and taping/pressing is also done manually. **Figures 16** and **17** show an actual application with tape for refrigerator; as shown in figures, many area of tape has detached area causing a low bonding strength and durability, and especially in the high-temperature, high-humidity aging condition, the bonding reliability decreases as the detached area is expanding due to temperature increase and moisture ingress over time. In general, the air pockets in the tape solution on the bondline are quite substantial such as 40–50% in some worst cases as the substrate surface is not always flat. On the other hand, the adhesive solution can prevent such air pocket possibilities as it is a liquid and it flows and fills all the

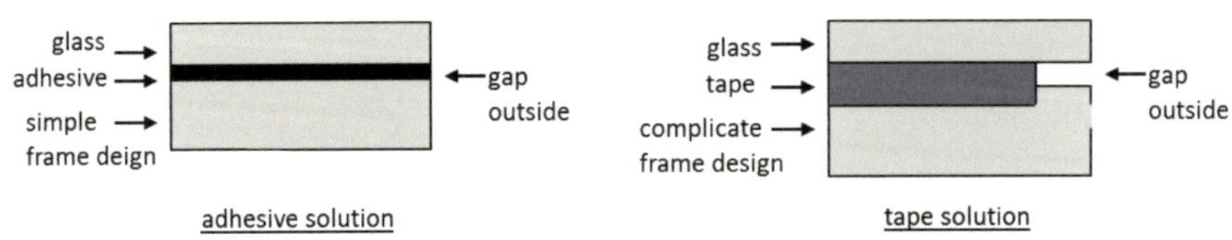

Figure 15.
Gap and design complexity between tape and adhesive.

Figure 16.
Air pocket for tape.

surface imperfections, meaning no air pocket, and provides a tough bonding resulting in better reliability and durability than the tape solution.

4.7 Bondline vs. environment condition

Tape solution when subjected to various environment conditions increases its thickness by expanding, due to high temperature and moisture ingress; it increases to 1 mm in some cases. The adhesive solution maintains quite consistently in terms of its thickness before and after various environment conditions. In case of high-temperature condition (80°C for 7 days), the delta is average 0.04 mm which is very minimal as shown in **Tables 1–3**. In case of thermal cycle condition (50°C, 95% RH for 8 h to 5°C, 95%RH for 8 h as 1 cycle for 1000 h), the delta is 0.01 mm. In case of bigger range of thermal cycle, (60°C, 95%RH for 16 h to −30°C for 4 h as 1 cycle for 1000 h), the delta is 0.05 mm. The above test results confirm the reliability and excellent environmental resistance of adhesive solution in achieving a durability and ultimately a better quality of appliances.

Figures 18 and **19** explain the gap generated between adhesive solution and tape solution. For adhesive solution, as shown in **Figure 18**, it is very narrow for a coin to penetrate, while for tape solution, as shown in **Figure 19**, there is a gap that is big enough for coin to penetrate.

Figure 17.
No air pocket for adhesive.

Point	Before aging	After aging	Δ
1	0.18 mm	0.15 mm	**- 0.03 mm**
2	0.20 mm	0.15 mm	**- 0.05 mm**
3	0.50 mm	0.50 mm	**0 mm**
4	0.30 mm	0.25 mm	**- 0.05 mm**
5	0.55 mm	0.50 mm	**- 0.05 mm**
6	0.60 mm	0.54 mm	**- 0.06 mm**
7	0.45 mm	0.45 mm	**0 mm**
8	0.15 mm	0.09 mm	**- 0.06 mm**
9	0.45 mm	0.40 mm	**- 0.05 mm**
		Avg	**- 0.04 mm**
		Max	**- 0.06 mm**

Table 1.
High temperature.

Point	Before aging	After aging	Δ
1	0.18 mm	0.18 mm	**0 mm**
2	0.02 mm	0 mm	**- 0.02 mm**
3	0.02 mm	0 mm	**- 0.02 mm**
4	0.10 mm	0.06 mm	**- 0.04 mm**
5	0.15 mm	0.14 mm	**- 0.01 mm**
6	0 mm	0 mm	**0 mm**
7	0.65 mm	0.66 mm	**0.01 mm**
8	0.63 mm	0.58 mm	**- 0.05 mm**
9	0.11 mm	0.10 mm	**- 0.01 mm**
10	0.03 mm	0 mm	**- 0.03 mm**
11	0.23 mm	0.23 mm	**0 mm**
12	0.40 mm	0.40 mm	**0 mm**
		Avg	- 0.01 mm
		Max	- 0.05 mm

Table 2.
Thermal cycle.

Point	Before aging	After aging	Δ
1	0.33 mm	0.38 mm	**0.05 mm**
2	0.07 mm	0.18 mm	**0.11 mm**
3	0.40 mm	0.50 mm	**0.10 mm**
4	0.47 mm	0.55 mm	**0.08 mm**
5	0.38 mm	0.45 mm	**0.07 mm**
6	0.40 mm	0.50 mm	**0.10 mm**
7	0.25 mm	0.30 mm	**0.05 mm**
8	0.42 mm	0.45 mm	**0.03 mm**
9	0.04 mm	0 mm	**- 0.04 mm**
10	0.20 mm	0.15 mm	**- 0.05 mm**
11	0.10 mm	0.06 mm	**- 0.04 mm**
12	0.12 mm	0.15 mm	**0.03 mm**
13	0.50 mm	0.60 mm	**0.10 mm**
		Avg	- 0.05 mm
		Max	0.11 mm

Table 3.
Higher thermal cycle.

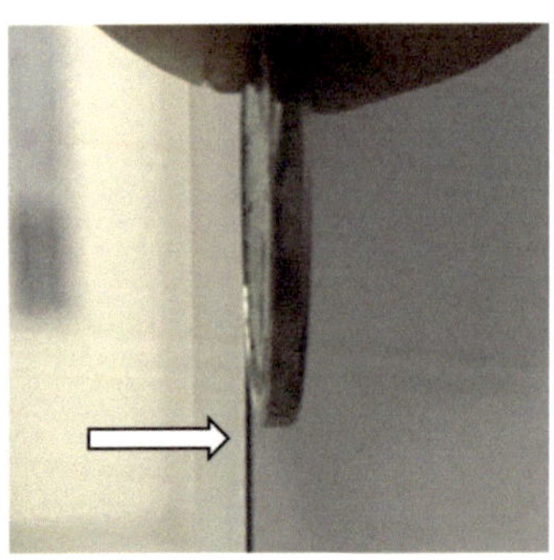

Figure 18.
No gap with adhesive.

4.8 Initial creep resistance vs. load

Testing has been done to measure a creep resistance of adhesive solution after applying onto substrate. The assembly test method was to hang 150 g weight to bonded specimen for max 10 min and measure the creep at each different fixture time. This weight and bond area are representations of actual glass bonding for refrigerator when the bonded assembly is put in upright position after bonding at horizontal position; then the glass, due to its weight and gravity, tends to move downward. The adhesive solution has to grip the glass in place after a certain fixture time at horizontal position. As shown in **Figure 20**, "1 time" means the same weight condition as actual glass load (glass weight/bonding area). "2 times" means 2 times more weight that was used to measure the creep. It has no creep in 5 min of fixture time and this is a severe condition. "5 times" means 5 times more weight that was used to measure the creep. It has no creep in 10 min of fixture time and this is a very severe condition. For "1 min" condition, 5 kg weight has a 2.5 mm creep at 15 s of fixture time, but as fixture time goes further, the creep reduces and at 4 min fixture time after assembly, it has no creep at all. This is the time that customers have and use in their assembly line. Anyway adhesive solution has a good fixture time with 4 mins of bonding after which bonded glass plate is placed vertically in such short amount of time and there is no creep that is adopted by various manufacturers, mostly their time is longer than 4 min. This explains the efficiency and durability of adhesive solution for process consideration when the refrigerator is put in upright position.

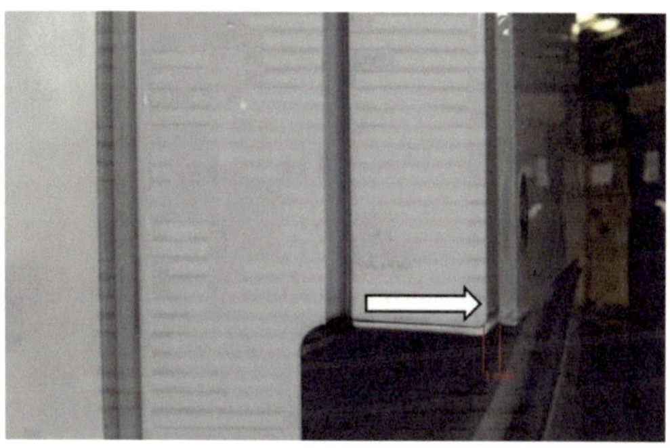

Figure 19.
Gap with tape.

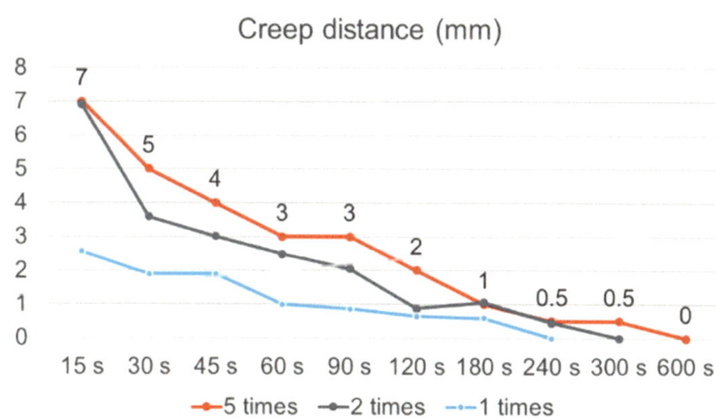

Figure 20.
Initial creep resistance vs. load.

Price for tape solution:	€1,100,000
Positive value for adhesive solution:	€798,000
Additional cost of adhesive solution:	€0
Total Value:	€1,898,000
Price for adhesive solution:	(€730,000)
Value to manufacturer:	**€1,168,000**

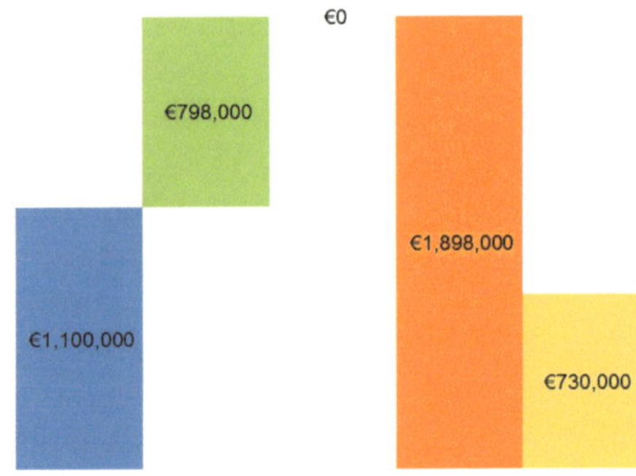

Figure 21.
Overall value analysis per year.

4.9 Automation equipment

While tape solution cannot be automated by using equipment, one of many benefits of adhesive solutions is to automate the assembly line. Adhesive solution in this case requires a melting machine and a robot. Melting machine is required to melt the solid-state adhesive into the molten state. The adhesive is then applied in a liquid state onto substrate and as soon as it touches the substrate, it loses its energy quickly and hardens into solid. Robot is required to position (control) a dispensing area and location; currently, a 6-axis robot is recommended to this application for manufacturers. Typical parameters of melting machine and robot are 10 s of adhesive dispensing for 650 × 750 mm refrigerator door, dispensing speed is 300 mm/s for 15 g/min, pump speed is 103 RPM, application temperature is 130°C, adhesive bead height is 2–2.5 mm, and bead width is 7–11 mm. These allow manufacturers to make the assembly line with less manpower, less space, and better work in process (WIP).

4.10 Overall value analysis

Cost comparison has been done for overall required cost between tape solution and adhesive solution for glass plate bonding for refrigerator door as shown in **Figure 21**.

The cost for tape was 1,100,000 Euro, which is positive value for adhesive solution, meaning the value they can save by using adhesive solution was 798,000 Euro; and additional cost of adhesive solution was 0 Euro. The total prospect value that the manufacturer can save by using adhesive solution was 1,898,000 Euro. The cost for total adhesive solution was 730,000 Euro, which includes adhesive cost 660,000 Euro + equipment cost (melting machine + robot) 210,000 Euro, and 1 year depreciation cost for equipment is 70,000 Euro. Therefore, the adhesive cost becomes 730,000 Euro (660,000 + 70,000). the ultimate value to manufacturer can be 1,168,000 Euro which was calculated based on 1-year production. The capital investment cost for melting machine and 6-axis robot required when using adhesive solution can be calculated and the pay-back time will be 0.4 (730,000/1,898,000) year.

5. Washing machine

In this section, we are going to compare polyurethane hot melt solution vs. 1 component room temperature vulcanizing (RTV) silicone solution for top loading type that used to use a folding plastic top cover which changed to an ink printed glass cover for better appearance and high-end image as a value sold in the market place. The application area is for bonding of glass plate to ABS plastic frame for top loading washing machine cover [2].

5.1 Assembly process

One-component room temperature vulcanizing solution is to apply silicone onto a plastic frame by dispenser with 3 axis robot that is an assembly of bare ABS and chrome-plated ABS that was assembled by mechanical screws (23 s) → place glass plate on top by 1 operator (2 s) → put several mechanical clamps or wrap the tape around the assembly for fixture by 1 operator (10 s) → transfer it to the rack (3 s) → rack placed in the corner of plant for curing (24 h, space) → move to next assembly process.

Polyurethane hot melt solution is to apply to a plastic frame that is an assembly of bare ABS and chrome-plated ABS that was assembled by mechanical screws by using an automatic dispensing machine as well as dispensing robot (23 s) → place glass plate on top by 1 operator and press for 30 s → assembly is fixtured and then goes to the next assembly process.

Comparing those two solutions, polyurethane hot melt adhesive solution takes a total of 1 or no operator and very compact assembly process time (total 1 min) while existing 1C silicone solution requires 2 workers and complicated and longer assembly process time such as taping 4 corners of cover, clamping or wrapping the assembled parts for curing, and a space for stacked assemblies to cure for 24 h before it moves to the next assembly process [2].

5.2 Initial fixture time vs. loading

Testing has been carried out to measure and compare an initial fixture strength per loading between polyurethane hot melt adhesive solution and 1 component silicone solution. As shown in **Figure 22**, adhesive is applied onto ABS plastic assembly; then, a glass plate is placed on it and pressed for 30 s, and the assembly is conditioned at different interval time, placed upside down; 25 kg weight is placed upon the glass part and then it is observed if the assembled glass remains bonded or falls apart. As shown in **Table 4**, polyurethane hot melt adhesive has no de-bonding even at 1 min curing while 1 component silicone takes a much longer curing time;

Figure 22.
Fixture time test condition.

Solution	1 min cure	3 min cure	5 min cure	10 min cure
PUR HM	No de-bonding	No de-bonding	No de-bonding	No de-bonding
1C silicone	De-bonding	De-bonding	De-bonding	De-bonding

Table 4.
Fixture time comparison.

therefore, it has a de-bonding even after 10 min of curing time. This explains more compact and process efficient assembly process with less time and less manpower for polyurethane hot melt adhesive compared to 1 component silicone solution.

5.3 Bonding strength vs. environmental condition 1 vs. loading

Testing has been done to measure the bonding strength, failure mode per different loading after a high temperature environmental condition (70°C for 1 week), and a high humidity condition (50°C/95%RH for 1 week). As shown in **Table 5**, polyurethane hot melt adhesive solution has no de-bonding when the weight loading of 25, 50, and 100 kg was placed on top of glass part; then, 3 times of 100 kg impact was applied to the glass part and the bond line remained intact between glass and mating substrates but the glass in the diddle was shattered instead. **Figures 23** and **24** explain the intact bond line area of both chrome-plated + glass and ABS + glass bonding.

5.4 Tensile shear strength vs. environmental condition

Testing has been conducted to measure and compare using test specimen the tensile shear strength at different high temperature (85°C) and high humidity (85% RH) for max 1000 h aging time. Specimen was glass and ABS, together with painted ABS substrates. After adhesive dispensing and glass part assembly, it was put into environment chamber immediately. As shown in **Figures 25** and **26**, we can see that as the aging time progresses, the tensile shear strength difference is widened explaining that the PUR HM adhesive increases as time goes by, while 1 component silicone has no curing within 1 h of aging and maintains a lower strength than polyurethane hot melt adhesive; especially for painted ABS substrate, it shows a

Aging condition	25 kg weight	50 kg weight	100 kg weight
70°C × 1 week	No de-bonding	No de-bonding	No de-bonding
50 °C × 95%RH × 1 week	No de-bonding	No de-bonding	No de-bonding

Table 5.
Bonding strength vs. loading.

Figure 23.
Chrome-plated + glass bonding.

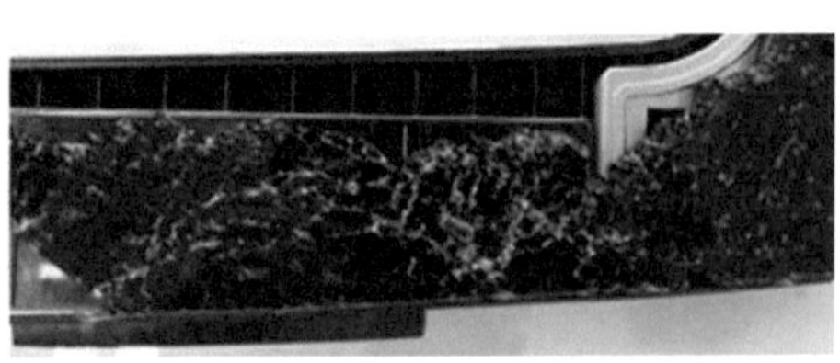

Figure 24.
ABS + glass bonding.

sharp decrease of bonding strength at 1000 h. Polyurethane hot melt adhesive shows an immediate certain strength initially due to pure hot melt fixturing function, but as time goes by, due to the chemical reaction, the adhesion strength increases to a higher reliable level. The assembly of glass to painted ABS shows a higher ultimate strength than the assembly of glass to ABS.

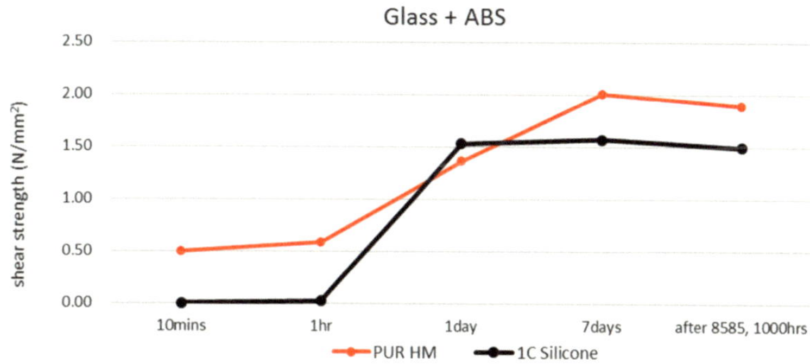

Figure 25.
Tensile shear strength comparison for glass + ABS.

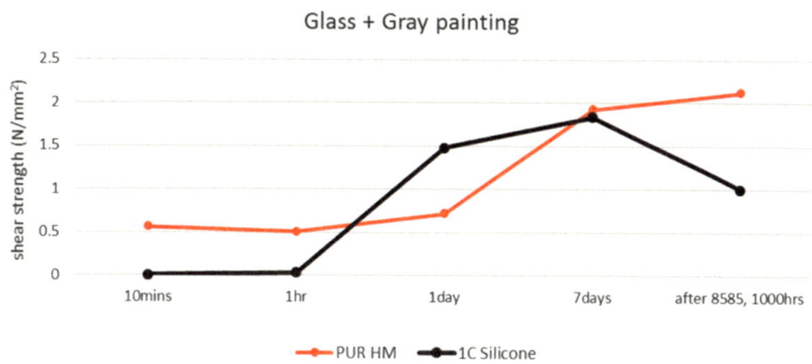

Figure 26.
Tensile shear strength comparison for glass + painted ABS.

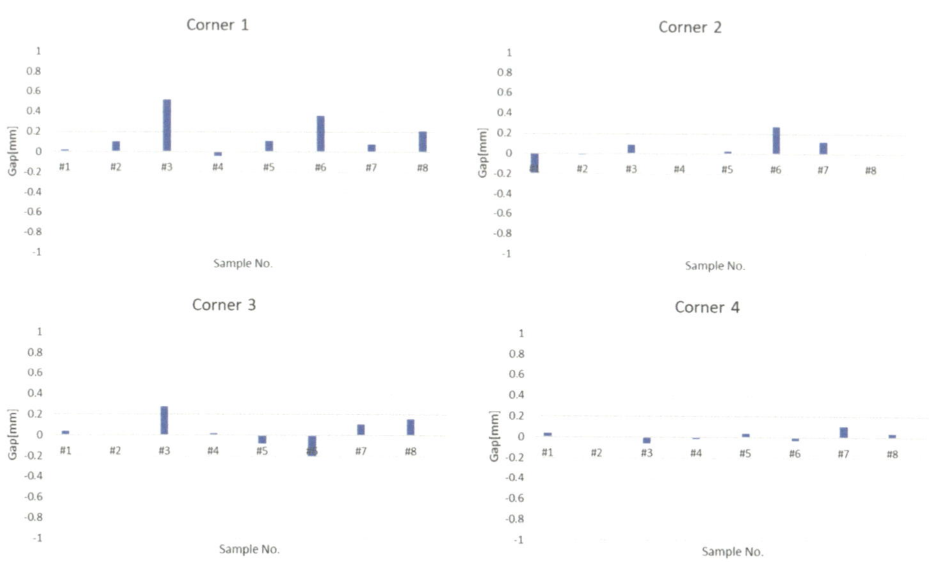

Figure 27.
Gap change at different corners vs. environment conditions.

5.5 Bond line vs. environmental condition 3

Testing has been done to measure the difference of bond line gap change before and after three different environmental conditions. As mentioned in the refrigerator section, polyurethane hot melt adhesive provides an excellent environment resistance. As shown in **Figure 27**, polyurethane hot melt adhesive exhibits a minimal gap change before and after three different environmental conditions. Specimens #1, #2, and #3 are the conditions of thermal cycle 4 h at 70°C/65%RH to 4 h at −30°C (10 cycles); specimens #4, #5, and #6 are on the condition of high temperature 168 h at 70°C; and specimens #7 and #8 are on the condition of high humidity 168 h at 40°C/95%RH. The gap change at 4 corner area of top cover that has a biggest tendency of gap change was measured.

6. Microwave

In this section, we are going to compare 2 component silicone vs. 1 component silicone in terms of process efficiency. Application is for bonding of glass plate to plastic case in the external three front door [3].

6.1 Production efficiency

The 1 component silicone solution is to apply a primer onto plastic case by 0.5 operator (15 s) → place double-sided tape on a few location by 0.5 operator for instant fixing (20 s) → silicone is dispensed by dispenser (15 s) → glass plate is placed on the plastic case and pressed (5 s) → clip and tape are used to fix the bonded assembly by one operator (20 s) → fixed assemblies go through a heating conveyor → after conveyor, tapes and clips are removed by 0.5 operators (15 s) → the assembly stacked onto rack by 0.5 operator (10 s) → rack is placed in the corner of plant for 24 h curing → cured assembly transfer to next process.

The 2 component silicone solution is to apply primer by machine (10 s) → primer drying by fan (10 s) → adhesive dispensing by machine (15 s) → glass plate placed onto plastic case by machine (15 s) → glass plate pressing by machine (45 s) → initial fixture of assembly achieved and transfer to next process.

The 1 component silicone solution takes three operators and many steps, and especially takes minimum 1 day for silicone curing before transferring to next process, while the 2 component silicone solution requires no operator and its processing time from primer application to initial fixture takes only 3 min which is a dramatic improvement of its assembly efficiency, and this solution required a minimal space for the whole process which is another benefit; it requires no heating conveyor, thus eliminates the concern on extra energy consumption and is operator

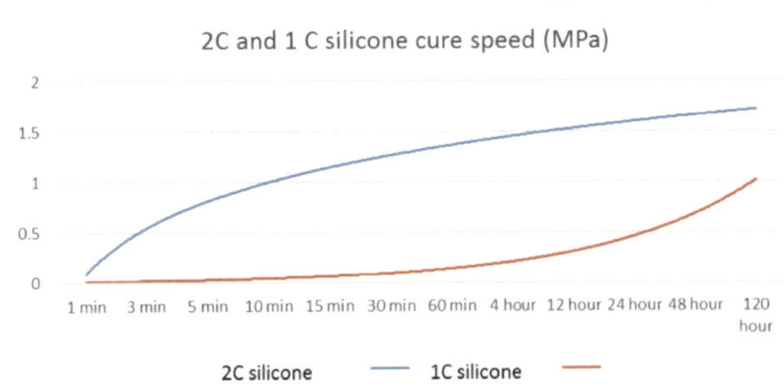

Figure 28.
Fixture time between 2C silicone and 1C silicone.

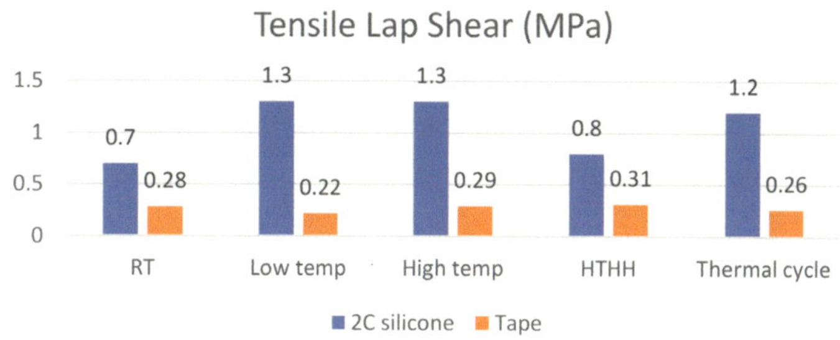

Figure 29.
Tensile shear strength vs. different aging conditions.

safety too; it also increases the quality of the assembly as the time to be exposed to the ambient environment contamination such as dirt sticking to silicone surface is minimal due to its fast cure while the 1 component silicone solution is difficult to avoid such appearance and contamination quality issue. **Figure 28** explains the fast fixture time of 2 component silicone within a short time (3 min) compared to 1 component silicone that takes 24 h to achieve the same initial strength. Using 2 component silicone solution, within a few minutes, the assembly can moves to the next process [3].

6.2 Tensile shear strength vs. aging condition

Testing has been carried out to measure a tensile shear strength to compare the bonding reliability before and after aging condition between 2 component silicone solution and tape solution as some manufacturers use tape only without 1 component silicone. Substrate was a ink printed glass and ABS or polycarbonate. Testing conditions were at room temperature (23°C), low temperature (−40°C for 2 h), high temperature (80°C for 2 h), high temperature and high humidity (80°C for 8 h, −40°C for 8 h as 1 cycle for 27 days) and thermal cycle tests (60°C for 4 h to −30°C for 4 h as 1 cycle for 27 days). As shown in **Figures 29**, 2 component silicone solution has at least 3–5 times higher bonding strength for all aging conditions compared to tape solution. Overall, it is very clear that as for environmental durability, 2 component silicone solution provides a far more reliable bonding than tape solution.

7. Conclusions

As reviewed and compared for three appliances on the above such as refrigerator, washing machine, and microwave, those new technologies such as polyurethane hot melt and 2 component silicone solution can help appliance manufacturers open their eyes much wider for new assembly solutions that can provide overall assembly process efficiency such as faster work in process (WIP), automated assembly that can reduce overall man powers and reduce assembly space, increase assembly reliability such as higher bonding strength after severe environment conditions and ultimately the assembly can receive a better reputation and high end image at the market place.

Acknowledgements

This work was supported by various lab engineers who have devoted their time for various and lengthy testing. Great thanks to them for their professional and technical support as well as administrative aid. Gratitude is also owed to those sales engineers who kindly have contacted manufacturers and received their consent to share with us their actual parts.

Author details

Chulsoo Woo[1*], Bin Liang[1] and Jonghyuk Oh[2]

1 Henkel (China) Investment Co., Ltd, Shanghai, China

2 Henkel Korea Innovation Center, Seoul, South Korea

*Address all correspondence to: cs.woo@henkel.com

References

[1] Woo C. Technology Training for Refrigerator. China: Henkel (China) Investment, Co.; 2014

[2] Woo C. Technology Training for Washing Machine. China: Henkel (China) Investment, Co.; 2014

[3] Woo C. Technology Training for Oven, Cook Top and Microwave. China: Henkel (China) Investment, Co.; 2014

[4] Woo C. Technology Training for Air Conditioner. China: Henkel (China) Investment, Co.; 2013

[5] Woo C. Technology Training for Dish Washer. China: Henkel (China) Investment, Co.; 2013

[6] Woo C. Functional Adhesive Trend for Assembly Industry. China: Henkel (China) Investment, Co.; 2019

[7] Alan L, Rudolf N. Loctite Worldwide Design Handbook. 2nd ed. Germany: Henkel Corporation; 1998

[8] Woo C, Michael L. Silicone Technology. China: Henkel (China) Investment, Co.; 2006

[9] Woo C, Michael G. Polyurethane Technology. China: Henkel (China) Investment, Co.; 2011